CITES CACTACEAE CHECKLIST

Second Edition

C·I·T·E·S
Cactaceae
CHECKLIST

Second Edition

compiled by

David Hunt
Honorary Research Fellow
Royal Botanic Gardens Kew

assisted by members of the
International Organization for Succulent Plant Study

First published 1992
Second edition 1999

ISBN 1 900347 45 8

Cover design by
Media Resources RBG Kew

Illustration: *Turbinicarpus beguinii*,
a species listed on CITES Appendix I,
in its Mexican habitat, 15 April 1992
(photo: David Hunt)

Compiled with the financial assistance of the
CITES Nomenclature Committee and the
US Scientific Authority for CITES

Typeset by David Hunt
Printed and bound by
Remous Limited, Milborne Port

CITES CACTACEAE CHECKLIST
Second Edition

CONTENTS

TABLE DES MATIERES

INDICE

FOREWORD

For many years the fate of endangered plant species in trade received little attention from the Parties to CITES, although the plant species included in the CITES appendices outnumber the animal species many times. In the past few years this situation has considerably changed for the better. In the case of the family Cactaceae, the change has become visible through a record of confiscated shipments of illegally traded wild-collected cacti. Also, the number of people arrested in countries of origin such as Mexico for illegal collecting of cacti has strongly increased, even resulting in jail sentences.

Improving the quality of CITES controls automatically calls for adequate information regarding identification, nomenclature and distribution of the species concerned. With an ever-increasing number of species of Cactaceae listed in Appendix I and the remainder in Appendix II of CITES, this Checklist is a first and very important step towards a better implementation of CITES for this group of plants. It will also form the basis on which the necessary identification sheets can be developed.

The CITES Secretariat expects that all who are involved in the implementation of this important Convention as well as scientists and cactus-lovers world-wide will use this second edition of the Cactaceae Checklist for the benefit of the conservation of this interesting group of succulent plants.

CITES Secretariat
Geneva, November 1998

AVANT-PROPOS

Pendant de nombreuses années, le sort des plantes commercialisées menacées d'extinction n'a guère retenu l'attention des Parties à la Convention alors qu'il y a bien plus d'espèces végétales que d'espèces animales inscrites aux annexes CITES. Ces dernières années, la situation a beaucoup évolué - et dans le bon sens. Dans le cas de la famille des Cactaceae, le nombre de confiscations d'envois de cactus prélevés dans la nature et commercialisés illégalement témoigne de ce changement. Le nombre de personnes arrêtées dans des pays d'origine - tels que le Mexique - pour prélèvement illégal de cactus, a lui aussi fortement augmenté. Certaines arrestations ont même entraîné des condamnations à des peines d'emprisonnement.

Améliorer la qualité des contrôles CITES implique de disposer d'informations adéquates sur l'identification, la nomenclature et la répartition géographique des espèces concernées. Avec un nombre toujours plus grand d'espèces de Cactaceae inscrites à l'Annexe I de la CITES et le reste à l'Annexe II, la présente Liste est un premier pas très important vers une meilleure mise en œuvre de la CITES pour ce groupe de plantes. C'est également sur la base de la Liste que les fiches d'identification nécessaires pourront être développées.

Le Secrétariat CITES s'attend à ce que tous ceux qui contribuent à l'application de cette importante convention ainsi que les scientifiques et les amateurs de cactus du monde entier utilisent la deuxième édition de la Liste des Cactaceae, pour le plus grand bien de la conservation de cet intéressant groupe de plantes succulentes.

Secrétariat CITES
Genève, novembre 1998

PROLOGO

Durante numerosos años las Partes en la CITES prestaron poca atención al sino de las especies de plantas en peligro objeto de comercio, a pesar de que el número de plantas incluidas en los Apéndices de la CITES excedía sobremanera al de especies animales. En los últimos años esta situación ha cambiado positivamente. En el caso de la familia de las cactáceas, el cambio se ha percibido claramente mediante el registro de envíos confiscados de cactus recolectados en el medio silvestre comercializados ilícitamente. Asimismo, ha incrementado el número de personas detenidas en los países de origen, como México, por la recolección ilegal de cactus, resultando incluso en penas de prisión.

La mejora de la calidad de los controles de la CITES exige automáticamente información adecuada sobre la identificación, la nomenclatura y la distribución de las especies en cuestión. Habida cuenta del cada día mayor número de especies de cactáceas incluidas en el Apéndice I y en el Apéndice II de la CITES, esta Lista constituye la primera y más importante medida en aras a una mejor aplicación de la Convención para este grupo de plantas. Constituirá, asimismo, la base para la preparación de las fichas de identificación necesarias.

La Secretaría de la CITES alberga la esperanza de que todas aquellas personas que participan en la aplicación de la Convención, así como los científicos y amantes de los cactus en todo el mundo, utilizarán esta segunda edición de la Lista de cactáceas en beneficio de la conservación de este interesante grupo de plantas suculentas.

La Secretaría de la CITES
Ginebra, noviembre de 1998

Acknowledgments / Remerciements / Reconocimientos

A Working Party of the International Organization for Succulent Plant Study (IUBS-IOS) was set up in 1984 to seek a consensus on the classification of the Cactaceae. A few years later the CITES Plants Committee suggested that IOS should oversee the compilation of what was to be the first edition of this publication. Many of the 30 or so collaborators on that edition have again provided input to the second edition and are listed below, along with other specialists who have joined the *ad hoc* group since 1992. The compiler is most grateful to all these contributors, and especially to Nigel Taylor, who has reviewed a substantial number of genera in addition to those of Brazil on which he has recently been working. Where compatible with the general approach set out in the Preamble, the listings have been revised to take account of the suggestions received, but should not be assumed to reflect the views of IOS or any individual contributor, or of the Royal Botanic Gardens, Kew, which has acted as publisher. The compiler accepts sole responsibility for the content of the Checklist and the taxonomic opinions implied.

En 1984, un Groupe de travail a été constitué par l'Organisation Internationale de Recherches sur les Plantes Succulentes (IUBS-IOS) pour parvenir à un consensus sur la classification des Cactaceae. Quelques années plus tard, le Comité CITES pour les plantes suggéra que l'IOS supervise la compilation de ce qui devait être la première édition de cette publication. Plusieurs parmi la trentaine de collaborateurs à la première édition ont également contribué à la deuxième et sont cités ci-dessous avec les autres spécialistes qui ont rejoint le groupe ad hoc depuis 1992. Le compilateur remercie vivement tous ceux qui ont contribué à la deuxième édition, en particulier Nigel Taylor, qui a revu un grand nombre de genres en plus de ceux du Brésil sur lesquels il avait récemment travaillé. Lorsque c'était compatible avec la démarche définie dans le préambule, la Liste a été révisée pour tenir compte des suggestions; toutefois, elle ne reflète pas forcément les vues de l'IOS ou celles des différents collaborateurs, ou encore celles des Royal Botanic Gardens, Kew, qui l'a publiée. Le compilateur assume l'entière responsabilité du contenu de la Liste et des opinions qu'elle implique quant à la taxonomie.

En 1984 se estableció un Grupo de trabajo de la Organización Internacional para el Estudio de Plantas Suculentas (IUBS-IOS) con miras a lograr un acuerdo general respecto de la clasificación de las cactáceas. Años más tarde, el Comité de Flora de la CITES propuso que la IOS supervisase la compilación de lo que habría de ser la primera edición de esta publicación. La mayor parte de las 30 personas que colaboraron en esa edición, cuyos nombres figuran a continuación, han vuelto a participar en esta segunda edición, junto con otros especialistas que se han unido al grupo desde 1992. El encargado de la compilación expresa su agradecimiento a todos los que han contribuido a la misma y, en particular, a Nigel Taylor, que ha revisado un considerable número de géneros, además de los correspondientes a Brasil, en los que ha trabajado personalmente en fecha reciente. Cuando ha sido posible con arreglo al enfoque general establecido en el preámbulo, se han revisado las inclusiones tomando en consideración las sugerencias formuladas, pero no debe considerarse que reflejan las opiniones de la IOS, de cualquier colaborador o del Royal Botanic Gardens, Kew, que ha publicado esta lista. La responsabilidad del contenido de la Lista y de las opiniones expresadas en lo que concierne a la taxonomía incumbe exclusivamente a la persona que se ha encargado de su compilación.

International panel of advisers and collaborators
Groupe international de conseillers et de collaborateurs
Grupo internacional de expertos y colaboradores

Dr Edward F. Anderson (USA), Eric Annal (UK), Salvador Arias Montes (Mexico), Rolando Barcénas (Mexico), Prof. Dr Wilhelm Barthlott (Germany), Karl Werner Beisel (Germany), Wolfgang Borgmann (Germany), Dr Pierre Braun (Germany), Dra Helia Bravo Hollis (Mexico), Drs Rob Bregman (Netherlands), John Brickwood (UK), Dr Hugo Cota Sánchez (Mexico), Dr Urs Eggli (Switzerland), Dr Thomas Engel (Germany), W.A. & B. Fitz Maurice (Mexico), the late Charles Glass (Mexico), Carlos Gomez (Mexico), Dr Héctor M. Hernandez Macias (Mexico), George S. Hinton (Mexico), Dr Mats Hjertson (Sweden), Andreas Hofacker (Germany), James Iliff (UK), Dr Roberto Kiesling (Argentina), Myron Kimnach (USA), Dr Beat Leuenberger (Germany), Dr Jonas Lüthy (Switzerland), Dr Bill Maddams (UK), Prof. James Mauseth (USA), Dr Massimo Meregalli (Italy), Dr Detlev Metzing (Germany), Alessandro Mosco (Italy), Roy Mottram (UK), Dr Reto Nyffeler (Switzerland), Dr Carlos Ostolaza (Peru), John Pilbeam (UK), Prof. Donald Pinkava (USA), Andy Powell (UK), Ken Preston-Mafham (UK), Alícia Rodríguez (Cuba), Gordon Rowley (UK), Dr Maurizio Sajeva (Italy), Dr Wolfgang Stuppy (Germany), Geoffrey Swales (UK), Nigel Taylor (UK), Prof. Baltasar Trujillo (Venezuela), Dr Robert Wallace (USA), Dra Daniela Zappi (UK).

CITES CACTACEAE CHECKLIST
Second Edition

Preamble

1. **Background**

1.1. The first edition of this Checklist was published in 1992 in response to a Resolution from the 6th Conference of the Parties to CITES (1987) calling for the development of a nomenclatorial reference for the Cactaceae. Compilation began in May 1989 and was completed in March 1992 as an official project at the Royal Botanic Gardens, Kew, under the auspices of the International Organization for Succulent Plant Study (IUBS-IOS). Funding assistance was provided by the CITES Nomenclature Committee and the US Scientific Authority for CITES.

1.2. Arrangements for the preparation of a new edition were concluded in September 1996 via a Consultancy Agreement between the Board of Trustees of the Royal Botanic Gardens Kew and the Compiler. Funding assistance was again provided by the CITES Nomenclature Committee.

1.3. The new edition follows the general plan of the first edition but has been expanded to include names of subspecies and a taxonomic or 'synonymized list' of preferred specific and subspecific names with their authorities and synonyms. For ease of use by CITES and Customs Officers – for whom the work is primarily intended – the general alphabetic listing has been placed directly after this Preamble, and the country checklists at the end of the volume, arranged alphabetically by countries. Reference material, including indexes of generic names, the general bibliography of works consulted, and the 'synonymized list' mentioned above, is sandwiched between the alphabetic and country listings.

2. **Computer aspects**

For this edition the hardware and software used was privately owned by the compiler. The main database was compiled in dBase from data tables used for the first edition and additional entries provided by Kew from *Index Kewensis* or by the compiler and members of the panel of advisers. In accordance with the terms of the Consultancy Agreement the data have been made available in ALICE Transfer Format for use in future editions.

3. **Compilation procedures**

3.1. ***Origination***. This edition of the CITES Cactaceae Checklist takes the first edition as its stepping-off point. Updating for the new edition has followed a number of routine procedures, supplemented by a taxonomic review of all entries by the compiler and members of the panel of advisers (see §4, below). The Bibliography (see Part II, sect. 4) has been extended to include various regional and taxonomic overviews which have become available since the first edition. As noted already, subspecific names are now included and a synonymized list, giving the full synonymy for all accepted taxa is provided.

3.2. ***List of genera and generic synonyms***. This has been brought up to date in the light of changes in the generic list adopted by the Cactaceae Working Party of the International Organization for Succulent Plant Study (IOS). The types of all generic names listed have been inserted.

3.3. ***Cactaceae names in current usage***

3.3.1. The scope of the list of binomials in the first edition was explained in the Preamble to that edition (pp. 10–11). Off the 6022 entries in that edition about 200 (nearly all synonyms) have been removed from the present one, for various reasons:

- Names under various disused or inadmissible generic names but included in the first edition because they were treated as accepted names in one or more of the works screened during its compilation;
- Names of species accepted in one or more of the works screened but known only from cultivation and of doubtful identity, usually due to inadequate or disputed typification;
- Other names treated as synonyms but identifiable to genus only;
- A few names from trade sources etc which appear to be undescribed.

3.3.2. Since the publication of the first edition, several incorrect entries have come to light and these have been corrected.

3.3.3. More than 500 additional names have been published at specific rank since the first edition, and these have been added to the database, with distribution data where appropriate.

3.3.4. The inclusion of names at subspecific rank in this second edition is a major change from the first. With few exceptions, all names previously published at this rank are listed here, and a percentage of taxa previously listed as species (mostly as 'provisionally accepted') have been reduced to this rank.

3.3.5. Synonymy at the species/subspecies level has been brought into line with changes in the generic list, current standard works and the views of advisers. Where it has not been possible to resolve differences of opinion, the compiler has made a personal judgment. New combinations have been routinely treated as synonyms if they are in genera not currently recognized by the IOS 'consensus' group or conflict with current views on generic synonymy.

3.3.6. CITES Appendix I listings have been brought into line with changes formally approved by Conferences of the Parties.

3.3.7. In the first edition, conservation ratings according to the old IUCN categories were included in some of the country lists. For most countries these have yet to be reassessed according to the new IUCN Conservation Categories and for this reason they have been deleted throughout.

4. **International Panel of Advisers**

4.1. So far as possible, it is the aim of the *CITES Cactaceae Checklist* to reflect a consensus of current opinion on the classification and nomenclature to be employed. Since 1984, meetings of an informal Working Party established under the auspices of the International Organization for Succulent Plant Study (IOS) have been held at regular intervals, usually annually, to discuss research in progress on the Cactaceae and its bearing on the classification of the family. Since arrangements for this edition of the Checklist were concluded, all proposed changes to entries in the first edition have been notified to members of the Working Party and others via a periodic newsletter published by the compiler (*Cactaceae Consensus Initiatives*; ISSN 1365-778X) and Technical Workshops have been convened in England (September 1996 and April 1998) and Germany (May 1997). Synonymized lists for individual countries have been submitted to various reviewers as follows:

Argentina	R. Kiesling	Guianas	B.E. Leuenberger
Brazil	N.P. Taylor & D.C Zappi	Mexico	H.M. Hernández Macias
Chile	U. Eggli & B.E. Leuenberger	Peru	C. Ostolaza
Cuba	A. Rodríguez	U.S.A.	E.F. Anderson

4.2. It must be stressed that the Checklist does not necessarily represent the views of the above reviewers, of the International Organization for Succulent Plant Study, its Executive Board or any individual contributor. The compiler accepts sole responsibility for the content of the Checklist and taxonomic opinions implied, but gratefully acknowledges the invaluable assistance of the advisers and collaborators listed above and on p. 6.

5. **How to use the Checklist**

5.1. It is intended that this Checklist be used as a quick reference for checking currently accepted names and synonymy, and the known distribution of accepted taxa. It is therefore divided into three parts:

> ***Part I*** is an alphabetic listing of all names of species and subspecies included in the Checklist and serves as an index to the work as a whole. The preferred names of accepted taxa are printed **bold** or ***bold-italic*** (provisionally accepted). Synonyms are printed *italic*, followed by an arrow (→) and the preferred name of the taxon to which they are considered to apply. Symbols (see §6 below) are used to denote various categories of names and accepted names and synonyms of taxa listed on CITES Appendix I.

5.2. ***Part II*** contains more detailed reference data under the following headings:

1. Alphabetic list of genera of Cactaceae and their principal synonyms – an alphabetic list of genera annotated with authorities, places of publication, nomenclatural types, numbers of taxa and geographical distribution.
2. Summary table of accepted genera with numbers of accepted taxa.
3. List of accepted taxa – still basically in alphabetic order, but with the preferred name of each of the currently accepted taxa followed by the synonyms referred to it in Part I and botanical authorities.
4. Selected bibliography – an index of works of reference consulted in the preparation of the Checklist, indicating those taken as the primary sources or 'benchmarks' for the treatment adopted.

5.3. ***Part III*** gives an alphabetic sublist for each country of the accepted taxa known to occur there. The countries are also listed in alphabetic order, using the United Nations Country Names as laid down in Terminology Bulletin no. 347: 1–41 (New York: United Nations, August 1995).

6. **Conventions employed in Parts I–III**

6.1. *Typography*. Preferred names of accepted taxa are printed **bold** and of provisionally accepted taxa and hybrids ***bold-italic***. Names regarded as synonyms are printed *italic*, followed by an arrow (→) and the preferred name of the taxon to which they are considered to apply.

6.2. *Symbols and codes*. The following are used to denote various categories of names and taxa:

→	Preceding name is regarded as a synonym of the one which follows;
→ ?	Preceding name *may be* a synonym of the one which follows;
×	Genus or species of hybrid origin;
♦	Name of taxon listed on CITES Appendix I;
ⓞ	Name outlawed by the International Code of Botanical Nomenclature;
¶	"as to type"; where a synonym is listed twice, this precedes the original and proper use of the name, but the name is commonly used in error for something else – see next entry;

® "misapplied"; the plant which is or has been known and/or traded under this name is the one to which it is here referred, not the one indicated or intended by the original author.

6.3. *Geographical distribution.* Two-letter ISO codes (except those marked * below, which are arbitrary) are used in Part II to indicate the countries of origin and distribution of accepted and provisionally accepted taxa:

Argentina	AR	El Salvador	SV	Peru	PE
Bahamas	BS	French Guiana	GF	Puerto Rico	PR
Belize	BZ	Guatemala	GT	Suriname	SR
Bolivia	BO	Guyana	GY	Trinidad & Tobago	TT
Brazil	BR	Haiti	HT	Uruguay	UY
Canada	CA	Honduras	HN	USA	US
Cayman Islands	KY	Jamaica	JM	Venezuela	VE
Chile	CL	Lesser Antilles	WI*	Virgin Islands	VI
Colombia	CO	Mexico	MX		
Costa Rica	CR	Neth. Antilles	AN	Old World	OW*
Cuba	CU	Nicaragua	NI		
Dominican Republic	DO	Panama	PA	Cult. origin	XC*
Ecuador	EC	Paraguay	PY		

7. **Cactaceae controlled by CITES**

7.1. Trade in all taxa of Cactaceae is controlled by the provisions of CITES Appendix II, including all parts and derivatives, except:

a) seeds, except those from Mexican cacti originating in Mexico, and pollen;
b) seedling or tissue cultures obtained *in vitro*, in solid or liquid media, transported in sterile containers;
c) cut flowers of artificially propagated plants;
d) fruits and parts and derivatives thereof of naturalized or artificially propagated plants; and
e) separate stem joints (pads) and parts and derivatives thereof of naturalized or artificially propagated plants of the genus *Opuntia* subgenus *Opuntia*;

and artificially propagated specimens of the following hybrids and/or cultivars: *Hatiora* × *graeseri* (*H. gaertneri* × *H. rosea*), *Schlumbergera* ×*buckleyi* (*S. russelliana* × *S. truncata*), *S.* ×*exotica* (*S. opuntioides* × *S. truncata*), *S.* ×*reginae* (*S. orssichiana* × *S. truncata*), *S. truncata* (cultivars), *Gymnocalycium mihanovichii* (cultivars) lacking chlorophyll, grafted on the following grafting stocks: *Harrisia* 'Jusbertii', *Hylocereus trigonus* or *Hylocereus undatus*, *Opuntia microdasys* (cultivars).

7.2. Trade in the taxa listed in Table 1 (page 23) is controlled by the provisions of CITES Appendix I.

LISTE DES CACTACEAE CITES

Deuxième Édition

Préambule

1. **Contexte**

1.1. La première édition de la Liste des Cactaceae CITES a été publiée en 1992 pour donner suite à une résolution de la sixième session de la Conférence des Parties à la CITES (1987) qui demandait l'élaboration d'une référence sur la nomenclature des Cactaceae. La compilation a commencé en mai 1989 et s'est achevée en mars 1992 en tant que projet officiel des Royal Botanic Gardens, Kew, sous les auspices de l'Organisation internationale de recherches sur les plantes succulentes (IUBS-IOS). Le Comité CITES de la nomenclature et l'autorité scientifique CITES des États-Unis d'Amérique ont fourni une assistance au financement du projet.

1.2. En septembre, 1996 la préparation de la deuxième édition a fait l'objet d'un accord de consultation entre le Conseil de direction des Royal Botanic Gardens, Kew, et le compilateur. Là encore, le Comité CITES de la nomenclature a alloué des fonds.

1.3.La nouvelle édition suit le plan général de la première mais a été augmentée des noms des sous-espèces et d'une liste taxonomique (liste des noms acceptés et de leurs synonymes), comportant les noms spécifiques et subspécifiques préférés avec les auteurs et les synonymes. Pour faciliter l'utilisation de la Liste par les organes CITES et les agents des douanes - auxquels elle s'adresse principalement - la liste alphabétique générale a été placée directement après le préambule, tandis que les listes des pays, dans l'ordre alphabétique, figurent à la fin de l'ouvrage. Les références – notamment les index des noms génériques, la bibliographie générale des travaux consultés et la liste des noms acceptés et de leurs synonymes – sont insérées entre la liste alphabétique et les listes par pays.

2. **Aspects informatiques**

Le compilateur a utilisé son propre matériel et ses logiciels pour préparer cette édition. Il a compilé la principale base de données en dBase à partir des tableaux de données utilisés pour la première édition; les entrées additionnelles ont été fournies par Kew à partir de l'Index Kewensis, par le compilateur, et par les membres du Groupe international de conseillers. Conformément à l'accord de consultation, les données ont été fournies en configuration ALICE pour pouvoir être utilisées dans de futures éditions.

3. **Protocole de compilation**

3.1. ***Point de départ***. La présente édition de la Liste des Cactaceae CITES part de la première édition, dont la mise à jour a suivi le protocole établi, complété par un examen taxonomique de toutes les entrées par le compilateur et les membres du Groupe international de conseillers (voir ci-dessous, §4). La bibliographie (voir deuxième partie, point 4) a été augmentée des diverses vues d'ensemble régionales et taxonomiques devenues disponibles après la première édition. Comme indiqué plus haut, les noms subspécifiques ont été inclus et une liste de tous les synonymes des taxons acceptés a été fournie.

3.2. ***Liste des genres et des synonymes génériques***. Cette liste a été mise à jour pour tenir compte des changements survenus dans la liste générique adoptée par le

Groupe de travail de l'IUBS-IOS. Les types de tous les noms génériques de la liste ont été inclus.

3.3. ***Noms des Cactaceae actuellement en usage***

3.3.1. Dans le préambule de la première édition, la portée de la liste des binômes est expliquée (p. 10-11). Sur les 6022 entrées de cette édition, environ 200 (presque toutes des synonymes) ont été retirées de la présente édition, pour diverses raisons:

- Divers noms génériques tombés en désuétude ou non valables mais néanmoins inclus dans la première édition parce qu'ils étaient traités comme des noms acceptés dans un ou plusieurs travaux examinés au cours de la compilation;
- Noms d'espèces acceptées dans un ou plusieurs des travaux examinés mais connues seulement en culture ou dont l'identité est contestable, habituellement en raison d'une typification contestée ou inadéquate;
- Noms traités comme des synonymes mais identifiables seulement au niveau du genre;
- Quelques noms de sources commerciales et autres n'ayant, semble-t-il, pas été décrits.

3.3.2. Depuis la publication de la première édition, plusieurs entrées incorrectes ont été décelées et corrigées.

3.3.3. Plus de 500 noms additionnels ont été publiés au rang spécifique depuis la publication de la première édition; ils ont été inclus dans la base de données avec des indications sur la répartition géographique, comme approprié.

3.3.4. L'inclusion des noms au rang subspécifique dans la deuxième édition constitue un changement important. A quelques exceptions près, tous les noms précédemment publiés à ce rang y figurent; une partie des taxons précédemment indiqués comme espèces (la plupart 'acceptées provisoirement') ont été réduits à ce rang.

3.3.5. La synonymie au niveau des espèces/sous-espèces suit les changements intervenus dans la liste générique, les travaux standard actuels et les vues des conseillers. Lorsqu'un consensus n'a pas pu être atteint, le compilateur s'est appuyé sur son jugement personnel. Les nouvelles combinaisons ont été automatiquement traitées comme des synonymes si elles apparaissaient dans des genres non actuellement acceptés par le groupe "de consensus" de l'IOS ou si elles ne correspondaient pas aux vues actuelles sur les synonymes génériques.

3.3.6. Les changements officiellement approuvés par la Conférence des Parties ont été intégrés dans la liste des espèces inscrites à l'Annexe I de la CITES.

3.3.7. Dans la première édition, l'état de conservation selon les anciennes catégories de l'UICN était indiqué dans certaines listes par pays. Pour la plupart des pays, l'état de conservation doit être réévalué en fonction des nouvelles catégories de l'UICN; c'est la raison pour laquelle cette indication a été supprimée partout.

4. **Groupe international de conseillers**

4.1. La Liste des Cactaceae CITES a pour but de refléter autant que possible un consensus de l'opinion actuelle sur la classification et la nomenclature à utiliser. Depuis 1984, le Groupe de travail constitué par l'IOS a tenu des réunions informelles à intervalles réguliers, habituellement chaque année, pour discuter des recherches en cours sur les Cactaceae et leur portée pour la classification de cette famille. Depuis que l'accord pour la préparation de la deuxième édition a été conclu, toutes les propositions de changements par rapport à la première édition ont été notifiées aux membres du Groupe de travail et aux autres collaborateurs par circulaires périodiques publiées par le compilateur (*Cactaceae Consensus Initiatives*; ISSN 1365-778X). Des ateliers techniques ont été convoqués en

Angleterre (septembre 1996 et avril 1998) et en Allemagne (mai 1997). Les listes de synonymes pour certains pays ont été soumises pour examen aux collaborateurs suivants:

Argentine	R. Kiesling	États-Unis d'Amérique	E.F. Anderson
Brésil	N.P. Taylor & D.C Zappi	Mexique	H.M. Hernández Macías
Chili	U. Eggli & B.E. Leuenberger	Pérou	C. Ostolaza
Cuba	A. Rodríguez	Région guyanaise	B.E. Leuenberger

4.2. Il convient de souligner que la Liste des Cactaceae CITES ne représente pas nécessairement les vues de ces collaborateurs, de l'Organisation internationale de recherches sur les plantes succulentes, de son Conseil de direction ou des autres collaborateurs. Le compilateur assume l'entière responsabilité du contenu de la Liste et des opinions taxonomiques qu'elle implique, tout en reconnaissant pleinement l'assistance précieuse fournie par les conseillers et les collaborateurs indiqués ci-dessus et à la page 6.

5. **Comment utiliser la liste**

5.1. La liste devrait permettre de trouver rapidement les noms et les synonymes actuellement acceptés des taxons et leur aire de répartition. Elle est divisée en trois parties:

5.2. La première partie est une liste alphabétique de tous les noms d'espèces et de sous-espèces inclus dans la Liste; elle sert d'index général. Les noms préférés des taxons acceptés figurent en **gras** et en ***gras-italique*** (= acceptés provisoirement). Les synonymes sont en *italique* et sont suivis de → et du nom préféré du taxon auquel ils s'appliquent. Les symboles (voir §6 ci-dessous) servent à distinguer les différentes catégories de noms et de noms et synonymes acceptés des taxons inscrits à l'Annexe I de la CITES.

5.3. La deuxième partie contient des données plus détaillées, sous les titres suivants:

1. Liste alphabétique des genres de Cactaceae et leurs principaux synonymes: liste alphabétique des genres indiquant les auteurs, les lieux de publication, les types nomenclaturaux, le nombre de taxons et la répartition géographique.
2. Tableau résumé des genres acceptés et nombre de taxons acceptés.
3. Liste des taxons acceptés: liste essentiellement alphabétique, comportant également le nom préféré de chaque taxon actuellement considéré comme accepté, suivi des synonymes indiqués dans la première partie et des autorités botaniques.
4. Choix bibliographique: index de quelque ouvrages de référence consultés au cours de la préparation de la Liste, indiquant ceux utilisés comme sources primaires pour le traitement adopté.

5.4. Dans la troisième partie, les noms acceptés de tous les genres sont donnés par ordre alphabétique des pays. Les noms des pays sont ceux utilisés dans le bulletin de terminologie des Nations Unies (Terminology Bulletin no. 347: 1-41, New York: United Nations, August 1995); ils sont indiqués dans l'ordre alphabétique.

6. **Conventions utilisées dans les première et deuxième parties**

6.1. *Typographie*. Les noms préférés des taxons acceptés sont en caractères **gras** et ceux des taxons provisoirement acceptés et des hybrides en ***gras-italique***. Les noms considérés comme des synonymes sont en *italique* suivis de → et du nom préféré du taxon auquel ils s'appliquent.

6.2. *Symboles et codes.* Les symboles et codes suivants distinguent les différentes catégories de noms et de taxons:

→	Le nom précèdent est considéré comme un synonyme du nom suivant;
→ ?	Le nom précèdent est peut-être un synonyme du nom suivant;
×	Genre ou espèce d'origine hybride;
♦	Nom d'un taxon inscrit à l'Annexe I de la CITES;
Ⓡ	Nom rejeté par le Code international de nomenclature botanique;
¶	"correspondant au type": lorsqu'un synonyme apparaît deux fois, ce code précède le nom original et correct si ce nom est couramment utilisé de manière erronée pour autre chose (voir le paragraphe suivant);
Ⓜ	"utilisé de manière erronée": la plante qui est ou a été connue ou commercialisée sous ce nom est celle à laquelle il est fait référence ici et non celle que l'auteur original a indiqué ou avait l'intention d'indiquer.

6.3. *Répartition géographique.* Les codes ISO à deux lettres (sauf ceux signalés ci-dessous par *, qui sont arbitraires) sont utilisés pour indiquer l'origine et la répartition géographique des taxons acceptés et provisoirement acceptés:

Antilles néerlandaises	AN	États-Unis d'Amérique	US	Paraguay	PY
Argentine	AR	Guatemala	GT	Pérou	PE
Bahamas	BS	Guyane française	GF	Petites Antilles	WI*
Belize	BZ	Guyana	GY	Porto Rico	PR
Bolivie	BO	Haïti	HT	République dominicaine	DO
Brésil	BR	Honduras	HN	Suriname	SR
Canada	CA	Iles Caïmanes	KY	Trinité-et-Tobago	TT
Chili	CL	Iles Vierges (US)	VI	Uruguay	UY
Colombie	CO	Jamaïque	JM	Venezuela	VE
Costa Rica	CR	Mexique	MX		
Cuba	CU	Nicaragua	NI	Ancien Monde	OW*
Equateur	EC	Panama	PA	Origine de la cult.	XC*
El Salvador	SV				

7. **Cactaceae soumises aux contrôles CITES**

7.1. Le commerce de tous les taxons de Cactaceae est contrôlé dans le cadre des dispositions relatives à l'Annexe II de la CITES, de même que leurs parties et produits, sauf:

a) les graines, sauf celles des cactus mexicains provenant du Mexique, et le pollen;
b) les cultures de plantules ou de tissus obtenues in vitro, en milieu solide ou liquide, et transportées en conteneurs stériles;
c) les fleurs coupées des plantes reproduites artificiellement;
d) les fruits, et leurs parties et produits, des plantes acclimatées ou reproduites artificiellement; et
e) les éléments de troncs (raquettes), et leurs parties et produits, de plantes du genre *Opuntia* sous-genre *Opuntia* acclimatées ou reproduites artificiellement;

et les spécimens reproduits artificiellement des hybrides et/ou cultivars suivants ne sont pas soumis aux dispositions de la Convention: *Hatiora* × *graeseri* (*H. gaertneri* × *H. rosea*), *Schlumbergera* ×*buckleyi* (*S. russelliana* × *S. truncata*), *S.* ×*exotica* (*S. opuntioides* × *S. truncata*), *S.* ×*reginae* (*S. orssichiana* × *S. truncata*), *S. truncata* (cultivars), *Gymnocalycium mihanovichii* (cultivars) formes sans chlorophylle, greffées sur les porte-greffes suivants: *Harrisia* 'Jusbertii', *Hylocereus trigonus* ou *Hylocereus undatus*, *Opuntia microdasys* (cultivars).

7.2. Le commerce des taxons figurant dans le tableau 1 (page 23) est soumis aux dispositions de l'Annexe I.

LISTA DE CACTACEAS CITES
Segunda Edición

Preámbulo

1. **Información general**

1.1. La primera edición de esta Lista se publicó en 1992, en respuesta a una resolución de la sexta reunión de la Conferencia de las Partes (1987), en la que se solicitaba que se preparase una referencia sobre las cactáceas. La compilación se inició en mayo de 1989 y se concluyó en marzo de 1992 como un proyecto oficial del Royal Botanic gardens, Kew, bajo los auspicios de la Organización Internacional para el Estudio de Plantas Suculentas (IUBS-IOS). Este proyecto contó con la asistencia financiera del Comité de Nomenclatura de la CITES y la Autoridad Científica CITES de Estados Unidos de América.

1.2. Los preparativos para la preparación de una nueva edición se concluyeron en septiembre de 1996, mediante un contrato de consultoría concertado entre la Junta Directiva del Royal Botanic Gardens, Kew, y la persona encargada de la compilación. El Comité de Nomenclatura de la CITES prestó nuevamente asistencia financiera.

1.3. En la nueva edición se ha seguido el plan general de la primera edición, pero se ha ampliado para incluir los nombres de las subespecies y una lista taxonómica o de sinónimos de los nombres específicos y subespecíficos preferidos, así como los autores. Para facilitar su utilización a los encargados de la CITES y los funcionarios de aduanas, a los que está destinada principalmente este lista, la inclusión por orden alfabético general se presenta inmediatamente después del preámbulo (Parte I), y las listas por países al final de la publicación (Parte III), presentados por orden alfabético. El material de referencia, inclusive los índices de los nombres genéricos, la bibliografía general de las obras consultadas y la lista de sinónimos precitada (Parte II), figura entre la Parte I y la Parte III.

2. **Programa informático**

El soporte lógico y físico utilizado para esta edición es propiedad del encargado de la compilación. La base de datos principal se compiló en dBase a partir de los datos utilizados en la primera edición y nuevas entradas proporcionadas por Kew a partir de Index Kewensis o por el compilador y los miembros del Grupo de expertos. Con arreglo al contrato de consultoría, los datos se han incluido en ALICE Transfer Format para su uso en futuras ediciones.

3. **Procedimientos para la compilación**

3.1. ***Origen***. Esta edición de la Lista de Cactáceas CITES es una continuación de la primera edición. La actualización de la nueva edición se ha realizado mediante diversos procedimientos habituales, además de una revisión taxonómica de todas las entradas realizada por el compilador y miembros del Grupo de expertos (véase el punto 4). La bibliografía (véase Parte II, sección 4) se ha ampliado para incluir varios análisis regionales y taxonómicos publicados después de la primera edición. Como ya se ha señalado, se han incluido los nombres subespecíficos y una lista de sinónimos, ofreciendo así la sinonimia completa para todos los taxa aceptados.

3.2. ***Lista de géneros y sinónimos genéricos***. Se ha actualizado a la luz de los cambios en la lista genérica adoptada por el Grupo de trabajo sobre cactáceas de la Organización Internacional para el Estudio de Plantas Suculentas (IOS). Se han insertado los tipos de todos los nombres genéricos incluidos.

3.3. ***Nombres de Cactáceas utilizados normalmente***

3.3.1. En el preámbulo de la primera edición se explicaba el alcance de la lista de binomiales (págs. 10-11). En la presente edición se han suprimido alrededor de 200 entradas (casi todas sinónimos) de las 6.022 entradas que figuran en la primera edición, por las razones siguientes:

- Nombres bajo varios nombres genéricos anticuados o no reconocidos que se habían incluido en la primera edición debido a que se consideraban como nombres aceptados en una o varias de las publicaciones consultadas durante la compilación;
- Nombres de las especies aceptadas en una o varias de las publicaciones consultadas, pero que se conocen únicamente en cultivo y cuya identidad es dudosa, normalmente debido a una tipificación inadecuada o impugnada;
- Otros nombres considerados como sinónimos, pero solamente reconocibles a nivel de género;
- Algunos nombres utilizados en el comercio que parece ser están erróneamente descritos.

3.3.2. Desde la publicación de la primera edición se han detectado varias entradas incorrectas y se han corregido.

3.3.3. Desde la primera edición se han publicado más de 500 nuevos nombres en categorías específicas, y se han incluido en la base de datos, adjuntando información sobre la distribución cuando ha sido posible.

3.3.4. La inclusión de nombres en categorías subespecíficas en esta segunda edición constituye un cambio importante en relación con la primera edición. Salvo en algunas excepciones, se incluyen los nombres previamente publicados en esta categoría, y un porcentaje de taxa anteriormente incluidos como especies (principalmente como 'provisionalmente aceptadas') se han clasificado en estas categorías.

3.3.5. Se ha alineado la sinonimia a nivel de especie/subespecie, con cambios en la lista genérica, las obras clásicas y las opiniones de los expertos. Cuando no ha sido posible resolver diferencias de opinión, el compilador ha emitido un juicio personal. Las nuevas combinaciones se han tratado como sinónimos si pertenecen a géneros que no son actualmente reconocidos por la ISO o están en pugna con las opiniones actuales sobre la sinonimia genérica.

3.3.6. Las especies incluidas en el Apéndice I de la CITES se han ajustado conforme a las modificaciones aprobadas por la Conferencia de las Partes.

3.3.7. En la primera edición se incluyeron niveles de conservación con arreglo a las antiguas categorías de la UICN en algunas de las listas por países. Para la mayoría de los países será preciso volver a evaluar dichos niveles tomando en consideración las nuevas Categorías de conservación de la UICN, y por esta razón se han suprimido completamente.

4. **Grupo internacional de expertos**

4.1. En la medida de lo posible, la Lista de Cactáceas CITES tiene por finalidad reflejar el acuerdo general en lo que concierne a la clasificación y nomenclatura que han de utilizarse. Desde 1984, un Grupo de trabajo, establecido por la Organización Internacional para el Estudio de Plantas Suculentas, se reunió de manera periódica, normalmente una vez al año, para examinar las investigaciones en curso sobre las cactáceas y su repercusión sobre la clasificación de las familias. Desde que se concluyeran los preparativos sobre esta edición, todos los cambios propuestos a las entradas de la primera edición se notificaron a los miembros del grupo de trabajo y otros interesados mediante un boletín periódico publicado por el compilador (Cactaceae Consensus Initiatives; ISSN 1365-778X), y se organizaron cursillos técnicos en Inglaterra (septiembre de 1996 y abril de 1998) y en Alemania (mayo de 1997). Las listas de sinónimos para países determinados y para cada género se sometieron a los expertos como sigue:

Argentina	R. Kiesling	E.U.A.	E.F. Anderson
Brasil	N.P. Taylor & D.C Zappi	Guyanas	B.E. Leuenberger
Chili	U. Eggli & B.E. Leuenberger	México	H.M. Hernández Macías
Cuba	A. Rodríguez	Perú	C. Ostolaza

4.2. Cabe señalar que la Lista no representa necesariamente la opinión de los expertos precitados, de la Organización Internacional para el Estudio de Plantas Suculentas, su Junta Directiva o cualquier otro colaborador. La responsabilidad del contenido de la Lista y de las opiniones expresadas sobre la taxonomía incumbe exclusivamente al compilador, el cual expresa su agradecimiento por la valiosa asistencia de los expertos y colaboradores precitados.

5. **¿Cómo emplear esta Lista?**

5.1. La finalidad consiste en que esta Lista se utilice como referencia rápida para controlar los nombres aceptados, los sinónimos y la distribución conocida de los taxa aceptados. Se divide en tres partes:

5.2. La ***Parte I*** es una enumeración por orden alfabético de todos los nombres de las especies y subespecies incluidas en la Lista y puede utilizarse como un índice general. Los nombres preferidos de los taxa aceptados se imprimen en **negritas** o en ***negritas-cursiva*** (= provisionalmente aceptados). Los sinónimos se imprimen en *cursiva*, seguido de → y el nombre preferido del taxón para los que supuestamente se aplican. Los símbolos (véase el párrafo 6 supra) se utilizan para denotar diversas categorías de nombres, nombres aceptados y sinónimos de los taxa incluidos en el Apéndice I de la CITES.

5.3. La ***Parte II*** contiene referencias detalladas bajo los apartados siguientes.

1. Lista por orden alfabético de los géneros de cactáceas y sus principales sinónimos – una lista por orden alfabético de los géneros anotados, con los autores, lugares de publicación, tipos de nomenclatura, números de taxa y distribución geográfica.
2. Cuadro sinóptico de los géneros aceptados con los números de los taxa aceptados.
3. Lista de taxa – presentada por orden alfabético, pero con el nombre preferido de cada uno de los taxa aceptados actualmente, seguido de los sinónimos a que se hace alusión en la Parte I y autoridades botánicas.
4. Bibliografía seleccionada – un índice de algunas obras de referencia consultadas en la preparación de la Lista, indicando las utilizadas como fuentes principales o "puntos de referencia", para la denominación adoptada.

5.4. En la ***Parte III*** figuran los nombres aceptados de todos los géneros por orden alfabético por país. Los países se presentan también por orden alfabético, tomando en consideración la publicación *Nombres de países* de las Naciones Unidas (Terminology Bulletin no. 347: 1-41, Nueva York: Naciones Unidas, agosto de 1995).

6. **Sistema de presentación utilizado en las Partes I–II**

6.1. *Tipografía.* Los nombres preferidos de los taxa aceptados figuran en negrita y los de los taxa aceptados provisionalmente (híbridos incluidos) en negritas-cursiva. Los nombres considerados como sinónimos aparecen en cursiva, seguido de → y el nombre preferido del taxón para los que supuestamente se aplican.

6.2. *Símbolos y códigos.* Se utilizan los siguientes para denotar distintas categorías de nombres y taxa:

→	El nombre anterior se considera como un sinónimo del nombre siguiente;
→ ?	El nombre anterior puede ser un sinónimo del nombre siguiente;
×	Género o especie de un híbrido;
♦	Nombre del taxón incluido en el Apéndice I de la CITES;
Ⓘ	Nombre declarado ilegal por el Código Internacional de Nomenclatura Botánica;
¶	"como tipo"; cuando un sinónimo se incluye dos veces, este precede al original y al uso apropiado del nombre, pero el nombre se utiliza en general erróneamente – véase la entrada siguiente;
Ⓜ	"utilizada erróneamente"; la planta que es, o ha sido, conocida o objeto de comercio bajo este nombre es a la que se hace referencia, no la indicada o pretendida por el autor original.

6.3. *Distribución geográfica.* Código ISO de dos letras (salvo los indicados con * debajo, que son arbitrarias) se utilizan para indicar los países de origen y distribución de los taxa definitiva o provisionalmente aceptados:

Antillas Holandesas	AN	El Salvador	SV	Panamá	PA
Antillas inferiores	WI*	Estados Unidos de América	US	Paraguay	PY
Argentina	AR			Perú	PE
Bahamas	BS	Guatemala	GT	Puerto Rico	PR
Belice	BZ	Guyana Francesa	GF	República Dominicana	DO
Bolivia	BO	Guyana	GY	Suriname	SR
Brasil	BR	Haití	HT	Trinidad y Tabago	TT
Canadá	CA	Honduras	HN	Uruguay	UY
Chile	CL	Islas Caimán	KY	Venezuela	VE
Colombia	CO	Islas Vírgenes (US)	VI		
Costa Rica	CR	Jamaica	JM	Viejo Mundo	OW*
Cuba	CU	México	MX		
Ecuador	EC	Nicaragua	NI	Origen: cultivo	XC*

7. **Cactáceas amparadas por la CITES**

7.1. El comercio de todos los taxa de Cactaceae está reglamentado por las disposiciones del Apéndice II de la CITES inclusive todas las partes y derivados, excepto:

a) las semillas, excepto las de las cactáceas mexicanas originarias de México, y el polen;
b) los cultivos de plántulas o de tejidos obtenidos in vitro, en medios sólidos o líquidos, que se transportan en envases estériles;
c) las flores cortadas de plantas reproducidas artificialmente;
d) los frutos, y sus partes y derivados, de plantas aclimatadas o reproducidas artificialmente; y
e) los elementos del tallo (ramificaciones), y sus partes y derivados, de plantas del género *Opuntia* subgénero *Opuntia* aclimatadas o reproducidas artificialmente;

y los especímenes reproducidos artificialmente de los siguientes híbridos y/o cultivares no están sujetos a las disposiciones de la Convención: *Hatiora* ×*graeseri* (*H. gaertneri* × *H. rosea*), *Schlumbergera* ×*buckleyi* (*S. russelliana* × *S. truncata*), *S.* ×*exotica* (*S. opuntioides* × *S. truncata*), *S.* ×*reginae* (*S. orssichiana* × *S. truncata*), *S. truncata* (cultivares), *Gymnocalycium mihanovichii* (cultivares) formas que carecen de clorofila, injertadas en los siguientes patrones: *Harrisia* 'Jusbertii', *Hylocereus trigonus* o *Hylocereus undatus*, *Opuntia microdasys* (cultivares).

7.2. El comercio de los taxa que figuran en el Cuadro 1 está reglamentado por las disposiciones del Apéndice I de la CITES.

Table / Tableau / Cuadro 1. Taxa controlled by the provisions of CITES Appendix I under their currently accepted names or any of their synonyms. Taxons soumis aux dispositions de l'Annexe I – noms actuellement acceptés ou synonymes. Taxa reglamentados por las disposiciones del Apéndice I de la CITES, según sus nombres aceptados normalmente o cualquiera de sus sinónimos.

Ariocarpus spp. (syn. *Neogomesia, Roseocactus*)
Astrophytum asterias (syn. *Echinocactus asterias*)
Aztekium ritteri
Coryphantha werdermannii (syn. *Mammillaria werdermannii*; *Coryphantha densispina*)
Discocactus spp.
Disocactus macdougallii (syn. *Lobeira macdougallii, Nopalxochia macdougallii*)
Echinocereus ferreirianus ssp. **lindsayi** (syn. *E. lindsayi*)
Echinocereus schmollii (syn. *Cereus schmollii, Wilcoxia schmollii*)
Escobaria minima (syn. *Coryphantha nellieae, Escobaria nellieae*)
Escobaria sneedii (syn. *Coryphantha sneedii, Escobaria sneedii* ssp. *leei*)
Mammillaria pectinifera (syn. *Solisia pectinata*)
Mammillaria solisioides
Melocactus conoideus
Melocactus deinacanthus
Melocactus glaucescens
Melocactus paucispinus
Obregonia denegrii
Pachycereus militaris (syn. *Backebergia militaris, Cephalocereus militaris, Mitrocereus militaris, Pachycereus chrysomallus*)
Pediocactus bradyi (syn. *Pediocactus bradyi* ssp. *despainii, P. bradyi* ssp. *winkleri, P. despainii, P. winkleri*)
Pediocactus knowltonii
Pediocactus paradinei
Pediocactus peeblesianus (syn. *P. peeblesianus* var. *fickeisenii, Navajoa peeblesiana, Toumeya peeblesiana, Utahia peeblesiana*)
Pediocactus sileri (syn. *Echinocactus sileri, Utahia sileri*)
Pelecyphora spp. (syn. *Encephalocarpus*)
Sclerocactus brevihamatus ssp. **tobuschii** (syn. *Ancistrocactus tobuschii, Ferocactus tobuschii*)
Sclerocactus erectocentrus (syn. *Echinomastus erectocentrus, Neolloydia erectocentra, Echinomastus acunensis, E. krausei*)
Sclerocactus glaucus (syn. *Ferocactus glaucus, S. brevispinus, S. wetlandicus, S. wetlandicus* ssp. *ilseae*)
Sclerocactus mariposensis (syn. *Echinocactus mariposensis, Echinomastus mariposensis, Neolloydia mariposensis*)
Sclerocactus mesae-verdae (syn. *Coloradoa mesae-verdae, Echinocactus mesae-verdae, Ferocactus mesae-verdae, Pediocactus mesae-verdae*)
Sclerocactus papyracanthus (syn. *Echinocactus papyracanthus, Mammillaria papyracantha, Pediocactus papyracanthus, Toumeya papyracantha*)
Sclerocactus pubispinus (syn. *Echinocactus pubispinus, Ferocactus pubispinus*)
Sclerocactus wrightiae (syn. *Pediocactus wrightiae*)
Strombocactus spp.
Turbinicarpus spp. (syn. *Gymnocactus, Normanbokea, Rapicactus*)
Uebelmannia spp.

Part I. Alphabetic List of Cactaceae Names in Current Usage

®*Acanthocalycium andreaeanum* → Eriosyce andreaeana
Acanthocalycium aurantiacum → Echinopsis glaucina
Acanthocalycium brevispinum → Echinopsis thionantha
Acanthocalycium catamarcense → Echinopsis thionantha
Acanthocalycium chionanthum → Echinopsis thionantha
Acanthocalycium ferrarii
Acanthocalycium glaucum → Echinopsis glaucina
®*Acanthocalycium griseum* → Echinopsis thionantha
Acanthocalycium klimpelianum
Acanthocalycium peitscherianum → Acanthocalycium klimpelianum
Acanthocalycium spiniflorum
Acanthocalycium thionanthum → Echinopsis thionantha
®*Acanthocalycium variiflorum* → Acanthocalycium ferrarii
Acanthocalycium violaceum → Acanthocalycium spiniflorum

Acanthocereus albicaulis → Cereus albicaulis
Acanthocereus baxaniensis
Acanthocereus brasiliensis → Pseudoacanthocereus brasiliensis
Acanthocereus chiapensis → Peniocereus sp.?
Acanthocereus colombianus
Acanthocereus floridanus → Acanthocereus tetragonus
®*Acanthocereus griseus* → Peniocereus sp.?
Acanthocereus horridus
Acanthocereus maculatus → Peniocereus maculatus
Acanthocereus occidentalis
Acanthocereus pentagonus → Acanthocereus tetragonus
Acanthocereus sicariguensis → Pseudoacanthocereus sicariguensis
Acanthocereus subinermis
Acanthocereus tetragonus
Acanthocereus undulosus → Dendrocereus undulosus

Acantholobivia incuiensis → Echinopsis tegeleriana
Acantholobivia tegeleriana → Echinopsis tegeleriana

®*Acanthorhipsalis brevispina* → Lepismium brevispinum
Acanthorhipsalis crenata → Lepismium crenatum
Acanthorhipsalis houlletiana → Lepismium houlletianum
Acanthorhipsalis incachacana → Lepismium incachacanum
Acanthorhipsalis incahuasina → Lepismium monacanthum
Acanthorhipsalis micrantha → Lepismium micranthum
Acanthorhipsalis monacantha → Lepismium monacanthum
Acanthorhipsalis paranganiensis → Lepismium paranganiense
Acanthorhipsalis samaipatana → Lepismium monacanthum

Acharagma aguirreana → Escobaria aguirreana

Ancistrocactus brevihamatus → Sclerocactus brevihamatus ssp. brevihamatus
Ancistrocactus crassihamatus → Sclerocactus uncinatus ssp. crassihamatus
Ancistrocactus megarhizus → Sclerocactus scheeri
Ancistrocactus scheeri → Sclerocactus scheeri
Ancistrocactus tobuschii♦ → Sclerocactus brevihamatus ssp. tobuschii♦
Ancistrocactus uncinatus → Sclerocactus uncinatus ssp. uncinatus

Anisocereus foetidus → Pachycereus gaumeri
Anisocereus gaumeri → Pachycereus gaumeri
Anisocereus lepidanthus → Pachycereus lepidanthus

Aporocactus conzattii → Disocactus martianus
Aporocactus flagelliformis → Disocactus flagelliformis
Aporocactus flagriformis → Disocactus flagelliformis
Aporocactus leptophis → Disocactus flagelliformis
Aporocactus martianus → Disocactus martianus

Arequipa australis → Oreocereus hempelianus
Arequipa erectocylindrica → Oreocereus hempelianus
Arequipa hempeliana → Oreocereus hempelianus
Arequipa leucotricha → Oreocereus leucotrichus
Arequipa mirabilis → Cleistocana mirabilis
Arequipa myriacantha → Matucana haynei ssp. myriacantha
Arequipa rettigii → Oreocereus hempelianus
Arequipa soehrensii → Oreocereus hempelianus
Arequipa spinosissima → Oreocereus hempelianus
Arequipa weingartiana → Oreocereus hempelianus

Ariocarpus agavoides♦
Ariocarpus bravoanus♦
Ariocarpus bravoanus ssp. **bravoanus♦**
Ariocarpus bravoanus ssp. **hintonii♦**
Ariocarpus confusus♦ → Ariocarpus retusus ssp. retusus♦
Ariocarpus elongatus♦ → Ariocarpus retusus♦
Ariocarpus fissuratus♦
Ariocarpus furfuraceus♦ → Ariocarpus retusus ssp. retusus♦
Ariocarpus kotschoubeyanus♦
Ariocarpus kotschoubeyanus ssp. *albiflorus♦* → Ariocarpus kotschoubeyanus♦
Ariocarpus retusus♦
Ariocarpus retusus ssp. **retusus♦**
Ariocarpus retusus ssp. *scapharostroides♦* → Ariocarpus retusus♦
Ariocarpus retusus ssp. **trigonus♦**
®*Ariocarpus scapharostrus♦* → Ariocarpus scaphirostris♦
Ariocarpus scaphirostris♦
Ariocarpus trigonus♦ → Ariocarpus retusus ssp. trigonus♦

Armatocereus arboreus → Armatocereus matucanensis
Armatocereus arduus
Armatocereus balsasensis → Armatocereus rauhii ssp. balsasensis
Armatocereus brevispinus
Armatocereus cartwrightianus
Armatocereus churinensis → Armatocereus matucanensis
Armatocereus ghiesbreghtii → Armatocereus sp.
Armatocereus godingianus
Armatocereus humilis
Armatocereus laetus
Armatocereus mataranus
Armatocereus mataranus ssp. **ancashensis**
Armatocereus mataranus ssp. **mataranus**
Armatocereus matucanensis
Armatocereus oligogonus
Armatocereus procerus
Armatocereus rauhii
Armatocereus rauhii ssp. **balsasensis**
Armatocereus rauhii ssp. **rauhii**
Armatocereus riomajensis
Armatocereus rupicola

Arrojadoa ×albiflora (A. dinae ssp. dinae × A. rhodantha)
Arrojadoa aureispina → Arrojadoa rhodantha
Arrojadoa bahiensis

Arrojadoa beateae → Arrojadoa dinae ssp. dinae
Arrojadoa canudosensis → Arrojadoa rhodantha
Arrojadoa dinae
Arrojadoa dinae ssp. **dinae**
Arrojadoa dinae ssp. **eriocaulis**
Arrojadoa dinae ssp. *nana* → Arrojadoa dinae ssp. dinae
Arrojadoa eriocaulis → Arrojadoa dinae ssp. eriocaulis
Arrojadoa eriocaulis ssp. *albicoronata* → Arrojadoa dinae ssp. eriocaulis
Arrojadoa horstiana → Arrojadoa rhodantha
Arrojadoa multiflora → Arrojadoa dinae ssp. dinae
Arrojadoa penicillata
Arrojadoa polyantha → Micranthocereus polyanthus
Arrojadoa rhodantha
Arrojadoa rhodantha ssp. *aureispina* → Arrojadoa rhodantha
Arrojadoa rhodantha ssp. *canudosensis* → Arrojadoa rhodantha
Arrojadoa rhodantha ssp. *reflexa* → Arrojadoa rhodantha
Arrojadoa theunisseniana → Arrojadoa rhodantha

Arthrocereus campos-portoi → Arthrocereus glaziovii
©*Arthrocereus damazioi* → Arthrocereus glaziovii
Arthrocereus glaziovii
Arthrocereus itabiriticola → Arthrocereus glaziovii
Arthrocereus melanurus
Arthrocereus melanurus ssp. *estevesii* → Arthrocereus melanurus ssp. melanurus
Arthrocereus melanurus ssp. **magnus**
Arthrocereus melanurus ssp. **melanurus**
Arthrocereus melanurus ssp. *mello-barretoi* → Arthrocereus melanurus ssp. melanurus
Arthrocereus melanurus ssp. **odorus**
Arthrocereus mello-barretoi → Arthrocereus melanurus ssp. melanurus
Arthrocereus microsphaericus → Schlumbergera microsphaerica
Arthrocereus odorus → Arthrocereus melanurus ssp. odorus
Arthrocereus rondonianus
Arthrocereus rowleyanus → Pygmaeocereus bylesianus
Arthrocereus spinosissimus

Astrophytum asterias♦
Astrophytum capricorne
Astrophytum coahuilense → Astrophytum myriostigma
Astrophytum columnare → Astrophytum myriostigma
Astrophytum myriostigma
Astrophytum myriostigma ssp. *potosinum* → Astrophytum myriostigma
Astrophytum myriostigma ssp. *tulense* → Astrophytum myriostigma
Astrophytum ornatum
Astrophytum senile → Astrophytum capricorne
Astrophytum tulense → Astrophytum myriostigma

Austrocactus bertinii
Austrocactus coxii
Austrocactus dusenii → Austrocactus bertinii
Austrocactus gracilis → Austrocactus bertinii
Austrocactus hibernus → Austrocactus philippii
Austrocactus patagonicus
Austrocactus philippii
Austrocactus spiniflorus

Austrocephalocereus albicephalus → Micranthocereus albicephalus
Austrocephalocereus dolichospermaticus → Micranthocereus dolichospermaticus
Austrocephalocereus dybowskii → Espostoopsis dybowskii
Austrocephalocereus estevesii → Micranthocereus estevesii
Austrocephalocereus estevesii ssp. *grandiflorus* → Micranthocereus estevesii

Austrocephalocereus estevesii ssp. *insigniflorus* → Micranthocereus estevesii
Austrocephalocereus fluminensis → Coleocephalocereus fluminensis
Austrocephalocereus lehmannianus → Micranthocereus purpureus
Austrocephalocereus purpureus → Micranthocereus purpureus
Austrocephalocereus salvadorensis → Pilosocereus catingicola ssp. salvadorensis

Austrocylindropuntia chuquisacana → Opuntia vestita
Austrocylindropuntia clavarioides → Opuntia clavarioides
Austrocylindropuntia colubrina → Opuntia colubrina
Austrocylindropuntia cylindrica → Opuntia cylindrica
Austrocylindropuntia exaltata → Opuntia subulata
Austrocylindropuntia floccosa → Opuntia floccosa
Austrocylindropuntia haematacantha → Opuntia verschaffeltii
Austrocylindropuntia humahuacana → Opuntia shaferi
Austrocylindropuntia inarmata → Opuntia verschaffeltii
Austrocylindropuntia intermedia → Opuntia cylindrica
Austrocylindropuntia ipatiana → Opuntia salmiana
®*Austrocylindropuntia lagopus* → Opuntia lagopus
Austrocylindropuntia lauliacoana → Opuntia floccosa
Austrocylindropuntia machacana → Opuntia floccosa
Austrocylindropuntia malyana → Opuntia lagopus
Austrocylindropuntia miquelii → Opuntia miquelii
Austrocylindropuntia pachypus → Opuntia pachypus
Austrocylindropuntia salmiana → Opuntia salmiana
Austrocylindropuntia schickendantzii → Opuntia schickendantzii
Austrocylindropuntia shaferi → Opuntia shaferi
Austrocylindropuntia steiniana → Opuntia shaferi
Austrocylindropuntia subulata → Opuntia subulata
Austrocylindropuntia tephrocactoides → Opuntia floccosa
Austrocylindropuntia teres → Opuntia vestita
Austrocylindropuntia verschaffeltii → Opuntia verschaffeltii
Austrocylindropuntia vestita → Opuntia vestita
Austrocylindropuntia weingartiana → Opuntia shaferi

Aylostera albiflora → Rebutia albiflora
Aylostera albipilosa → Rebutia fiebrigii
Aylostera deminuta → Rebutia deminuta
Aylostera fiebrigii → Rebutia fiebrigii
Aylostera fulviseta → Rebutia fulviseta
Aylostera heliosa → Rebutia heliosa
Aylostera jujuyana → Rebutia fiebrigii
Aylostera kupperiana → ?Rebutia pseudodeminuta
Aylostera muscula → Rebutia fiebrigii
Aylostera narvaecensis → Rebutia narvaecensis
Aylostera padcayensis → Rebutia padcayensis
Aylostera pseudodeminuta → Rebutia pseudodeminuta
Aylostera pseudominuscula → Rebutia deminuta
Aylostera pulchella → Rebutia fiebrigii
Aylostera pulvinosa → Rebutia pulvinosa
Aylostera rubiginosa → Rebutia spegazziniana
Aylostera spegazziniana → Rebutia spegazziniana
Aylostera spinosissima → Rebutia spinosissima
Aylostera steinmannii → Rebutia steinmannii
Aylostera tuberosa → Rebutia spegazziniana
Aylostera zavaletae → Rebutia canigueralii

Aztekium hintonii
Aztekium ritteri♦

Azureocereus deflexispinus → Echinopsis knuthiana
Azureocereus hertlingianus → Browningia hertlingiana
Azureocereus viridis → Browningia viridis

Backebergia militaris♦ → Pachycereus militaris♦

Bartschella schumannii → Mammillaria schumannii

Bergerocactus emoryi

Blossfeldia atroviridis → Blossfeldia liliputana
Blossfeldia campaniflora → Blossfeldia liliputana
Blossfeldia fechseri → Blossfeldia liliputana
Blossfeldia liliputana
Blossfeldia minima → Blossfeldia liliputana
Blossfeldia pedicellata → Blossfeldia liliputana

Bolivicereus brevicaulis → Cleistocactus samaipatanus
Bolivicereus croceus → Cleistocactus samaipatanus
Bolivicereus pisacensis → Corryocactus erectus
Bolivicereus rufus → Cleistocactus samaipatanus
Bolivicereus samaipatanus → Cleistocactus samaipatanus
Bolivicereus serpens → Cleistocactus serpens
Bolivicereus soukupii → Corryocactus erectus
Bolivicereus tenuiserpens → Cleistocactus tenuiserpens

Borzicactus acanthurus → Cleistocactus acanthurus ssp. acanthurus
Borzicactus aequatorialis → Cleistocactus sepium
Borzicactus aurantiacus → Matucana aurantiaca ssp. aurantiaca
Borzicactus aurivillus → Cleistocactus icosagonus
Borzicactus cajamarcensis → Cleistocactus fieldianus ssp. fieldianus
Borzicactus calocephalus → Matucana haynei ssp. myriacantha
Borzicactus calvescens → Matucana aurantiaca ssp. aurantiaca
Borzicactus calviflorus → Cleistocactus fieldianus ssp. fieldianus
Borzicactus celsianus → Oreocereus celsianus
Borzicactus decumbens → Haageocereus decumbens
Borzicactus doelzianus → Oreocereus doelzianus
Borzicactus fieldianus → Cleistocactus fieldianus ssp. fieldianus
Borzicactus formosus → Matucana formosa
Borzicactus fossulatus → Oreocereus celsianus
Borzicactus fruticosus → Matucana fruticosa
Borzicactus haynei → Matucana haynei ssp. haynei
Borzicactus hendriksenianus → Oreocereus leucotrichus
Borzicactus huagalensis → Matucana huagalensis
Borzicactus icosagonus → Cleistocactus icosagonus
Borzicactus intertextus → Matucana intertexta
Borzicactus krahnii → Matucana krahnii
Borzicactus leucotrichus → Oreocereus leucotrichus
Borzicactus madisoniorum → Matucana madisoniorum
Borzicactus mirabilis → Cleistocana mirabilis
Borzicactus morleyanus → Cleistocactus sepium
Borzicactus myriacanthus → Matucana haynei ssp. myriacantha
Borzicactus neoroezlii → Cleistocactus neoroezlii
Borzicactus oreodoxus → Matucana oreodoxa
Borzicactus paucicostatus → Matucana paucicostata
Borzicactus pisacensis → Corryocactus erectus
Borzicactus plagiostoma → Cleistocactus plagiostoma
Borzicactus pseudothelegonus → Cleistocactus serpens
Borzicactus purpureus → Cleistocactus plagiostoma

Borzicactus ritteri → Matucana ritteri
Borzicactus samaipatanus → Cleistocactus samaipatanus
Borzicactus samnensis → Cleistocactus fieldianus ssp. samnensis
Borzicactus sepium → Cleistocactus sepium
Borzicactus serpens → Cleistocactus serpens
Borzicactus sextonianus → Cleistocactus sextonianus
Borzicactus soukupii → Corryocactus erectus
Borzicactus sulcifer → Cleistocactus serpens
Borzicactus tenuiserpens → Cleistocactus tenuiserpens
Borzicactus trollii → Oreocereus trollii
Borzicactus tuberculatus → Matucana tuberculata
Borzicactus variabilis → Matucana haynei ssp. haynei
Borzicactus ventimigliae → Cleistocactus sepium
Borzicactus weberbaueri → Matucana weberbaueri
Borzicactus websterianus → Cleistocactus sepium

Brachycalycium tilcarense → Gymnocalycium saglionis ssp. tilcarense

Brachycereus nesioticus

Brasilicactus elachisanthus → Parodia haselbergii ssp. graessneri
Brasilicactus graessneri → Parodia haselbergii ssp. graessneri
Brasilicactus haselbergii → Parodia haselbergii ssp. haselbergii

Brasilicereus breviflorus → Brasilicereus phaeacanthus
Brasilicereus markgrafii
Brasilicereus phaeacanthus
Brasilicereus phaeacanthus ssp. *breviflorus* → Brasilicereus phaeacanthus

Brasiliopuntia bahiensis
Brasiliopuntia brasiliensis → Opuntia brasiliensis
Brasiliopuntia neoargentina → Opuntia brasiliensis
Brasiliopuntia schulzii → Opuntia brasiliensis
Brasiliopuntia subacarpa

Brasiliparodia alacriportana → Parodia alacriportana ssp. alacriportana
Brasiliparodia brevihamata → Parodia alacriportana ssp. brevihamata
Brasiliparodia buenekeri → Parodia alacriportana ssp. buenekeri
Brasiliparodia catarinensis → Parodia alacriportana ssp. buenekeri
Brasiliparodia rechensis → Parodia rechensis

Browningia albiceps
Browningia altissima
Browningia amstutziae
Browningia caineana
Browningia candelaris
Browningia chlorocarpa
Browningia columnaris
Browningia hertlingiana
Browningia icaensis → Browningia candelaris
Browningia microsperma
Browningia pilleifera
Browningia riosaniensis → Rauhocereus riosaniensis ssp. riosaniensis
Browningia viridis

Buiningia aurea → Coleocephalocereus aureus
Buiningia brevicylindrica → Coleocephalocereus aureus
Buiningia purpurea → Coleocephalocereus purpureus

Calymmanthium fertile → Calymmanthium substerile
Calymmanthium substerile

Carnegiea euphorbioides → Neobuxbaumia euphorbioides
Carnegiea fulviceps → Pachycereus fulviceps
Carnegiea gigantea
Carnegiea laui → Neobuxbaumia laui
Carnegiea macrocephala → Neobuxbaumia macrocephala
Carnegiea mezcalaensis → Neobuxbaumia mezcalaensis
Carnegiea nova → Neobuxbaumia mezcalaensis
Carnegiea polylopha → Neobuxbaumia polylopha
Carnegiea scoparia → Neobuxbaumia scoparia
Carnegiea squamulosa → Neobuxbaumia squamulosa
Carnegiea tetetzo → Neobuxbaumia tetetzo

Castellanosia caineana → Browningia caineana

Cephalocereus alensis → Pilosocereus alensis
Cephalocereus apicicephalium
Cephalocereus arrabidae → Pilosocereus arrabidae
Cephalocereus bahamensis → Pilosocereus polygonus
Cephalocereus barbadensis → Pilosocereus royenii
Cephalocereus brasiliensis → Pilosocereus brasiliensis
Cephalocereus brooksianus → Pilosocereus royenii
Cephalocereus catingicola → Pilosocereus catingicola ssp. catingicola
Cephalocereus chrysacanthus → Pilosocereus chrysacanthus
Cephalocereus collinsii → ?Pilosocereus purpusii
Cephalocereus colombianus → Pilosocereus lanuginosus
Cephalocereus columna-trajani
Cephalocereus cometes → ?Pilosocereus leucocephalus
Cephalocereus dybowskii → Espostoopsis dybowskii
Cephalocereus fluminensis → Coleocephalocereus fluminensis
Cephalocereus fulviceps → Pachycereus fulviceps
Cephalocereus gaumeri → Pilosocereus royenii
Cephalocereus gounellei → Pilosocereus gounellei ssp. gounellei
Cephalocereus guentheri → Espostoa guentheri
Cephalocereus guerreronis → Pilosocereus alensis
Cephalocereus hoppenstedtii → Cephalocereus apicicephalium
Cephalocereus keyensis → Pilosocereus polygonus
Cephalocereus lanuginosus → Pilosocereus lanuginosus
Cephalocereus leucocephalus → Pilosocereus leucocephalus
Cephalocereus leucostele → Stephanocereus leucostele
Cephalocereus macrocephalus → Neobuxbaumia macrocephala
Cephalocereus maxonii → Pilosocereus leucocephalus
Cephalocereus militaris♦ → Pachycereus militaris♦
Cephalocereus millspaughii → Pilosocereus royenii
Cephalocereus monoclonos → Pilosocereus royenii
Cephalocereus moritzianus → Pilosocereus lanuginosus
Cephalocereus nizandensis
®*Cephalocereus nobilis* → Pilosocereus royenii
Cephalocereus palmeri → Pilosocereus leucocephalus
Cephalocereus pentaedrophorus → Pilosocereus pentaedrophorus ssp. pentaedrophorus
Cephalocereus phaeacanthus → Brasilicereus phaeacanthus
Cephalocereus piauhyensis → Pilosocereus piauhyensis
Cephalocereus polygonus → Pilosocereus polygonus
Cephalocereus polylophus → Neobuxbaumia polylopha
Cephalocereus purpusii → Pilosocereus purpusii
Cephalocereus quadricentralis → Pilosocereus quadricentralis
Cephalocereus robinii → Pilosocereus polygonus
Cephalocereus robustus → Pilosocereus ulei
Cephalocereus royenii → Pilosocereus royenii
Cephalocereus russel(l)ianus → Cereus fricii

Cephalocereus sartorianus → Pilosocereus leucocephalus
Cephalocereus scoparius → Neobuxbaumia scoparia
Cephalocereus senilis
Cephalocereus smithianus → Praecereus euchlorus ssp. smithianus
Cephalocereus swartzii → Pilosocereus royenii
Cephalocereus totolapensis
Cephalocereus tweedyanus → Pilosocereus lanuginosus
Cephalocereus zehntneri → Pilosocereus gounellei ssp. zehntneri

Cephalocleistocactus chrysocephalus
Cephalocleistocactus pallidus → Cleistocactus palhuayensis
Cephalocleistocactus ritteri → Cleistocactus ritteri
Cephalocleistocactus schattatianus → Cleistocactus varispinus

Cereus adelmarii
Cereus aethiops
Cereus alacriportanus → Cereus sp.
Cereus albicaulis
Cereus alticostatus → Praecereus euchlorus ssp. euchlorus
Cereus amazonicus → Praecereus euchlorus ssp. amazonicus
Cereus anisitsii → Cereus spegazzinii
Cereus apoloensis → Praecereus euchlorus ssp. amazonicus
Cereus argentinensis
Cereus atroviridis → Cereus repandus
Cereus azureus → Cereus aethiops
Cereus ballivianii → Praecereus euchlorus ssp. amazonicus
Cereus baumannii → Cleistocactus baumannii ssp. baumannii
Cereus bertinii → Austrocactus bertinii
Cereus bicolor
Cereus boeckmannii → Selenicereus boeckmannii
Cereus braunii
Cereus calcirupicola → Cereus jamacaru ssp. calcirupicola
Cereus calcirupicola ssp. *cabralensis* → Cereus jamacaru ssp. calcirupicola
Cereus calcirupicola ssp. *cipoensis* → Cereus jamacaru ssp. calcirupicola
Cereus campinensis → Praecereus euchlorus ssp. euchlorus
Cereus campinensis ssp. *piedadensis* → Praecereus euchlorus ssp. euchlorus
Cereus chacoanus → Stetsonia coryne
Cereus chalybaeus → Cereus aethiops
Cereus cochabambensis
Cereus comarapanus
Cereus crassisepalus → Cipocereus crassisepalus
Cereus dayami → Cereus stenogonus
Cereus diffusus → Praecereus euchlorus ssp. diffusus
Cereus emoryi → Bergerocactus emoryi
Cereus eriophorus → Harrisia eriophora
Cereus euchlorus → Praecereus euchlorus ssp. euchlorus
Cereus euchlorus ssp. *alticostatus* → Praecereus euchlorus ssp. euchlorus
Cereus euchlorus ssp. *leucanthus* → Praecereus euchlorus ssp. euchlorus
Cereus euchlorus ssp. *rhodoleucanthus* → Praecereus saxicola
¶*Cereus extensus* → ?Selenicereus setaceus
⊗*Cereus extensus* → ?Hylocereus lemairei
Cereus fernambucensis
Cereus fernambucensis ssp. **fernambucensis**
Cereus fernambucensis ssp. **sericifer**
Cereus forbesii → Cereus validus
Cereus fricii
Cereus giganteus → Carnegiea gigantea
Cereus goiasensis → Cereus jamacaru ssp. goiasensis
Cereus gracilis → Harrisia gracilis
Cereus grandiflorus → Selenicereus grandiflorus
Cereus greggii → Peniocereus greggii

Cereus grenadensis → Cereus repandus
Cereus haageanus
Cereus hankeanus
Cereus hassleri → Selenicereus setaceus
Cereus hexagonus
Cereus hildmannianus
Cereus hildmannianus ssp. **hildmannianus**
Cereus hildmannianus ssp. **uruguayanus**
Cereus hildmannianus ssp. *xanthocarpus* → Cereus hildmannianus
Cereus horrispinus
Cereus huilunchu
Cereus insularis
Cereus jamacaru
Cereus jamacaru ssp. **calcirupicola**
Cereus jamacaru ssp. ***goiasensis***
Cereus jamacaru ssp. **jamacaru**
Cereus kroenleinii
Cereus lamprospermus
Cereus lamprospermus ssp. *colosseus* → ?C. lamprospermus
Cereus lanosus
Cereus lauterbachii → Praecereus euchlorus ssp. euchlorus
Cereus lindbergianus → Selenicereus setaceus
Cereus lindmanii → Selenicereus setaceus
Cereus longiflorus → Cereus hexagonus
Cereus margaritensis → Cereus repandus
Cereus markgrafii → Brasilicereus markgrafii
Cereus martinii → Harrisia martinii
Cereus milesimus → Cereus hildmannianus ssp. hildmannianus
Cereus mirabella
Cereus mortensenii
Cereus neonesioticus → Cereus hildmannianus
ⓤ*Cereus neotetragonus* → ?Cereus fernambucensis ssp. fernambucensis
Cereus nudiflorus → Dendrocereus nudiflorus
Cereus orcuttii → Pacherocactus orcuttii
Cereus pachyrrhizus
Cereus pentagonus → Acanthocereus tetragonus
Cereus perlucens → Cereus hexagonus
ⓤ*Cereus pernambucensis* → Cereus fernambucensis ssp. fernambucensis
Cereus peruvianus → Cereus repandus
ⓤ*Cereus peruvianus* → Cereus hildmannianus
Cereus phatnospermus
Cereus phatnospermus ssp. *adelmarii* → Cereus adelmarii
Cereus phatnospermus ssp. *kroenleinii* → Cereus kroenleinii
Cereus poselgeri → Echinocereus poselgeri
Cereus pteranthus → Selenicereus pteranthus
Cereus repandus
Cereus rhodoleucanthus → Praecereus saxicola
Cereus ridleii
Cereus ritteri → Praecereus saxicola
Cereus ritteri ssp. *parapetiensis* → Praecereus saxicola
Cereus robinii → Pilosocereus polygonus
Cereus roseiflorus
ⓤ*Cereus russelianus* → Cereus fricii
Cereus saddianus
Cereus saxicola → Praecereus saxicola
Cereus schottii → Pachycereus schottii
Cereus sericifer → Cereus fernambucensis ssp. sericifer
Cereus smithianus → Praecereus euchlorus ssp. smithianus
Cereus spachianus → Echinopsis spachiana
Cereus spegazzinii
Cereus spinulosus → Selenicereus spinulosus

Cereus stenogonus
Cereus striatus → Peniocereus striatus
Cereus tacaquirensis → Echinopsis tacaquirensis
Cereus tacuaralensis
⊗*Cereus tetragonus* → Cereus fernambucensis ssp. fernambucensis
Cereus thurberi → Stenocereus thurberi
Cereus tortuosus → Harrisia tortuosa
Cereus trigonodendron
Cereus undatus → Hylocereus undatus
Cereus uruguayanus → Cereus hildmannianus ssp. uruguayanus
Cereus validus
Cereus vargasianus
Cereus variabilis → Acanthocereus tetragonus
Cereus xanthocarpus → Cereus hildmannianus
Cereus yunckeri → Stenocereus yunckeri

Chamaecereus silvestrii → Echinopsis chamaecereus

Chiapasia nelsonii → Disocactus nelsonii

Chileorebutia duripulpa → Eriosyce napina ssp. lembckei
Chileorebutia esmeraldana → Eriosyce esmeraldana
Chileorebutia malleolata → Eriosyce krausii

Cintia knizei

Cipocereus bradei
Cipocereus crassisepalus
Cipocereus laniflorus
Cipocereus minensis
Cipocereus minensis ssp. **minensis**
Cipocereus minensis ssp. **pleurocarpus**
Cipocereus pleurocarpus → Cipocereus minensis ssp. pleurocarpus
Cipocereus pusilliflorus

Cleistocactus acanthurus
Cleistocactus acanthurus ssp. **acanthurus**
Cleistocactus acanthurus ssp. **faustianus**
Cleistocactus acanthurus ssp. **pullatus**
Cleistocactus angosturensis → Cleistocactus buchtienii
Cleistocactus anguinus → Cleistocactus baumannii ssp. anguinus
Cleistocactus apurimacensis → Cleistocactus morawetzianus
Cleistocactus areolatus → Cleistocactus parviflorus
Cleistocactus aureispinus → Cleistocactus baumannii ssp. baumannii
Cleistocactus ayopayanus → Cleistocactus buchtienii
Cleistocactus azerensis → Cleistocactus parapetiensis
Cleistocactus baumannii
Cleistocactus baumannii ssp. ***anguinus***
Cleistocactus baumannii ssp. **baumannii**
Cleistocactus baumannii ssp. ***chacoanus***
Cleistocactus baumannii ssp. ***croceiflorus***
Cleistocactus baumannii ssp. **horstii**
Cleistocactus baumannii ssp. ***santacruzensis***
Cleistocactus brevispinus → Cleistocactus peculiaris
Cleistocactus brookeae
Cleistocactus bruneispinus → Cleistocactus baumannii ssp. baumannii
Cleistocactus buchtienii
Cleistocactus candelilla
Cleistocactus capadalensis → Cleistocactus tominensis
Cleistocactus chacoanus → Cleistocactus baumannii ssp. chacoanus

Cleistocactus chotaensis
Cleistocactus clavicaulis → Cleistocactus tominensis
Cleistocactus clavispinus
Cleistocactus compactus → Cleistocactus tarijensis
Cleistocactus crassicaulis → Cleistocactus tominensis
Cleistocactus ×crassiserpens (C. icosagonus × C. serpens)
Cleistocactus croceiflorus → Cleistocactus baumannii ssp. croceiflorus
Cleistocactus dependens
Cleistocactus ferrarii
Cleistocactus fieldianus
Cleistocactus fieldianus ssp. **fieldianus**
Cleistocactus fieldianus ssp. **samnensis**
Cleistocactus fieldianus ssp. **tessellatus**
Cleistocactus flavispinus → Cleistocactus baumannii ssp. baumannii
Cleistocactus fossulatus → Oreocereus pseudofossulatus
Cleistocactus fusiflorus → Cleistocactus parviflorus
Cleistocactus glaucus → Cleistocactus luribayensis
Cleistocactus granjaensis → Cleistocactus luribayensis
Cleistocactus grossei
Cleistocactus herzogianus → Cleistocactus parviflorus
Cleistocactus hildegardiae
Cleistocactus horstii → Cleistocactus baumannii ssp. horstii
Cleistocactus humboldtii → Cleistocactus icosagonus
Cleistocactus hyalacanthus
Cleistocactus hystrix
Cleistocactus ianthinus → Cleistocactus candelilla
Cleistocactus icosagonus
Cleistocactus jujuyensis → Cleistocactus hyalacanthus
Cleistocactus laniceps
Cleistocactus leonensis → Cleistocactus serpens
®*Cleistocactus luminosus* → Cleistocactus morawetzianus
Cleistocactus luribayensis
Cleistocactus mendozae → Cleistocactus tominensis
Cleistocactus micropetalus
Cleistocactus morawetzianus
Cleistocactus muyurinensis
Cleistocactus neoroezlii
Cleistocactus orthogonus
Cleistocactus pachycladus
Cleistocactus palhuayensis
Cleistocactus paraguariensis
Cleistocactus parapetiensis
Cleistocactus parviflorus
Cleistocactus peculiaris
Cleistocactus piraymirensis
Cleistocactus plagiostoma
Cleistocactus pojoensis → Cleistocactus candelilla
Cleistocactus pungens
Cleistocactus pycnacanthus → Cleistocactus morawetzianus
Cleistocactus reae
Cleistocactus ressinianus → Cleistocactus buchtienii
Cleistocactus ritteri
Cleistocactus roezlii
Cleistocactus samaipatanus
Cleistocactus santacruzensis → Cleistocactus baumannii ssp. santacruzensis
Cleistocactus sepium
Cleistocactus serpens
Cleistocactus sextonianus
Cleistocactus smaragdiflorus
Cleistocactus strausii

Cleistocactus sucrensis → Cleistocactus buchtienii
Cleistocactus tarijensis
Cleistocactus tenuiserpens
Cleistocactus tominensis
Cleistocactus tupizensis
Cleistocactus vallegrandensis → Cleistocactus candelilla
Cleistocactus varispinus
Cleistocactus villaazulensis → Cleistocactus morawetzianus
Cleistocactus viridialabastri → Cleistocactus tominensis
Cleistocactus viridiflorus → Cleistocactus palhuayensis
Cleistocactus vulpis-cauda
Cleistocactus wendlandiorum → Cleistocactus brookeae
Cleistocactus winteri
Cleistocactus xylorhizus

×***Cleistocana mirabilis*** (Cleistocactus fieldianus × Matucana supertexta)

Clistanthocereus calviflorus → Cleistocactus fieldianus ssp. fieldianus
Clistanthocereus fieldianus → Cleistocactus fieldianus ssp. fieldianus
Clistanthocereus samnensis → Cleistocactus fieldianus ssp. samnensis
Clistanthocereus tessellatus → Cleistocactus fieldianus ssp. tessellatus

Cochemiea halei → Mammillaria halei
Cochemiea maritima → Mammillaria pondii ssp. maritima
Cochemiea pondii → Mammillaria pondii ssp. pondii
Cochemiea poselgeri → Mammillaria poselgeri
Cochemiea setispina → Mammillaria pondii ssp. setispina

Cochiseia robbinsorum → Escobaria robbinsorum

Coleocephalocereus albicephalus → Micranthocereus albicephalus
Coleocephalocereus aureispinus → Pilosocereus aureispinus
Coleocephalocereus aureus
Coleocephalocereus aureus ssp. *brevicylindricus* → Coleocephalocereus aureus
Coleocephalocereus aureus ssp. *elongatus* → Coleocephalocereus aureus
Coleocephalocereus aureus ssp. *longispinus* → Coleocephalocereus aureus
Coleocephalocereus braunii → Coleocephalocereus buxbaumianus ssp. buxbaumianus
Coleocephalocereus brevicylindricus → Coleocephalocereus aureus
Coleocephalocereus buxbaumianus
Coleocephalocereus buxbaumianus ssp. **buxbaumianus**
Coleocephalocereus buxbaumianus ssp. **flavisetus**
Coleocephalocereus decumbens → Coleocephalocereus fluminensis ssp. decumbens
Coleocephalocereus diersianus → Coleocephalocereus fluminensis ssp. fluminensis
Coleocephalocereus dybowskii → Espostoopsis dybowskii
Coleocephalocereus elongatus → Coleocephalocereus aureus
Coleocephalocereus estevesii → Coleocephalocereus buxbaumianus ssp. flavisetus
Coleocephalocereus flavisetus → Coleocephalocereus buxbaumianus ssp. flavisetus
Coleocephalocereus fluminensis
Coleocephalocereus fluminensis ssp. *braamhaarii*
→ Coleocephalocereus fluminensis ssp. fluminensis
Coleocephalocereus fluminensis ssp. **decumbens**
Coleocephalocereus fluminensis ssp. **fluminensis**
Coleocephalocereus fluminensis ssp. *paulensis*
→ Coleocephalocereus fluminensis ssp. fluminensis
Coleocephalocereus goebelianus
Coleocephalocereus lehmannianus → Micranthocereus purpureus
Coleocephalocereus luetzelburgii → Stephanocereus luetzelburgii
Coleocephalocereus pachystele → Coleocephalocereus goebelianus
Coleocephalocereus paulensis → Coleocephalocereus fluminensis ssp. fluminensis
Coleocephalocereus pluricostatus

Coleocephalocereus pluricostatus ssp. *uebelmanniorum* → Coleocephalocereus pluricostatus
Coleocephalocereus purpureus

Coloradoa mesae-verdae♦ → Sclerocactus mesae-verdae♦

Consolea corallicola → Opuntia spinosissima
Consolea falcata → Opuntia falcata
Consolea macracantha → Opuntia macracantha
Consolea millspaughii → Opuntia millspaughii
Consolea moniliformis → Opuntia moniliformis
Consolea moniliformis ssp. *guantanamana* → Opuntia moniliformis
Consolea nashii → Opuntia nashii
Consolea nashii ssp. *gibarensis* → Opuntia nashii
Consolea rubescens → Opuntia rubescens
Consolea spinosissima → Opuntia spinosissima

Copiapoa alticostata → Copiapoa coquimbana
®*Copiapoa applanata* → Copiapoa cinerascens
Copiapoa atacamensis → Copiapoa calderana ssp. calderana
®*Copiapoa barquitensis* → Copiapoa hypogaea
Copiapoa boliviana → ?Copiapoa calderana ssp. calderana
Copiapoa bridgesii
Copiapoa brunnescens → Copiapoa megarhiza
Copiapoa calderana
Copiapoa calderana ssp. **calderana**
Copiapoa calderana ssp. **longistaminea**
Copiapoa carrizalensis → Copiapoa malletiana
Copiapoa castanea → Copiapoa serpentisulcata
Copiapoa chan(i)aralensis
Copiapoa cinerascens
Copiapoa cinerea
Copiapoa cinerea ssp. **cinerea**
®*Copiapoa cinerea* ssp. *columna-alba* → ?Copiapoa cinerea ssp. cinerea
Copiapoa cinerea ssp. *dealbata* → Copiapoa malletiana
Copiapoa cinerea ssp. *gigantea* → Copiapoa cinerea ssp. haseltoniana
Copiapoa cinerea ssp. **haseltoniana**
Copiapoa cinerea ssp. **krainziana**
®*Copiapoa cinerea* ssp. *longistaminea* → Copiapoa calderana ssp. longistaminea
®*Copiapoa conglomerata* → Copiapoa solaris
®***Copiapoa copiapensis***
Copiapoa coquimbana
Copiapoa cuprea → Copiapoa echinoides
Copiapoa cupreata → Copiapoa echinoides
Copiapoa dealbata → Copiapoa malletiana
Copiapoa desertorum
Copiapoa dura → Copiapoa echinoides
Copiapoa echinata → Copiapoa fiedleriana
Copiapoa echinoides
Copiapoa eremophila → Copiapoa cinerea ssp. haseltoniana
Copiapoa esmeraldana → Copiapoa humilis
®*Copiapoa ferox* → Copiapoa solaris
Copiapoa fiedleriana
Copiapoa gigantea → Copiapoa cinerea ssp. haseltoniana
Copiapoa grandiflora → Copiapoa montana ssp. grandiflora
Copiapoa haseltoniana → Copiapoa cinerea ssp. haseltoniana
Copiapoa hornilloensis
Copiapoa humilis
Copiapoa hypogaea
Copiapoa intermedia → Copiapoa fiedleriana
Copiapoa krainziana → Copiapoa cinerea ssp. krainziana

Copiapoa laui
®*Copiapoa lembckei* → Copiapoa calderana ssp. calderana
Copiapoa longispina → Copiapoa humilis
Copiapoa longistaminea → Copiapoa calderana ssp. longistaminea
®Copiapoa macracantha
Copiapoa malletiana
Copiapoa marginata
Copiapoa megarhiza
Copiapoa melanohystrix → Copiapoa cinerea ssp. haseltoniana
Copiapoa mollicula → Copiapoa montana ssp. montana
Copiapoa montana
Copiapoa montana ssp. **grandiflora**
Copiapoa montana ssp. **montana**
Copiapoa olivana → Copiapoa montana ssp. montana
Copiapoa paposoensis → Copiapoa humilis
Copiapoa pendulina → Copiapoa coquimbana
Copiapoa pepiniana → Copiapoa coquimbana
Copiapoa pseudocoquimbana → Copiapoa fiedleriana
Copiapoa rarissima → Copiapoa montana ssp. montana
Copiapoa rubriflora → Copiapoa rupestris
Copiapoa rupestris
Copiapoa serenana → Copiapoa coquimbana
Copiapoa serpentisulcata
Copiapoa solaris
Copiapoa streptocaulon → Copiapoa marginata
Copiapoa taltalensis → Copiapoa humilis
Copiapoa tenebrosa → Copiapoa cinerea ssp. haseltoniana
®Copiapoa tenuissima
Copiapoa tocopillana
Copiapoa totoralensis → Copiapoa fiedleriana
Copiapoa vallenarensis → Copiapoa coquimbana
Copiapoa varispinata
®*Copiapoa wagenknechtii* → Copiapoa coquimbana

Corryocactus acervatus
Corryocactus apiciflorus
Corryocactus aureus
Corryocactus ayacuchoensis
Corryocactus ayopayanus
Corryocactus brachycladus
Corryocactus brachypetalus
Corryocactus brevispinus
Corryocactus brevistylus
Corryocactus brevistylus ssp. **brevistylus**
Corryocactus brevistylus ssp. **puquiensis**
Corryocactus chachapoyensis
Corryocactus charazanensis
Corryocactus chavinilloensis
Corryocactus cuajonesensis
Corryocactus erectus
Corryocactus gracilis
Corryocactus heteracanthus
Corryocactus huincoensis
Corryocactus krausii → Corryocactus brevistylus ssp. brevistylus
Corryocactus matucanensis
Corryocactus maximus → Corryocactus apiciflorus
Corryocactus megarhizus
Corryocactus melaleucus
Corryocactus melanotrichus
Corryocactus odoratus

Corryocactus otuyensis
Corryocactus pachycladus
Corryocactus perezianus
Corryocactus pilispinus
Corryocactus prostratus
Corryocactus pulquinensis
Corryocactus puquiensis → Corryocactus brevistylus ssp. puquiensis
Corryocactus pyroporphyranthus
Corryocactus quadrangularis
Corryocactus quivillanus
Corryocactus serpens
Corryocactus solitarius
Corryocactus spiniflorus → Austrocactus spiniflorus
Corryocactus squarrosus
Corryocactus tarijensis
Corryocactus tenuiculus

Corynopuntia agglomerata → Opuntia agglomerata
Corynopuntia bulbispina → Opuntia bulbispina
Corynopuntia clavata → Opuntia clavata
Corynopuntia dumetorum → Opuntia dumetorum
Corynopuntia grahamii → Opuntia grahamii
Corynopuntia invicta → Opuntia invicta
Corynopuntia moelleriana → Opuntia moelleri
Corynopuntia pulchella → Opuntia pulchella
Corynopuntia reflexispina → Opuntia reflexispina
Corynopuntia schottii → Opuntia schottii
Corynopuntia stanlyi → Opuntia emoryi
Corynopuntia vilis → Opuntia vilis

®*Coryphantha aggregata* → Escobaria vivipara
Coryphantha albicolumnaria → Escobaria albicolumnaria
Coryphantha alversonii → Escobaria alversonii
Coryphantha andreae → Coryphantha pycnacantha
Coryphantha arizonica → Escobaria vivipara
Coryphantha asterias → Coryphantha ottonis
Coryphantha bergeriana → Coryphantha glanduligera
Coryphantha bernalensis → Coryphantha radians
Coryphantha borwigii → Coryphantha salinensis
Coryphantha bumamma → Coryphantha elephantidens
Coryphantha bussleri → Coryphantha ottonis
Coryphantha calipensis
Coryphantha calochlora
Coryphantha chlorantha → Escobaria deserti
Coryphantha clava → Coryphantha octacantha
Coryphantha clavata
Coryphantha columnaris → Escobaria vivipara
Coryphantha compacta
Coryphantha conimamma → Coryphantha sulcolanata
Coryphantha connivens → Coryphantha pycnacantha
Coryphantha cornifera
Coryphantha cornuta
Coryphantha cubensis → Escobaria cubensis
Coryphantha cuencamensis → Coryphantha delaetiana
Coryphantha delaetiana
Coryphantha delicata → Coryphantha radians
Coryphantha densispina♦ → Coryphantha werdermannii♦
Coryphantha deserti → Escobaria deserti
Coryphantha difficilis
Coryphantha duncanii → Escobaria duncanii

Coryphantha durangensis
Coryphantha echinoidea
Coryphantha echinus
Coryphantha elephantidens
Coryphantha erecta
Coryphantha exsudans → Coryphantha ottonis
Coryphantha garessii → Coryphantha elephantidens
Coryphantha georgii
Coryphantha gladiispina → Coryphantha delaetiana
Coryphantha glanduligera
Coryphantha gracilis
Coryphantha grandis → Coryphantha longicornis
Coryphantha grata
Coryphantha greenwoodii → Coryphantha elephantidens
Coryphantha guerkeana
Coryphantha henricksonii → Escobaria chihuahuensis ssp. henricksonii
Coryphantha indensis
Coryphantha jalpanensis
®***Coryphantha jaumavei***
Coryphantha laredoi → Escobaria laredoi
Coryphantha laui → Coryphantha delaetiana
Coryphantha longicornis
Coryphantha macromeris
Coryphantha macromeris ssp. **macromeris**
Coryphantha macromeris ssp. **runyonii**
Coryphantha maiz-tablasensis
Coryphantha maliterrarum
Coryphantha melleospina
Coryphantha minima♦ → Escobaria minima♦
Coryphantha missouriensis → Escobaria missouriensis ssp. missouriensis
®*Coryphantha muehlenpfordtii* → Coryphantha robustispina ssp. scheeri
®*Coryphantha muehlenpfordtii* ssp. *robustispina*
→ Coryphantha robustispina ssp. robustispina
®*Coryphantha muehlenpfordtii* ssp. *uncinata* → Coryphantha robustispina ssp. uncinata
Coryphantha neglecta
Coryphantha neoscheeri → Coryphantha robustispina ssp. scheeri
Coryphantha nickelsiae
Coryphantha obscura → Coryphantha sulcata
Coryphantha octacantha
Coryphantha odorata
Coryphantha orcuttii → Escobaria orcuttii
Coryphantha organensis → Escobaria organensis
Coryphantha ottonis
Coryphantha pallida
¶*Coryphantha palmeri* → Coryphantha compacta
®*Coryphantha palmeri* → Coryphantha jaumavei
Coryphantha pectinata → Coryphantha echinus
Coryphantha pirtlei → Coryphantha macromeris
Coryphantha poselgeriana
Coryphantha potosiana
Coryphantha pseudoechinus
Coryphantha pseudonickelsiae → Coryphantha delaetiana
Coryphantha pseudoradians
Coryphantha pulleineana
Coryphantha pusilliflora
Coryphantha pycnacantha
Coryphantha radians
Coryphantha ramillosa
Coryphantha recurvata
Coryphantha recurvispina → Coryphantha elephantidens

Coryphantha reduncispina
Coryphantha retusa
Coryphantha robbinsorum → Escobaria robbinsorum
Coryphantha robustispina
Coryphantha robustispina ssp. **robustispina**
Coryphantha robustispina ssp. **scheeri**
Coryphantha robustispina ssp. **uncinata**
Coryphantha roederiana → Coryphantha salinensis
Coryphantha runyonii → Coryphantha macromeris ssp. runyonii
Coryphantha salinensis
Ⓜ*Coryphantha salm-dyckiana* → Coryphantha delaetiana
Coryphantha scheeri → Coryphantha robustispina ssp. scheeri
Coryphantha schwarziana → Coryphantha echinoidea
Coryphantha scolymoides → Coryphantha sp.
Coryphantha sneedii♦ → Escobaria sneedii♦
Coryphantha speciosa → Coryphantha sulcata
Coryphantha sulcata
Coryphantha sulcolanata
Coryphantha tripugionacantha
Coryphantha unicornis
Coryphantha valida → Coryphantha poselgeriana
Coryphantha vaupeliana
Coryphantha villarensis → Coryphantha georgii
Coryphantha vivipara → Escobaria vivipara
Coryphantha vogtherriana
Coryphantha werdermannii♦
Coryphantha wohlschlageri

Cryptocereus anthonyanus → Selenicereus anthonyanus
Cryptocereus imitans → Weberocereus imitans
Cryptocereus rosei → Weberocereus rosei

Cumarinia odorata → Coryphantha odorata

Cumulopuntia alboareolata → Opuntia zehnderi
Ⓜ*Cumulopuntia berteri* → Opuntia sphaerica
Ⓜ*Cumulopuntia boliviana* → Opuntia chichensis
¶*Cumulopuntia boliviana* → Opuntia boliviana
Ⓤ*Cumulopuntia crassicylindrica* → Opuntia crassicylindrica
Cumulopuntia echinacea → Opuntia boliviana
Cumulopuntia famatinensis → Opuntia boliviana
Cumulopuntia frigida → Opuntia frigida
Cumulopuntia galerasensis → Opuntia galerasensis
Cumulopuntia hystrix → Opuntia sanctae-barbarae
Cumulopuntia ignescens → Opuntia ignescens
Ⓤ*Cumulopuntia ignota* → Opuntia corotilla
Ⓤ*Cumulopuntia kuehnrichiana* → Opuntia sphaerica
Cumulopuntia multiareolata → Opuntia sphaerica
Cumulopuntia pampana → Opuntia boliviana
Cumulopuntia pentlandii → Opuntia pentlandii
Cumulopuntia pyrrhacantha → Opuntia pyrrhacantha
Cumulopuntia rauppiana → Opuntia sphaerica
Cumulopuntia rossiana → Opuntia rossiana
Ⓤ*Cumulopuntia subterranea* → Opuntia subterranea
Cumulopuntia ticnamarensis → Opuntia ticnamarensis
Cumulopuntia tortispina → Opuntia guatinensis
Cumulopuntia tubercularis → Opuntia sphaerica
Cumulopuntia tumida → Opuntia tumida
Cumulopuntia unguispina → Opuntia unguispina
Cumulopuntia zehnderi → Opuntia zehnderi

Cylindropuntia abyssi → Opuntia abyssi
Cylindropuntia acanthocarpa → Opuntia acanthocarpa
Cylindropuntia alamosensis → Opuntia thurberi
Cylindropuntia alcahes → Opuntia prolifera
Cylindropuntia arbuscula → Opuntia arbuscula
Cylindropuntia bigelowii → Opuntia bigelovii
Cylindropuntia brevispina → Opuntia alcahes
Cylindropuntia brittonii → Opuntia leptocaulis
Cylindropuntia burrageana → Opuntia burrageana
Cylindropuntia californica → Opuntia californica
Cylindropuntia calmalliana → Opuntia molesta
Cylindropuntia caribaea → Opuntia caribaea
Cylindropuntia cholla → Opuntia cholla
Cylindropuntia ciribe → Opuntia bigelovii
Cylindropuntia clavellina → Opuntia molesta
Cylindropuntia davisii → Opuntia davisii
Cylindropuntia echinocarpa → Opuntia echinocarpa
Cylindropuntia fulgida → Opuntia fulgida
Cylindropuntia haematacantha → Opuntia verschaffeltii
Cylindropuntia hualpaensis → Opuntia whipplei
Cylindropuntia humahuacana → Opuntia shaferi
Cylindropuntia imbricata → Opuntia imbricata
Cylindropuntia intermedia → Opuntia cylindrica
Cylindropuntia kleiniae → Opuntia kleiniae
Cylindropuntia leptocaulis → Opuntia leptocaulis
Cylindropuntia lloydii → Opuntia lloydii
Cylindropuntia metuenda → Opuntia caribaea
Cylindropuntia molesta → Opuntia molesta
Cylindropuntia multigeniculata → Opuntia multigeniculata
Cylindropuntia munzii → Opuntia munzii
Cylindropuntia parryi → Opuntia parryi
Cylindropuntia prolifera → Opuntia prolifera
Cylindropuntia ramosissima → Opuntia ramosissima
Cylindropuntia recondita → Opuntia kleiniae
Cylindropuntia rosarica → Opuntia rosarica
Cylindropuntia rosea → Opuntia rosea
Cylindropuntia spinosior → Opuntia spinosior
Cylindropuntia teres → Opuntia vestita
Cylindropuntia tesajo → Opuntia tesajo
Cylindropuntia tetracantha → Opuntia tetracantha
Cylindropuntia thurberi → Opuntia thurberi
Cylindropuntia tunicata → Opuntia tunicata
Cylindropuntia versicolor → Opuntia versicolor
Cylindropuntia viridiflora → Opuntia viridiflora
Cylindropuntia vivipara → Opuntia arbuscula
Cylindropuntia weingartiana → Opuntia shaferi
Cylindropuntia whipplei → Opuntia whipplei
Cylindropuntia wigginsii → Opuntia wigginsii

Deamia diabolica → Selenicereus testudo
Deamia testudo → Selenicereus testudo

®*Delaetia woutersiana* → Eriosyce taltalensis ssp. paucicostata

Dendrocereus nudiflorus
Dendrocereus undulosus

Denmoza erythrocephala → Denmoza rhodacantha
Denmoza rhodacantha

Discocactus albispinus♦ → Discocactus zehntneri ssp. zehntneri♦
Discocactus alteolens♦ → Discocactus placentiformis♦
Discocactus araneispinus♦ → Discocactus zehntneri ssp. boomianus♦
Discocactus bahiensis♦
Discocactus bahiensis ssp. *subviridigriseus*♦ → Discocactus bahiensis♦
Discocactus boliviensis♦ → Discocactus heptacanthus ssp. heptacanthus♦
Discocactus boomianus♦ → Discocactus zehntneri ssp. boomianus♦
Discocactus buenekeri♦ → Discocactus zehntneri ssp. boomianus♦
Discocactus cangaensis♦ → Discocactus heptacanthus ssp. heptacanthus♦
Discocactus catingicola♦ → Discocactus heptacanthus ssp. catingicola♦
Discocactus catingicola ssp. *griseus*♦ → Discocactus heptacanthus ssp. heptacanthus♦
Discocactus catingicola ssp. *rapirhizus*♦ → Discocactus heptacanthus ssp. heptacanthus♦
Discocactus cephaliaciculosus♦ → Discocactus heptacanthus ssp. heptacanthus♦
Ⓤ*Discocactus cephaliaciculosus* ssp. *nudicephalus*♦
→ Discocactus heptacanthus ssp. heptacanthus♦
Discocactus crassispinus♦ → Discocactus heptacanthus ssp. heptacanthus♦
Discocactus crassispinus ssp. *araguaiensis*♦
→ Discocactus heptacanthus ssp. heptacanthus♦
Discocactus crystallophilus♦ → Discocactus placentiformis♦
Discocactus diersianus♦ → Discocactus heptacanthus ssp. heptacanthus♦
Discocactus diersianus ssp. *goianus*♦ → Discocactus heptacanthus ssp. heptacanthus♦
Discocactus estevesii♦ → Discocactus heptacanthus ssp. heptacanthus♦
Discocactus ferricola♦
Discocactus flavispinus♦ → Discocactus heptacanthus ssp. heptacanthus♦
Discocactus goianus♦ → Discocactus heptacanthus ssp. heptacanthus♦
Discocactus griseus♦ → Discocactus heptacanthus ssp. heptacanthus♦
Discocactus hartmannii♦ → Discocactus heptacanthus ssp. magnimammus♦
Discocactus hartmannii ssp. *giganteus*♦ → Discocactus heptacanthus ssp. magnimammus♦
Discocactus hartmannii ssp. *magnimammus*♦
→ Discocactus heptacanthus ssp. magnimammus♦
Discocactus hartmannii ssp. *patulifolius*♦ → Discocactus heptacanthus ssp. magnimammus♦
Discocactus hartmannii ssp. *setosiflorus*♦ → Discocactus heptacanthus ssp. heptacanthus♦
Discocactus heptacanthus♦
Discocactus heptacanthus ssp. **catingicola**♦
Discocactus heptacanthus ssp. **heptacanthus**♦
Discocactus heptacanthus ssp. **magnimammus**♦
Discocactus heptacanthus ssp. *melanochlorus*♦
→ Discocactus heptacanthus ssp. heptacanthus♦
Discocactus horstii♦
Discocactus insignis♦ → Discocactus placentiformis♦
Discocactus latispinus♦ → Discocactus placentiformis♦
Discocactus latispinus ssp. *pseudolatispinus*♦ → Discocactus placentiformis♦
Discocactus latispinus ssp. *pulvinicapitatus*♦ → Discocactus placentiformis♦
Ⓤ*Discocactus lindaianus* → Discocactus heptacanthus ssp. heptacanthus
Discocactus lindanus♦ → Discocactus heptacanthus ssp. heptacanthus♦
Discocactus magnimammus♦ → Discocactus heptacanthus ssp. magnimammus♦
Discocactus magnimammus ssp. *bonitoensis*♦
→ Discocactus heptacanthus ssp. magnimammus♦
Discocactus mamillosus♦ → Discocactus heptacanthus ssp. magnimammus♦
Discocactus melanochlorus♦ → Discocactus heptacanthus ssp. heptacanthus♦
Discocactus multicolorispinus♦ → Discocactus placentiformis♦
Discocactus nigrisaetosus♦ → Discocactus heptacanthus ssp. catingicola♦
Discocactus pachythele♦ → Discocactus heptacanthus ssp. magnimammus♦
Discocactus paranaensis♦ → Discocactus heptacanthus♦
Discocactus patulifolius♦ → Discocactus heptacanthus ssp. magnimammus♦
Discocactus piauiensis♦ → Discocactus heptacanthus ssp. catingicola♦
Discocactus placentiformis♦
Discocactus placentiformis ssp. *alteolens*♦ → Discocactus placentiformis♦
Discocactus placentiformis ssp. *multicolorispinus*♦ → Discocactus placentiformis♦
Discocactus placentiformis ssp. *pugionacanthus*♦ → Discocactus placentiformis♦

Discocactus prominentigibbus♦ → Discocactus heptacanthus ssp. heptacanthus♦
Discocactus pseudoinsignis♦
Discocactus pseudolatispinus♦ → Discocactus placentiformis♦
Discocactus pugionacanthus♦ → Discocactus placentiformis♦
Discocactus pulvinicapitatus♦ → Discocactus placentiformis♦
Discocactus rapirhizus♦ → Discocactus heptacanthus ssp. heptacanthus♦
Discocactus semicampaniflorus♦ → Discocactus heptacanthus ssp. heptacanthus♦
Discocactus silicicola♦ → Discocactus heptacanthus ssp. heptacanthus♦
Discocactus silvaticus♦ → Discocactus heptacanthus ssp. heptacanthus♦
Discocactus spinosior♦ → Discocactus heptacanthus ssp. catingicola♦
Discocactus squamibaccatus♦ → Discocactus heptacanthus ssp. heptacanthus♦
Discocactus subterraneo-proliferans♦ → Discocactus heptacanthus ssp. heptacanthus♦
Discocactus subviridigriseus♦ → Discocactus bahiensis♦
Discocactus tricornis♦ → Discocactus placentiformis♦
Discocactus woutersianus♦ → Discocactus horstii♦
Discocactus zehntneri♦
Discocactus zehntneri ssp. *albispinus*♦ → Discocactus zehntneri ssp. zehntneri♦
Discocactus zehntneri ssp. *araneispinus*♦ → Discocactus zehntneri ssp. boomianus♦
Discocactus zehntneri ssp. **boomianus♦**
Discocactus zehntneri ssp. *buenekeri*♦ → Discocactus zehntneri ssp. boomianus♦
Discocactus zehntneri ssp. *horstiorum*♦ → Discocactus zehntneri ssp. boomianus♦
Discocactus zehntneri ssp. **zehntneri♦**

Disocactus ackermannii
Disocactus acuminatus → Pseudorhipsalis acuminata
Disocactus alatus → Pseudorhipsalis alata
Disocactus amazonicus
Disocactus aurantiacus
Disocactus biformis
Disocactus cinnabarinus
Disocactus eichlamii
Disocactus flagelliformis
Disocactus himantocladus → Pseudorhipsalis himantoclada
Disocactus horichii → Pseudorhipsalis horichii
Disocactus kimnachii
Disocactus lankesteri → Pseudorhipsalis lankesteri
Disocactus macdougallii♦
Disocactus macranthus
Disocactus martianus
Disocactus nelsonii
Disocactus phyllanthoides
Disocactus quezaltecus
Disocactus ramulosus → Pseudorhipsalis ramulosa
Disocactus schrankii
Disocactus speciosus

Dolichothele albescens → Mammillaria decipiens ssp. albescens
Dolichothele balsasoides → Mammillaria beneckei
Dolichothele baumii → Mammillaria baumii
Dolichothele beneckei → Mammillaria beneckei
Dolichothele camptotricha → Mammillaria decipiens ssp. camptotricha
Dolichothele decipiens → Mammillaria decipiens ssp. decipiens
Dolichothele longimamma → Mammillaria longimamma
Dolichothele longimamma ssp. *uberiformis* → Mammillaria longimamma
Dolichothele melaleuca → Mammillaria melaleuca
Dolichothele nelsonii → Mammillaria beneckei
Dolichothele sphaerica → Mammillaria sphaerica
Dolichothele surculosa → Mammillaria surculosa
Dolichothele uberiformis → Mammillaria longimamma
Dolichothele zephyranthoides → Mammillaria zephyranthoides

Eccremocactus bradei → Weberocereus bradei
Eccremocactus imitans → Weberocereus imitans
Eccremocactus rosei → Weberocereus rosei

Echinocactus acanthodes → Ferocactus sp.
Echinocactus asterias♦ → Astrophytum asterias♦
Echinocactus grandis → Echinocactus platyacanthus
Echinocactus grusonii
Echinocactus horizonthalonius
Echinocactus ingens → Echinocactus platyacanthus
Echinocactus palmeri → Echinocactus platyacanthus
Echinocactus parryi
Echinocactus platyacanthus
Echinocactus polycephalus
Echinocactus polycephalus ssp. **polycephalus**
Echinocactus polycephalus ssp. **xeranthemoides**
Echinocactus texensis
Echinocactus visnaga → Echinocactus platyacanthus

Echinocereus abbeae → Echinocereus fasciculatus ssp. fasciculatus
Echinocereus acifer → Echinocereus polyacanthus ssp. acifer
Echinocereus acifer ssp. *huitcholensis* → Echinocereus polyacanthus ssp. huitcholensis
Echinocereus acifer ssp. *tubiflorus* → Echinocereus polyacanthus ssp. acifer
Echinocereus adustus
Echinocereus adustus ssp. **adustus**
Echinocereus adustus ssp. **bonatzii**
Echinocereus adustus ssp. **schwarzii**
⊗*Echinocereus albatus* → Echinocereus nivosus
Echinocereus albispinus → Echinocereus reichenbachii ssp. baileyi
Echinocereus amoenus → Echinocereus pulchellus ssp. pulchellus
Echinocereus angusticeps → Echinocereus papillosus
Echinocereus apachensis
Echinocereus arizonicus → Echinocereus coccineus ssp. coccineus
Echinocereus arizonicus ssp. *matudae* → Echinocereus coccineus ssp. coccineus
Echinocereus arizonicus ssp. *nigrihorridispinus* → Echinocereus coccineus ssp. coccineus
Echinocereus armatus → Echinocereus reichenbachii ssp. armatus
Echinocereus baileyi → Echinocereus reichenbachii ssp. baileyi
Echinocereus barthelowanus
Echinocereus berlandieri
⊗*Echinocereus blanckii* → Echinocereus berlandieri
Echinocereus bonatzii → Echinocereus adustus ssp. bonatzii
Echinocereus bonkerae
Echinocereus boyce-thompsonii
Echinocereus brandegeei
Echinocereus bristolii
Echinocereus bristolii ssp. *floresii* → Echinocereus sciurus ssp. floresii
Echinocereus caespitosus → Echinocereus reichenbachii
Echinocereus canyonensis → Echinocereus coccineus ssp. coccineus
Echinocereus carmenensis → ?Echinocereus viridiflorus ssp. chloranthus
Echinocereus chisoensis
Echinocereus chloranthus → Echinocereus viridiflorus ssp. chloranthus
Echinocereus chloranthus ssp. *cylindricus* → Echinocereus viridiflorus ssp. cylindricus
Echinocereus chloranthus ssp. *neocapillus* → ?Echinocereus viridiflorus ssp. chloranthus
Echinocereus chloranthus ssp. *rhyolithensis* → ?Echinocereus viridiflorus ssp. chloranthus
Echinocereus chlorophthalmus → Echinocereus cinerascens ssp. cinerascens
Echinocereus cinerascens
Echinocereus cinerascens ssp. **cinerascens**
Echinocereus cinerascens ssp. *ehrenbergii* → Echinocereus cinerascens ssp. cinerascens
Echinocereus cinerascens ssp. **septentrionalis**
Echinocereus cinerascens ssp. **tulensis**

Echinocereus coccineus
Echinocereus coccineus ssp. *aggregatus* → Echinocereus coccineus ssp. coccineus
Echinocereus coccineus ssp. **coccineus**
Echinocereus coccineus ssp. *mojavensis* → Echinocereus mojavensis
Echinocereus coccineus ssp. *paucispinus* → Echinocereus coccineus ssp. coccineus
Echinocereus coccineus ssp. *roemeri* → Echinocereus coccineus ssp. coccineus
Echinocereus coccineus ssp. *rosei* → Echinocereus coccineus ssp. coccineus
Echinocereus conglomeratus → Echinocereus stramineus
Echinocereus ctenoides → Echinocereus dasyacanthus
Echinocereus dasyacanthus
Echinocereus dasyacanthus ssp. ***rectispinus***
Echinocereus davisii → Echinocereus viridiflorus ssp. davisii
Echinocereus decumbens → Echinocereus coccineus ssp. coccineus
Echinocereus delaetii → Echinocereus longisetus ssp. delaetii
Echinocereus dubius → Echinocereus enneacanthus ssp. enneacanthus
Echinocereus durangensis → Echinocereus polyacanthus ssp. polyacanthus
Echinocereus ehrenbergii → Echinocereus cinerascens ssp. cinerascens
Echinocereus engelmannii
Echinocereus engelmannii ssp. ***decumbens***
Echinocereus engelmannii ssp. *fasciculatus* → Echinocereus fasciculatus ssp. fasciculatus
Echinocereus enneacanthus
Echinocereus enneacanthus ssp. **brevispinus**
Echinocereus enneacanthus ssp. **enneacanthus**
Echinocereus fasciculatus
Echinocereus fasciculatus ssp. *bonkerae* → Echinocereus bonkerae
Echinocereus fasciculatus ssp. *boyce-thompsonii* → Echinocereus boyce-thompsonii
Echinocereus fendleri
Echinocereus fendleri ssp. **fendleri**
Echinocereus fendleri ssp. ***hempelii***
Echinocereus fendleri ssp. **rectispinus**
Echinocereus ferreirianus
Echinocereus ferreirianus ssp. **ferreirianus**
Echinocereus ferreirianus ssp. **lindsayi♦**
Echinocereus fitchii → Echinocereus reichenbachii ssp. fitchii
Echinocereus fitchii ssp. *albertii* → Echinocereus reichenbachii ssp. fitchii
Echinocereus fitchii ssp. *armatus* → Echinocereus reichenbachii ssp. armatus
Echinocereus floresii → Echinocereus sciurus ssp. floresii
Echinocereus fobeanus → Echinocereus chisoensis
Echinocereus fobeanus ssp. *metornii* → Echinocereus chisoensis
Echinocereus freudenbergeri
Echinocereus gentryi → Echinocereus scheeri ssp. gentryi
Echinocereus glycimorphus → Echinocereus cinerascens ssp. cinerascens
Echinocereus gonacanthus → Echinocereus triglochidiatus
Echinocereus grandis
Echinocereus hancockii → Echinocereus maritimus ssp. hancockii
Echinocereus hempelii → Echinocereus fendleri ssp. hempelii
Echinocereus hexaedrus → Echinocereus coccineus ssp. coccineus
Echinocereus hildmannii → Echinocereus dasyacanthus
Echinocereus huitcholensis → Echinocereus polyacanthus ssp. huitcholensis
Echinocereus klapperi
Echinocereus knippelianus
Echinocereus knippelianus ssp. *kruegeri* → Echinocereus knippelianus
Echinocereus knippelianus ssp. *reyesii* → Echinocereus knippelianus
Echinocereus krausei → Echinocereus coccineus ssp. coccineus
Echinocereus kuenzleri → Echinocereus fendleri ssp. fendleri
Echinocereus kunzei → Echinocereus coccineus ssp. coccineus
Echinocereus laui
Echinocereus ledingii
Echinocereus leeanus → Echinocereus polyacanthus ssp. polyacanthus
Echinocereus leonensis → Echinocereus pentalophus ssp. leonensis

Echinocereus leptacanthus → Echinocereus pentalophus ssp. pentalophus
Echinocereus leucanthus
Echinocereus lindsayi♦ → Echinocereus ferreirianus ssp. lindsayi♦
Echinocereus ×lloydii → Echinocereus ×roetteri
Echinocereus longisetus
Echinocereus longisetus ssp. **delaetii**
Echinocereus longisetus ssp. *freudenbergeri* → Echinocereus freudenbergeri
Echinocereus longisetus ssp. **longisetus**
Echinocereus luteus → Echinocereus subinermis ssp. subinermis
Echinocereus mamillatus → Echinocereus brandegeei
Echinocereus mapimiensis
Echinocereus mariae → Echinocereus reichenbachii
Echinocereus maritimus
Echinocereus maritimus ssp. ***hancockii***
Echinocereus marksianus → Echinocereus polyacanthus ssp. acifer
®*Echinocereus matthesianus* → Echinocereus polyacanthus ssp. huitcholensis
Echinocereus matudae → Echinocereus coccineus ssp. coccineus
®*Echinocereus melanocentrus* → Echinocereus reichenbachii ssp. fitchii
Echinocereus merkeri → Echinocereus enneacanthus ssp. enneacanthus
Echinocereus metornii → Echinocereus chisoensis
Echinocereus mojavensis
Echinocereus mombergerianus → Echinocereus polyacanthus ssp. pacificus
Echinocereus morricalii → Echinocereus viereckii ssp. morricalii
Echinocereus munzii → Echinocereus engelmannii
Echinocereus neocapillus → ?Echinocereus viridiflorus ssp. chloranthus
Echinocereus neomexicanus → ?Echinocereus coccineus ssp. coccineus
Echinocereus nicholii
Echinocereus nicholii ssp. ***llanuraensis***
Echinocereus nivosus
Echinocereus ochoterenae → Echinocereus subinermis ssp. ochoterenae
Echinocereus ortegae
Echinocereus ortegae ssp. *koehresianus* → Echinocereus ortegae
Echinocereus pacificus → Echinocereus polyacanthus ssp. pacificus
Echinocereus pacificus ssp. *mombergerianus* → Echinocereus polyacanthus ssp. pacificus
Echinocereus palmeri
Echinocereus pamanesiorum
Echinocereus pamanesiorum ssp. *bonatzii* → Echinocereus adustus ssp. bonatzii
Echinocereus papillosus
Echinocereus parkeri
Echinocereus parkeri ssp. ***arteagensis***
Echinocereus parkeri ssp. **gonzalezii**
Echinocereus parkeri ssp. **mazapilensis**
Echinocereus parkeri ssp. **parkeri**
Echinocereus pectinatus
Echinocereus pectinatus ssp. *ctenoides* → Echinocereus dasyacanthus
Echinocereus pectinatus ssp. ***wenigeri***
Echinocereus pensilis
Echinocereus pentalophus
Echinocereus pentalophus ssp. **leonensis**
Echinocereus pentalophus ssp. **pentalophus**
Echinocereus pentalophus ssp. **procumbens**
Echinocereus perbellus → Echinocereus reichenbachii ssp. perbellus
Echinocereus polyacanthus
Echinocereus polyacanthus ssp. **acifer**
Echinocereus polyacanthus ssp. **huitcholensis**
Echinocereus polyacanthus ssp. *mombergerianus*
→ Echinocereus polyacanthus ssp. pacificus
Echinocereus polyacanthus ssp. **pacificus**
Echinocereus polyacanthus ssp. **polyacanthus**
Echinocereus poselgeri

Echinocereus poselgeri ssp. *kroenleinii* → Echinocereus poselgeri
Echinocereus primolanatus
Echinocereus pseudopectinatus
Echinocereus pulchellus
Echinocereus pulchellus ssp. **acanthosetus**
Echinocereus pulchellus ssp. **pulchellus**
Echinocereus pulchellus ssp. **sharpii**
Echinocereus pulchellus ssp. *venustus* → Echinocereus pulchellus ssp. weinbergii
Echinocereus pulchellus ssp. **weinbergii**
Echinocereus purpureus → Echinocereus reichenbachii
Echinocereus radians → Echinocereus adustus ssp. adustus
Echinocereus rayonesensis
Echinocereus rectispinus → Echinocereus fendleri ssp. rectispinus
Echinocereus reichenbachii
Echinocereus reichenbachii ssp. **armatus**
Echinocereus reichenbachii ssp. **baileyi**
©*Echinocereus reichenbachii* ssp. *caespitosus* → Echinocereus reichenbachii ssp. perbellus
Echinocereus reichenbachii ssp. **fitchii**
Echinocereus reichenbachii ssp. **perbellus**
Echinocereus reichenbachii ssp. **reichenbachii**
Echinocereus rigidissimus
Echinocereus rigidissimus ssp. **rigidissimus**
Echinocereus rigidissimus ssp. **rubispinus**
Echinocereus roemeri → Echinocereus coccineus ssp. coccineus
Echinocereus* ×*roetteri (E. coccineus × E. dasyacanthus)
Echinocereus rosei → Echinocereus coccineus ssp. coccineus
Echinocereus rufispinus → Echinocereus adustus ssp. adustus
Echinocereus russanthus
Echinocereus russanthus ssp. *fiehnii* → Echinocereus russanthus
Echinocereus russanthus ssp. *weedinii* → ?Echinocereus russanthus
Echinocereus salm-dyckianus → ?Echinocereus scheeri ssp. scheeri
Echinocereus salm-dyckianus ssp. *obscuriensis* → Echinocereus scheeri ssp. scheeri
Echinocereus salmianus → ?Echinocereus scheeri ssp. scheeri
Echinocereus sanpedroensis → Echinocereus scheeri ssp. scheeri
Echinocereus santaritensis → Echinocereus polyacanthus ssp. polyacanthus
Echinocereus sarissophorus → Echinocereus enneacanthus ssp. enneacanthus
Echinocereus scheeri
Echinocereus scheeri ssp. **gentryi**
Echinocereus scheeri ssp. **scheeri**
Echinocereus schereri
Echinocereus schmollii♦
Echinocereus schwarzii → Echinocereus adustus ssp. schwarzii
Echinocereus sciurus
Echinocereus sciurus ssp. **floresii**
Echinocereus sciurus ssp. **sciurus**
Echinocereus scopulorum
Echinocereus scopulorum ssp. *pseudopectinatus* → Echinocereus pseudopectinatus
Echinocereus spinigemmatus
Echinocereus standleyi → Echinocereus viridiflorus ssp. viridiflorus
Echinocereus steereae → Echinocereus dasyacanthus
Echinocereus stoloniferus
Echinocereus stoloniferus ssp. **stoloniferus**
Echinocereus stoloniferus ssp. **tayopensis**
Echinocereus stramineus
Echinocereus stramineus ssp. **occidentalis**
Echinocereus stramineus ssp. **stramineus**
Echinocereus subinermis
Echinocereus subinermis ssp. **ochoterenae**
Echinocereus subinermis ssp. **subinermis**
©*Echinocereus subterraneus* → ?Echinocereus sciurus

Echinocereus tamaulipensis ssp. *deherdtii* → Echinocereus poselgeri
Echinocereus tamaulipensis ssp. *waldeisii* → Echinocereus poselgeri
Echinocereus tayopensis → Echinocereus stoloniferus ssp. tayopensis
Echinocereus toroweapensis → Echinocereus coccineus ssp. coccineus
Echinocereus triglochidiatus
Echinocereus triglochidiatus ssp. *mojavensis* → Echinocereus mojavensis
Echinocereus tulensis → Echinocereus cinerascens ssp. tulensis
Echinocereus uspenskii → Echinocereus enneacanthus ssp. enneacanthus
Echinocereus viereckii
Echinocereus viereckii ssp. ***huastecensis***
Echinocereus viereckii ssp. **morricalii**
Echinocereus viereckii ssp. **viereckii**
Echinocereus viridiflorus
Echinocereus viridiflorus ssp. **chloranthus**
Echinocereus viridiflorus ssp. ***correllii***
Echinocereus viridiflorus ssp. **cylindricus**
Echinocereus viridiflorus ssp. **davisii**
Echinocereus viridiflorus ssp. **viridiflorus**
Echinocereus waldeisii → Echinocereus poselgeri
Echinocereus websterianus
Echinocereus weinbergii → Echinocereus pulchellus ssp. sharpii

Echinofossulocactus albatus → ?Stenocactus vaupelianus
Echinofossulocactus arrigens → Stenocactus crispatus
Echinofossulocactus caespitosus → Stenocactus obvallatus
Echinofossulocactus confusus → Stenocactus crispatus
Echinofossulocactus coptonogonus → Stenocactus coptonogonus
Echinofossulocactus crispatus → Stenocactus crispatus
Echinofossulocactus densispinus → Stenocactus ochoterenanus
Echinofossulocactus dichroacanthus → Stenocactus crispatus
®*Echinofossulocactus erectocentrus* → Stenocactus multicostatus
Echinofossulocactus flexispinus → Stenocactus crispatus
Echinofossulocactus guerraianus → Stenocactus crispatus
Echinofossulocactus hastatus → Stenocactus hastatus
Echinofossulocactus heteracanthus → ?Stenocactus ochoterenanus
Echinofossulocactus kellerianus → Stenocactus crispatus
Echinofossulocactus lamellosus → Stenocactus crispatus
Echinofossulocactus lancifer → Stenocactus crispatus
Echinofossulocactus lexarzae → Stenocactus ochoterenanus
Echinofossulocactus lloydii → Stenocactus multicostatus
Echinofossulocactus multiareolatus → Stenocactus crispatus
Echinofossulocactus multicostatus → Stenocactus multicostatus
Echinofossulocactus obvallatus → Stenocactus obvallatus
®*Echinofossulocactus ochoterenaus* → Stenocactus ochoterenanus
Echinofossulocactus parksianus → Stenocactus ochoterenanus
Echinofossulocactus pentacanthus → Stenocactus obvallatus
Echinofossulocactus phyllacanthus → Stenocactus phyllacanthus
Echinofossulocactus rosasianus → Stenocactus ochoterenanus
Echinofossulocactus sulphureus → Stenocactus sulphureus
Echinofossulocactus tricuspidatus → Stenocactus phyllacanthus
Echinofossulocactus vaupelianus → Stenocactus vaupelianus
Echinofossulocactus violaciflorus → Stenocactus crispatus
Echinofossulocactus zacatecasensis → Stenocactus multicostatus

Echinomastus acunensis♦ → Sclerocactus erectocentrus♦
Echinomastus dasyacanthus → Sclerocactus intertextus
Echinomastus durangensis → Sclerocactus unguispinus
Echinomastus erectocentrus♦ → Sclerocactus erectocentrus♦
Echinomastus gautii → ?Sclerocactus warnockii
Echinomastus intertextus → Sclerocactus intertextus

Echinomastus johnsonii → Sclerocactus johnsonii
Echinomastus krausei♦ → Sclerocactus erectocentrus♦
Echinomastus laui → Sclerocactus unguispinus
Echinomastus macdowellii → Thelocactus macdowellii
Echinomastus mapimiensis → Sclerocactus unguispinus
Echinomastus mariposensis♦ → Sclerocactus mariposensis♦
Echinomastus unguispinus → Sclerocactus unguispinus
Echinomastus unguispinus ssp. *laui* → Sclerocactus unguispinus
Echinomastus warnockii → Sclerocactus warnockii

Echinopsis adolfofriedrichii
Echinopsis albispinosa → Echinopsis tubiflora
Echinopsis amblayensis → Echinopsis haematantha
Echinopsis ancistrophora
Echinopsis ancistrophora ssp. ***arachnacantha***
Echinopsis ancistrophora ssp. ***cardenasiana***
Echinopsis ancistrophora ssp. ***pojoensis***
Echinopsis angelesiae
Echinopsis antezanae
Echinopsis apiculata → Echinopsis sp.
Echinopsis arachnacantha → Echinopsis ancistrophora ssp. arachnacantha
Echinopsis arboricola
Echinopsis arebaloi
Echinopsis atacamensis
Echinopsis atacamensis ssp. **atacamensis**
Echinopsis atacamensis ssp. **pasacana**
Echinopsis aurantiaca → Echinopsis glaucina
Echinopsis aurea
Echinopsis ayopayana → Echinopsis comarapana
Echinopsis backebergii
Echinopsis baldiana
Echinopsis bertramiana
Echinopsis boedekeriana → Echinopsis backebergii
Echinopsis boyuibensis
Echinopsis brasiliensis
Echinopsis brevispina → Echinopsis thionantha
Echinopsis bridgesii
Echinopsis bridgesii ssp. ***yungasensis***
Echinopsis bruchii
Echinopsis ×cabrerae (E. terscheckii × E. strigosus)
Echinopsis caineana
Echinopsis cajasensis
Echinopsis calliantholilacina
Echinopsis callichroma
Echinopsis calochlora
Echinopsis calochlora ssp. ***glaetzleana***
Echinopsis calorubra → Echinopsis obrepanda ssp. calorubra
Echinopsis camarguensis
Echinopsis candicans
Echinopsis cardenasiana → Echinopsis ancistrophora ssp. cardenasiana
®*Echinopsis carmineiflora* → Echinopsis obrepanda
Echinopsis cephalomacrostibas
Echinopsis cerdana
Echinopsis chacoana → Echinopsis rhodotricha ssp. chacoana
Echinopsis chalaensis
Echinopsis chamaecereus
Echinopsis chiloensis
Echinopsis chrysantha
Echinopsis chrysochete
Echinopsis cinnabarina

Echinopsis clavatus
Echinopsis cochabambensis
Echinopsis comarapana
Echinopsis conaconensis
Echinopsis coquimbana
Echinopsis cordobensis → Echinopsis leucantha
Echinopsis coronata
Echinopsis cotacajesii
Echinopsis courantii → Echinopsis candicans
Echinopsis crassicaulis
Echinopsis cuzcoensis
Echinopsis cylindracea → Echinopsis aurea
Echinopsis densispina
Echinopsis derenbergii
Echinopsis deserticola
Echinopsis elongata → Echinopsis haematantha
Echinopsis escayachensis
Echinopsis eyriesii
Echinopsis fabrisii
Echinopsis fallax → Echinopsis aurea
Echinopsis famatinensis
Echinopsis ferox
Echinopsis fiebrigii → Echinopsis obrepanda
Echinopsis forbesii → ?Echinopsis rhodotricha
Echinopsis formosa
Echinopsis formosissima → ?Echinopsis atacamensis ssp. pasacana
Echinopsis fricii → Echinopsis tiegeliana
Echinopsis friedrichii
Echinopsis fulvilana → Echinopsis deserticola
Echinopsis glauca
Echinopsis glaucina
Echinopsis graciliflora
Echinopsis grandiflora → ?Echinopsis calochlora
Echinopsis grandis → Echinopsis bruchii
Echinopsis haematantha
Echinopsis hahniana
Echinopsis hamatacantha → Echinopsis ancistrophora
Echinopsis hammerschmidii
Echinopsis hardeniana → Echinopsis pentlandii
Echinopsis herbasii → ?Echinopsis mamillosa
Echinopsis hertrichiana
Echinopsis herzogiana → Echinopsis tarijensis ssp. herzogianus
Echinopsis hualfinensis → Echinopsis haematantha
Echinopsis huascha
Echinopsis huotii
Echinopsis huotii ssp. ***vallegrandensis***
Echinopsis hystrichoides
Echinopsis ibicuatensis
Echinopsis ingens → Echinopsis bruchii
Echinopsis intricatissima → Echinopsis leucantha
Echinopsis kermesina → Echinopsis mamillosa
Echinopsis kladiwana
Echinopsis klingleriana
Echinopsis knuthiana
Echinopsis korethroides
Echinopsis kratochviliana → Echinopsis ancistrophora ssp. arachnacantha
Echinopsis kuehnrichii → Echinopsis haematantha
Echinopsis lageniformis
Echinopsis lamprochlora
Echinopsis lateritia

Echinopsis lecoriensis → Echinopsis ferox
Echinopsis leucantha
Echinopsis leucomalla → Echinopsis aurea
Echinopsis litoralis
Echinopsis longispina → Echinopsis ferox
Echinopsis macrogona
Echinopsis mamillosa
Echinopsis mamillosa ssp. ***silvatica***
Echinopsis manguinii → Echinopsis schickendantzii
Echinopsis marsoneri
Echinopsis mataranensis
Echinopsis maximiliana
Echinopsis melanopotamica → Echinopsis leucantha
Echinopsis meyeri
Echinopsis mieckleyi
Echinopsis minuana
Echinopsis mirabilis
Echinopsis molesta
Echinopsis multiplex → Echinopsis oxygona
Echinopsis narvaecensis → ?Echinopsis tarijensis
Echinopsis nealeana → Echinopsis saltensis
Echinopsis nigra
Echinopsis obrepanda
Echinopsis obrepanda ssp. ***calorubra***
Echinopsis obrepanda ssp. ***tapecuana***
Echinopsis orurensis → ?Echinopsis ferox
Echinopsis oxygona
Echinopsis pachanoi
Echinopsis pampana
Echinopsis pamparuizii
Echinopsis paraguayensis → Echinopsis oxygona
Echinopsis pasacana → Echinopsis atacamensis ssp. pasacana
Echinopsis pecheretiana → Echinopsis huascha
Echinopsis peitscheriana → Acanthocalycium klimpelianum
Echinopsis pentlandii
Echinopsis pentlandii ssp. *hardeniana* → Echinopsis pentlandii
Echinopsis pentlandii ssp. *larae* → Echinopsis pentlandii
Echinopsis pereziensis → Echinopsis comarapana
Echinopsis peruviana
Echinopsis peruviana ssp. **peruviana**
Echinopsis peruviana ssp. **puquiensis**
Echinopsis poco → Echinopsis tarijensis
Echinopsis pojoensis
Echinopsis polyancistra
Echinopsis potosina → Echinopsis ferox
Echinopsis pseudocachensis → Echinopsis saltensis
®*Echinopsis pseudocandicans* → Echinopsis candicans
Echinopsis pseudomamillosa
Echinopsis pudantii → Echinopsis eyriesii
Echinopsis pugionacantha
Echinopsis pugionacantha ssp. ***rossii***
Echinopsis puquiensis → Echinopsis peruviana ssp. puquiensis
Echinopsis purpureopilosa → Echinopsis lamprochlora
Echinopsis quadratiumbonatus
Echinopsis quinesensis → Echinopsis aurea
Echinopsis randallii → Echinopsis formosa
Echinopsis rauschii → Echinopsis obrepanda ssp. calorubra
Echinopsis rebutioides → Echinopsis densispina
Echinopsis rhodotricha
Echinopsis rhodotricha ssp. ***chacoana***

Echinopsis ritteri → Echinopsis mamillosa
Echinopsis riviere-de-caraltii
Echinopsis rivierei → Echinopsis atacamensis ssp. pasacana
Echinopsis rojasii → ?Echinopsis obrepanda
Echinopsis roseolilacina → ?Echinopsis mamillosa
Echinopsis rowleyi → Echinopsis huascha
©*Echinopsis rubinghiana* → Echinopsis thelegonoides
Echinopsis saltensis
Echinopsis sanguiniflora
Echinopsis santaensis
Echinopsis santiaguensis → Echinopsis sp.
Echinopsis schickendantzii
Echinopsis schieliana
Echinopsis schoenii
Echinopsis schreiteri
Echinopsis schwantesii → Echinopsis oxygona
Echinopsis scopulicola
©*Echinopsis semidenudata*
Echinopsis shaferi → Echinopsis leucantha
Echinopsis silvatica → Echinopsis mamillosa ssp. silvatica
Echinopsis silvestrii
Echinopsis skottsbergii
Echinopsis smrziana
Echinopsis spachiana
Echinopsis spegazziniana → Echinopsis leucantha
Echinopsis spinibarbis
Echinopsis spiniflora → Acanthocalycium spiniflorum
Echinopsis stollenwerkiana → Echinopsis pugionacantha
Echinopsis strigosa
Echinopsis subdenudata
Echinopsis sucrensis
Echinopsis tacaquirensis
Echinopsis tacaquirensis ssp. ***taquimbalensis***
Echinopsis tapecuana → Echinopsis obrepanda ssp. tapecuana
Echinopsis taquimbalensis → Echinopsis tacaquirensis ssp. taquimbalensis
Echinopsis taratensis
Echinopsis tarijensis
Echinopsis tarijensis ssp. ***herzogianus***
Echinopsis tarijensis ssp. ***totorensis***
Echinopsis tarmaensis
Echinopsis tegeleriana
Echinopsis terscheckii
Echinopsis thelegona
Echinopsis thelegonoides
Echinopsis thionantha
Echinopsis tiegeliana
Echinopsis toralapana → ?Echinopsis obrepanda
Echinopsis torrecillasensis → Echinopsis ancistrophora ssp. arachnacantha
Echinopsis trichosa
Echinopsis tubiflora
Echinopsis tulhuayacensis
Echinopsis tunariensis
Echinopsis turbinata → Echinopsis eyriesii
©*Echinopsis uebelmanniana* → ?Echinopsis formosa
Echinopsis uyupampensis
Echinopsis valida → Echinopsis sp.
Echinopsis vallegrandensis → Echinopsis huotii ssp. vallegrandensis
Echinopsis vasquezii
Echinopsis vatteri
Echinopsis volliana

Echinopsis walteri
Echinopsis werdermanniana → Echinopsis terscheckii
Echinopsis werdermannii
Echinopsis yungasensis → Echinopsis bridgesii ssp. yungasensis
Echinopsis yuquina

Emorycactus parryi → Echinocactus parryi
Emorycactus polycephalus → Echinocactus polycephalus
Emorycactus xeranthemoides → Echinocactus polycephalus ssp. xeranthemoides

Encephalocarpus strobiliformis♦ → Pelecyphora strobiliformis♦

Eomatucana madisoniorum → Matucana madisoniorum
Eomatucana oreodoxa → Matucana oreodoxa

Epiphyllanthus candidus → Schlumbergera microsphaerica ssp. candida
Epiphyllanthus microsphaericus → Schlumbergera microsphaerica
®*Epiphyllanthus obovatus* → Schlumbergera opuntioides
Epiphyllanthus obtusangulus → Schlumbergera microsphaerica

Epiphyllopsis gaertneri → Hatiora gaertneri

Epiphyllum ackermannii → Disocactus ackermannii
Epiphyllum anguliger
Epiphyllum biforme → Disocactus biformis
Epiphyllum bridgesii → Schlumbergera truncata
Epiphyllum cartagense
Epiphyllum caudatum
Epiphyllum caulorhizum → Epiphyllum crenatum
Epiphyllum chrysocardium → Selenicereus chrysocardium
Epiphyllum columbiense
Epiphyllum costaricense
Epiphyllum crenatum
Epiphyllum darrahii → Epiphyllum anguliger
Epiphyllum eichlamii → Disocactus eichlamii
Epiphyllum floribundum
Epiphyllum gigas → Epiphyllum grandilobum
Epiphyllum grandilobum
Epiphyllum guatemalense
Epiphyllum hookeri
Epiphyllum laui
Epiphyllum lepidocarpum
Epiphyllum macrocarpum → Epiphyllum costaricense
Epiphyllum nelsonii → Disocactus nelsonii
Epiphyllum oxypetalum
Epiphyllum phyllanthus
Epiphyllum pittieri
Epiphyllum pumilum
Epiphyllum quezaltecum → Disocactus quezaltecus
Epiphyllum rubrocoronatum
Epiphyllum stenopetalum → Epiphyllum hookeri
Epiphyllum steyermarkii → Selenicereus inermis
Epiphyllum strictum → Epiphyllum hookeri
Epiphyllum thomasianum
Epiphyllum trimetrale

Epithelantha bokei
Epithelantha densispina → Epithelantha micromeris ssp. greggii
Epithelantha greggii → Epithelantha micromeris ssp. greggii
Epithelantha micromeris

Epithelantha micromeris ssp. **greggii**
Epithelantha micromeris ssp. **micromeris**
Epithelantha micromeris ssp. **pachyrhiza**
Epithelantha micromeris ssp. **polycephala**
Epithelantha micromeris ssp. **unguispina**
Epithelantha pachyrhiza → Epithelantha micromeris ssp. pachyrhiza
Epithelantha polycephala → Epithelantha micromeris ssp. polycephala
Epithelantha rufispina → Epithelantha micromeris ssp. greggii

Erdisia apiciflora → Corryocactus apiciflorus
Erdisia aureispina → Corryocactus erectus
Erdisia erecta → Corryocactus erectus
Erdisia fortalezensis → Corryocactus tenuiculus
Erdisia maxima → Corryocactus apiciflorus
Erdisia meyenii → Corryocactus aureus
Erdisia philippii → Austrocactus philippii
Erdisia quadrangularis → Corryocactus quadrangularis
Erdisia ruthae → Corryocactus erectus
Erdisia spiniflora → Austrocactus spiniflorus
Erdisia squarrosa → Corryocactus squarrosus
Erdisia tenuicula → Corryocactus tenuiculus
Eriocactus ampliocostatus → Parodia schumanniana

Eriocactus claviceps → Parodia schumanniana ssp. claviceps
Eriocactus leninghausii → Parodia leninghausii
Eriocactus magnificus → Parodia magnifica
Eriocactus nigrispinus → Parodia nigrispina
Eriocactus schumannianus → Parodia schumanniana ssp. schumanniana
Eriocactus warasii → Parodia warasii

Eriocereus adscendens → Harrisia adscendens
Eriocereus arendtii → Harrisia tortuosa
®*Eriocereus bonplandii* → Harrisia pomanensis
Eriocereus guelichii → Harrisia balansae
Eriocereus jusbertii → Harrisia 'Jusbertii'
Eriocereus martinii → Harrisia martinii
Eriocereus polyacanthus → Harrisia pomanensis
Eriocereus pomanensis → Harrisia pomanensis
Eriocereus regelii → Harrisia pomanensis ssp. regelii
Eriocereus spinosissimus → Arthrocereus spinosissimus
Eriocereus tarijensis → Harrisia pomanensis
Eriocereus tephracanthus → Harrisia tetracantha
Eriocereus tortuosus → Harrisia tortuosa

Eriosyce aerocarpa
Eriosyce algarrobensis → Eriosyce aurata
Eriosyce andreaeana
Eriosyce aspillagae
Eriosyce aurata
Eriosyce bulbocalyx
®*Eriosyce ceratistes* → Eriosyce aurata
Eriosyce chilensis
Eriosyce confinis
Eriosyce crispa
Eriosyce crispa ssp. **atroviridis**
Eriosyce crispa ssp. **crispa**
Eriosyce curvispina
Eriosyce engleri
Eriosyce esmeraldana
Eriosyce garaventae

Eriosyce heinrichiana
Eriosyce heinrichiana ssp. **heinrichiana**
Eriosyce heinrichiana ssp. **intermedia**
Eriosyce heinrichiana ssp. **simulans**
Eriosyce ihotzkyanae → Eriosyce aurata
Eriosyce islayensis
Eriosyce krausii
Eriosyce kunzei
Eriosyce lapampaensis → Eriosyce aurata
Eriosyce laui
Eriosyce limariensis
Eriosyce marksiana
Eriosyce megacarpa → Eriosyce rodentiophila
Eriosyce napina
Eriosyce napina ssp. **lembckei**
Eriosyce napina ssp. **napina**
Eriosyce occulta
Eriosyce odieri
Eriosyce odieri ssp. **fulva**
Eriosyce odieri ssp. **glabrescens**
Eriosyce odieri ssp. **odieri**
Eriosyce omasensis
Eriosyce recondita
Eriosyce recondita ssp. **iquiquensis**
Eriosyce recondita ssp. **recondita**
Eriosyce rodentiophila
Eriosyce sandillon → Eriosyce aurata
Eriosyce senilis
Eriosyce senilis ssp. **coimasensis**
Eriosyce senilis ssp. **elquiensis**
Eriosyce senilis ssp. **senilis**
Eriosyce sociabilis
Eriosyce spinibarbis → Eriosyce aurata
Eriosyce strausiana
Eriosyce subgibbosa
Eriosyce subgibbosa ssp. **clavata**
Eriosyce subgibbosa ssp. **subgibbosa**
Eriosyce taltalensis
Eriosyce taltalensis ssp. **echinus**
Eriosyce taltalensis ssp. **paucicostata**
Eriosyce taltalensis ssp. **pilispina**
Eriosyce taltalensis ssp. **taltalensis**
Eriosyce tenebrica
Eriosyce umadeave
Eriosyce vertongenii
Eriosyce villicumensis
Eriosyce villosa

Erythrorhipsalis burchellii → Rhipsalis burchellii
Erythrorhipsalis campos-portoana → Rhipsalis campos-portoana
Erythrorhipsalis cereuscula → Rhipsalis cereuscula
®*Erythrorhipsalis cribrata* → Rhipsalis teres
Erythrorhipsalis pilocarpa → Rhipsalis pilocarpa

Escobaria aguirreana
Escobaria albicolumnaria
Escobaria alversonii
Escobaria asperispina → Escobaria missouriensis ssp. asperispina
Escobaria bella → Escobaria emskoetteriana
Escobaria bisbeeana → Escobaria vivipara

Escobaria chaffeyi → Escobaria dasyacantha ssp. chaffeyi
Escobaria chihuahuensis
Escobaria chihuahuensis ssp. **chihuahuensis**
Escobaria chihuahuensis ssp. **henricksonii**
Escobaria cubensis
Escobaria dasyacantha
Escobaria dasyacantha ssp. **chaffeyi**
Escobaria dasyacantha ssp. **dasyacantha**
Escobaria deserti
Escobaria duncanii
Escobaria emskoetteriana
Escobaria guadalupensis
Escobaria henricksonii → Escobaria chihuahuensis ssp. henricksonii
Escobaria hesteri
Escobaria laredoi
Escobaria leei♦ → Escobaria sneedii ssp. leei♦
Escobaria lloydii
Escobaria minima♦
Escobaria missouriensis
Escobaria missouriensis ssp. **asperispina**
Escobaria missouriensis ssp. **missouriensis**
Escobaria missouriensis ssp. *navajoensis* → Escobaria missouriensis ssp. missouriensis
Escobaria muehlbaueriana → Escobaria emskoetteriana
Escobaria nellieae♦ → Escobaria minima♦
Escobaria neomexicana → Escobaria vivipara
Escobaria orcuttii
Escobaria organensis
Escobaria radiosa → Escobaria vivipara
®*Escobaria rigida* → Escobaria laredoi
Escobaria robbinsorum
Escobaria roseana
Escobaria roseana ssp. ***galeanensis***
Escobaria runyonii → Escobaria emskoetteriana
Escobaria sandbergii
Escobaria sneedii♦
Escobaria sneedii ssp. **leei**♦
Escobaria sneedii ssp. **sneedii**♦
¶*Escobaria strobiliformis* → ?Escobaria chihuahuensis
®*Escobaria strobiliformis* → Escobaria tuberculosa
Escobaria tuberculosa
Escobaria varicolor → Escobaria tuberculosa
Escobaria villardii
Escobaria vivipara
Escobaria zilziana

Escontria chiotilla
Escontria lepidantha → Pachycereus lepidanthus

Espostoa baumannii
Espostoa blossfeldiorum
Espostoa calva
Espostoa dautwitzii → Espostoa lanata
Espostoa frutescens
Espostoa guentheri
®*Espostoa haagei* → Espostoa melanostele
Espostoa huanucoensis
Espostoa hylaea
Espostoa lanata
Espostoa lanianuligera
Espostoa laticornua → Espostoa lanata

Espostoa melanostele
Espostoa mirabilis
Espostoa nana
Espostoa procera → Espostoa lanata
Espostoa ritteri
Espostoa ruficeps
Espostoa senilis
Espostoa superba

×***Espostocactus mirabilis*** (Cleistocactus icosagonus × Espostoa lanata)

Espostoopsis dybowskii

Eulychnia acida
Eulychnia aricensis
Eulychnia barquitensis → Eulychnia breviflora
Eulychnia breviflora
Eulychnia castanea
Eulychnia iquiquensis
Eulychnia morromorenoensis → Eulychnia iquiquensis
Ⓤ***Eulychnia procumbens***
Eulychnia ritteri
Eulychnia saint-pieana → Eulychnia breviflora
Eulychnia spinibarbis → Echinopsis spinibarbis

Facheiroa cephaliomelana
Facheiroa cephaliomelana ssp. **cephaliomelana**
Facheiroa cephaliomelana ssp. **estevesii**
Facheiroa chaetacantha → Facheiroa squamosa
Facheiroa estevesii → Facheiroa cephaliomelana ssp. estevesii
Facheiroa pilosa → Facheiroa cephaliomelana ssp. cephaliomelana
Facheiroa pubiflora → Facheiroa ulei
Facheiroa squamosa
Facheiroa squamosa ssp. *polygona*
Facheiroa tenebrosa → Facheiroa cephaliomelana ssp. cephaliomelana
Facheiroa ulei

Ⓜ*Ferocactus acanthodes* → Ferocactus cylindraceus
Ferocactus alamosanus
Ferocactus alamosanus ssp. **alamosanus**
Ferocactus alamosanus ssp. **reppenhagenii**
Ferocactus bicolor → Thelocactus bicolor ssp. bicolor
Ferocactus chrysacanthus
Ferocactus chrysacanthus ssp. **chrysacanthus**
Ferocactus chrysacanthus ssp. **grandiflorus**
Ferocactus coloratus → Ferocactus gracilis ssp. coloratus
Ferocactus coptonogonus → Stenocactus coptonogonus
Ferocactus covillei → Ferocactus emoryi ssp. emoryi
Ferocactus crassihamatus → Sclerocactus uncinatus ssp. crassihamatus
Ferocactus crispatus → Stenocactus crispatus
Ferocactus cylindraceus
Ferocactus cylindraceus ssp. **cylindraceus**
Ferocactus cylindraceus ssp. **lecontei**
Ferocactus cylindraceus ssp. **tortulispinus**
Ferocactus diguetii
Ferocactus eastwoodiae
Ferocactus echidne
Ferocactus emoryi
Ferocactus emoryi ssp. **emoryi**
Ferocactus emoryi ssp. **rectispinus**

Ferocactus flavovirens
Ferocactus fordii
Ferocactus gatesii → Ferocactus gracilis ssp. gatesii
Ferocactus glaucescens
Ferocactus glaucus♦ → Sclerocactus glaucus♦
Ferocactus gracilis
Ferocactus gracilis ssp. **coloratus**
Ferocactus gracilis ssp. **gatesii**
Ferocactus gracilis ssp. **gracilis**
Ferocactus haematacanthus
Ferocactus hamatacanthus
Ferocactus hamatacanthus ssp. **hamatacanthus**
Ferocactus hamatacanthus ssp. **sinuatus**
Ferocactus hastifer → Thelocactus hastifer
Ferocactus herrerae
Ferocactus hertrichii → Ferocactus cylindraceus ssp. lecontei
Ferocactus heterochromus → Thelocactus heterochromus
Ferocactus histrix
Ferocactus horridus → Ferocactus peninsulae
Ferocactus johnsonii → Sclerocactus johnsonii
Ferocactus johnstonianus
Ferocactus latispinus
Ferocactus latispinus ssp. **latispinus**
Ferocactus latispinus ssp. **spiralis**
Ferocactus lecontei → Ferocactus cylindraceus ssp. lecontei
Ferocactus leucacanthus → Thelocactus leucacanthus
Ferocactus lindsayi
Ferocactus macrodiscus
Ferocactus macrodiscus ssp. **macrodiscus**
Ferocactus macrodiscus ssp. **septentrionalis**
Ferocactus mathssonii → Sclerocactus uncinatus ssp. crassihamatus
Ferocactus melocactiformis → Ferocactus histrix
Ferocactus mesae-verdae♦ → Sclerocactus mesae-verdae♦
Ferocactus nobilis → ?Ferocactus latispinus
Ferocactus orcuttii → Ferocactus viridescens
Ferocactus parviflorus → Sclerocactus parviflorus ssp. parviflorus
Ferocactus peninsulae
Ferocactus pfeifferi → Ferocactus glaucescens
Ferocactus phyllacanthus → Stenocactus phyllacanthus
Ⓤ*Ferocactus piliferus* → Ferocactus pilosus
Ferocactus pilosus
Ferocactus polyancistrus → Sclerocactus polyancistrus
Ferocactus pottsii
Ferocactus pringlei → Ferocactus pilosus
Ferocactus pubispinus♦ → Sclerocactus pubispinus♦
Ferocactus rectispinus → Ferocactus emoryi ssp. rectispinus
Ⓜ*Ferocactus recurvus* → Ferocactus latispinus
Ferocactus reppenhagenii → Ferocactus alamosanus ssp. reppenhagenii
Ⓤ*Ferocactus rhodanthus* → Ferocactus echidne
Ferocactus robustus
Ferocactus rostii → Ferocactus cylindraceus
Ferocactus santa-maria
Ferocactus scheeri → Sclerocactus scheeri
Ferocactus schwarzii
Ferocactus setispinus → Thelocactus setispinus
Ferocactus spinosior
Ferocactus stainesii → Ferocactus pilosus
Ferocactus tiburonensis
Ferocactus tobuschii♦ → Sclerocactus brevihamatus ssp. tobuschii♦
Ferocactus tortulispinus → Ferocactus cylindraceus ssp. tortulispinus

©*Ferocactus tortulospinus* → Ferocactus cylindraceus ssp. tortulispinus
Ferocactus townsendianus
Ferocactus uncinatus → Sclerocactus uncinatus ssp. uncinatus
Ferocactus vaupelianus → Stenocactus vaupelianus
Ferocactus victoriensis → Ferocactus echidne
Ferocactus viridescens
Ferocactus viscainensis → Ferocactus gracilis ssp. coloratus
Ferocactus whipplei → Sclerocactus whipplei
Ferocactus wislizeni
Ferocactus wrightiae♦ → Sclerocactus wrightiae♦

Floribunda bahiensis → Arrojadoa bahiensis
Floribunda pusilliflora → Cipocereus pusilliflorus

Frailea alacriportana → Frailea gracillima ssp. gracillima
Frailea albiareolata → Frailea pumila
Frailea albicolumnaris → Frailea pygmaea ssp. albicolumnaris
Frailea albifusca → Frailea gracillima ssp. gracillima
Frailea asperispina → Frailea pygmaea ssp. pygmaea
Frailea aureinitens → Frailea pygmaea ssp. pygmaea
Frailea aureispina → Frailea pygmaea ssp. pygmaea
Frailea buenekeri
Frailea buenekeri ssp. ***densispina***
Frailea buiningiana
Frailea caespitosa → Parodia concinna
Frailea carminifilamentosa → Frailea pumila
Frailea castanea
Frailea castanea ssp. ***harmoniana***
Frailea cataphracta
Frailea cataphracta ssp. **cataphracta**
Frailea cataphracta ssp. **duchii**
Frailea cataphracta ssp. *melitae* → Frailea cataphracta ssp. duchii
Frailea cataphracta ssp. *tuyensis* → Frailea cataphracta ssp. cataphracta
Frailea cataphractoides → Frailea cataphracta ssp. duchii
Frailea chiquitana
©*Frailea chrysacantha* → Frailea pumila
Frailea colombiana → Frailea pumila
Frailea concepcionensis → Frailea schilinzkyana
Frailea curvispina
Frailea deminuta → Frailea pumila ssp. deminuta
Frailea friedrichii
Frailea fulviseta → Frailea pygmaea ssp. fulviseta
Frailea gracillima
Frailea gracillima ssp. *alacriportana* → Frailea gracillima ssp. gracillima
Frailea gracillima ssp. *albifusca* → Frailea gracillima ssp. gracillima
Frailea gracillima ssp. **gracillima**
Frailea gracillima ssp. **horstii**
Frailea grahliana
Frailea grahliana ssp. *concepcionensis* → Frailea schilinzkyana
Frailea grahliana ssp. ***moseriana***
Frailea grahliana ssp. *ybatensis* → Frailea schilinzkyana
Frailea hlineckyana → Frailea pumila
Frailea horstii → Frailea gracillima ssp. horstii
Frailea horstii ssp. *fecotrigensis* → Frailea gracillima ssp. horstii
Frailea ignacionensis → Frailea schilinzkyana
Frailea jajoiana → Frailea pumila
Frailea knippeliana
Frailea larae → Frailea chiquitana
Frailea lepida → Frailea gracillima ssp. gracillima
©*Frailea magnifica* → Frailea mammifera

Frailea mammifera
Frailea matoana → Frailea cataphracta ssp. duchii
Frailea melitae → Frailea cataphracta ssp. duchii
Frailea moseriana → Frailea grahliana ssp. moseriana
Frailea perbella → Frailea phaeodisca
Frailea perumbilicata
Frailea phaeodisca
Frailea pseudogracillima → Frailea gracillima ssp. gracillima
Frailea pseudopulcherrima
Frailea pulcherrima → Frailea pygmaea
Frailea pullispina → Frailea chiquitana
Frailea pumila
Frailea pumila ssp. *albiareolata* → Frailea pumila
Frailea pumila ssp. *colombiana* → Frailea pumila
Frailea pumila ssp. ***deminuta***
Frailea pumila ssp. *hlineckyana* → Frailea pumila
Frailea pumila ssp. *jajoiana* → Frailea pumila
Frailea pumila ssp. *maior* → Frailea pumila
Frailea pygmaea
Frailea pygmaea ssp. **albicolumnaris**
Frailea pygmaea ssp. *altigibbera* → Frailea pygmaea ssp. pygmaea
Frailea pygmaea ssp. *asperispina* → Frailea pygmaea ssp. pygmaea
Frailea pygmaea ssp. *aureinitens* → Frailea pygmaea ssp. pygmaea
Frailea pygmaea ssp. *aureispina* → Frailea pygmaea ssp. pygmaea
Frailea pygmaea ssp. ***fulviseta***
Frailea pygmaea ssp. *lilalunula* → Frailea pygmaea ssp. pygmaea
Frailea pygmaea ssp. **pygmaea**
Frailea schilinzkyana
Frailea schilinzkyana ssp. *concepcionensis* → Frailea schilinzkyana
Frailea uhligiana → Frailea cataphracta ssp. duchii
Frailea ybatensis → Frailea schilinzkyana

Geohintonia mexicana

Gerocephalus dybowskii → Espostoopsis dybowskii

Glandulicactus crassihamatus → Sclerocactus uncinatus ssp. crassihamatus
Glandulicactus mathssonii → Sclerocactus uncinatus ssp. crassihamatus
Glandulicactus uncinatus → Sclerocactus uncinatus ssp. uncinatus
Glandulicactus wrightii → Sclerocactus uncinatus ssp. wrightii

Grusonia bradtiana → Opuntia bradtiana
Grusonia bulbispina → Opuntia bulbispina
Grusonia hamiltonii → Opuntia rosarica
Grusonia santamaria → Opuntia santamaria
Grusonia wrightiana → Opuntia kunzei

Gymnanthocereus macracanthus → Browningia pilleifera
Gymnanthocereus pilleifer → Browningia pilleifera

Gymnocactus aguirreanus → Escobaria aguirreana
ⓂⓊ*Gymnocactus beguinii*♦ → Sclerocactus erectocentrus♦
Ⓤ*Gymnocactus beguinii* → Turbinicarpus beguinii
Gymnocactus conothelos → Thelocactus conothelos ssp. conothelos
Gymnocactus gielsdorfianus♦ → Turbinicarpus gielsdorfianus♦
Gymnocactus goldii♦ → Turbinicarpus horripilus♦
Gymnocactus horripilus♦ → Turbinicarpus horripilus♦
Gymnocactus knuthianus♦ → Turbinicarpus knuthianus♦
Gymnocactus mandragora♦ → Turbinicarpus mandragora♦
Gymnocactus roseanus → Escobaria roseana

Gymnocactus roseiflorus♦ → Turbinicarpus sp.♦
Gymnocactus saueri♦ → Turbinicarpus saueri♦
Gymnocactus subterraneus♦ → Turbinicarpus subterraneus♦
Gymnocactus valdezianus♦ → Turbinicarpus valdezianus♦
Gymnocactus viereckii♦ → Turbinicarpus viereckii♦
Gymnocactus ysabelae♦ → Turbinicarpus ysabelae♦

Gymnocalycium achirasense → Gymnocalycium monvillei ssp. achirasense
Gymnocalycium acorrugatum → Gymnocalycium castellanosii
Gymnocalycium albispinum → Gymnocalycium bruchii
Gymnocalycium ambatoense
Gymnocalycium amerhauseri
Gymnocalycium andreae
Gymnocalycium andreae ssp. ***carolinense***
Gymnocalycium angelae
Gymnocalycium anisitsii
Gymnocalycium anisitsii ssp. ***multiproliferum***
Gymnocalycium antherostele → Gymnocalycium schickendantzii
Gymnocalycium armatum → Gymnocalycium spegazzinii ssp. cardenasianum
Gymnocalycium artigas → Gymnocalycium uruguayense
Gymnocalycium asterium → Gymnocalycium stellatum
Gymnocalycium baldianum
Gymnocalycium bayrianum
Gymnocalycium berchtii
Gymnocalycium bicolor → Gymnocalycium mostii
Gymnocalycium bodenbenderianum
Gymnocalycium bodenbenderianum ssp. ***intertextum***
Gymnocalycium borthii
Gymnocalycium bozsingianum → Gymnocalycium castellanosii
Gymnocalycium brachyanthum → Gymnocalycium monvillei ssp. brachyanthum
Gymnocalycium brachypetalum → Gymnocalycium gibbosum
®*Gymnocalycium brevistylum* → Gymnocalycium marsoneri ssp. matoense
Gymnocalycium bruchii
Gymnocalycium buenekeri
Gymnocalycium calochlorum
Gymnocalycium capillaense
Gymnocalycium cardenasianum → Gymnocalycium spegazzinii ssp. cardenasianum
Gymnocalycium carminanthum
Gymnocalycium castellanosii
Gymnocalycium catamarcense
Gymnocalycium catamarcense ssp. ***acinacispinum***
Gymnocalycium catamarcense ssp. ***schmidianum***
Gymnocalycium chiquitanum
Gymnocalycium chubutense → Gymnocalycium gibbosum
Gymnocalycium chuquisacanum → Gymnocalycium pflanzii
Gymnocalycium comarapense → Gymnocalycium pflanzii
Gymnocalycium curvispinum
Gymnocalycium damsii → Gymnocalycium anisitsii
Gymnocalycium deeszianum
Gymnocalycium delaetii
Gymnocalycium denudatum
Gymnocalycium erinaceum
Gymnocalycium eurypleurum
Gymnocalycium eytianum
®*Gymnocalycium ferox* → Gymnocalycium castellanosii
Gymnocalycium ferrarii → Gymnocalycium mucidum
®***Gymnocalycium fleischerianum***
Gymnocalycium fricianum → Gymnocalycium marsoneri
®*Gymnocalycium friedrichii* → Gymnocalycium stenopleurum
Gymnocalycium gerardii → Gymnocalycium gibbosum

Gymnocalycium gibbosum
Gymnocalycium gibbosum ssp. ***ferox***
Gymnocalycium glaucum → Gymnocalycium mucidum
Gymnocalycium grandiflorum → Gymnocalycium mostii
Gymnocalycium griseopallidum → Gymnocalycium anisitsii
Gymnocalycium guanchinense → Gymnocalycium hossei
Gymnocalycium guerkeanum → Gymnocalycium uruguayense
Gymnocalycium hamatum → Gymnocalycium marsoneri
Gymnocalycium hammerschmidii → Gymnocalycium chiquitanum
Gymnocalycium horizonthalonium → Gymnocalycium spegazzinii
Gymnocalycium horridispinum → Gymnocalycium monvillei ssp. horridispinum
Gymnocalycium horstii
Gymnocalycium hossei
Gymnocalycium hybopleurum
Gymnocalycium hyptiacanthum
Gymnocalycium immemoratum → Gymnocalycium mostii
Gymnocalycium intertextum → Gymnocalycium bodenbenderianum ssp. intertextum
Gymnocalycium izozogsii → Gymnocalycium pflanzii
Gymnocalycium joossensianum → Gymnocalycium anisitsii
Gymnocalycium kieslingii
Gymnocalycium kurtzianum → Gymnocalycium mostii
Gymnocalycium lagunillasense → Gymnocalycium pflanzii
Gymnocalycium leeanum
Gymnocalycium leptanthum
®*Gymnocalycium loricatum* → Gymnocalycium spegazzinii
Gymnocalycium mackieanum
Gymnocalycium marquezii → Gymnocalycium pflanzii
Gymnocalycium marsoneri
Gymnocalycium marsoneri ssp. ***matoense***
Gymnocalycium matoense → Gymnocalycium marsoneri ssp. matoense
Gymnocalycium mazanense → Gymnocalycium hossei
Gymnocalycium megalothelon → ?Gymnocalycium monvillei
Gymnocalycium megatae → Gymnocalycium marsoneri ssp. matoense
Gymnocalycium melanocarpum → Gymnocalycium uruguayense
Gymnocalycium mesopotamicum
Gymnocalycium michoga → Gymnocalycium schickendantzii
Gymnocalycium mihanovichii
Gymnocalycium millaresii → Gymnocalycium pflanzii
Gymnocalycium monvillei
Gymnocalycium monvillei ssp. ***achirasense***
Gymnocalycium monvillei ssp. ***brachyanthum***
Gymnocalycium monvillei ssp. ***horridispinum***
Gymnocalycium moserianum → Gymnocalycium bodenbenderianum ssp. intertextum
Gymnocalycium mostii
Gymnocalycium mucidum
Gymnocalycium multiflorum → Gymnocalycium monvillei
Gymnocalycium netrelianum
Gymnocalycium neuhuberi
Gymnocalycium neumannianum → Rebutia neumanniana
Gymnocalycium nidulans → Gymnocalycium hossei
Gymnocalycium nigriareolatum → Gymnocalycium hybopleurum
Gymnocalycium obductum
®*Gymnocalycium occultum* → Gymnocalycium stellatum ssp. occultum
Gymnocalycium ochoterenae
Gymnocalycium ochoterenae ssp. ***herbsthoferianum***
Gymnocalycium ochoterenae ssp. **ochoternae**
Gymnocalycium ochoterenae ssp. **vatteri**
Gymnocalycium oenanthemum
Gymnocalycium onychacanthum → Gymnocalycium marsoneri
Gymnocalycium ourselianum → Gymnocalycium monvillei

®*Gymnocalycium paediophilum* → Gymnocalycium pediophilum
Gymnocalycium paraguayense
Gymnocalycium parvulum
Gymnocalycium pediophilum
Gymnocalycium pflanzii
Gymnocalycium pflanzii ssp. *argentinense* → Gymnocalycium pflanzii
Gymnocalycium piltziorum → Gymnocalycium riojense ssp. piltziorum
Gymnocalycium platense
®*Gymnocalycium platygonum* → Gymnocalycium riojense
Gymnocalycium proliferum → Gymnocalycium calochlorum
Gymnocalycium pseudomalacocarpus → Gymnocalycium marsoneri ssp. matoense
Gymnocalycium pugionacanthum
Gymnocalycium quehlianum
Gymnocalycium ragonesei
Gymnocalycium rauschii
Gymnocalycium reductum → Gymnocalycium gibbosum
®*Gymnocalycium rhodantherum* → Gymnocalycium hossei
Gymnocalycium riograndense → Gymnocalycium pflanzii
Gymnocalycium riojense
Gymnocalycium riojense ssp. ***kozelskyanum***
Gymnocalycium riojense ssp. ***paucispinum***
Gymnocalycium riojense ssp. ***piltziorum***
Gymnocalycium ritterianum
Gymnocalycium rosae
®*Gymnocalycium saglione* → Gymnocalycium saglionis
Gymnocalycium saglionis
Gymnocalycium saglionis ssp. ***tilcarense***
Gymnocalycium sanguiniflorum → Gymnocalycium baldianum
Gymnocalycium schatzlianum → Gymnocalycium mackieanum
Gymnocalycium schickendantzii
Gymnocalycium schroederianum
Gymnocalycium schroederianum ssp. ***bayense***
Gymnocalycium schroederianum ssp. ***paucicostatum***
Gymnocalycium schuetzianum → Gymnocalycium monvillei
Gymnocalycium sigelianum → Gymnocalycium capillaense
Gymnocalycium spegazzinii
Gymnocalycium spegazzinii ssp. **cardenasianum**
Gymnocalycium stellatum
Gymnocalycium stellatum ssp. ***occultum***
Gymnocalycium stenopleurum
Gymnocalycium striglianum
Gymnocalycium stuckertii
Gymnocalycium sutterianum → Gymnocalycium capillaense
Gymnocalycium taningaense
Gymnocalycium terweemeanum
®*Gymnocalycium tilcarense* → Gymnocalycium saglionis ssp. tilcarense
Gymnocalycium tillianum
Gymnocalycium tobuschianum → Gymnocalycium mostii
Gymnocalycium tortuga → Gymnocalycium marsoneri ssp. matoense
Gymnocalycium triacanthum → Gymnocalycium riojense
Gymnocalycium tudae → Gymnocalycium marsoneri ssp. matoense
Gymnocalycium uebelmannianum
Gymnocalycium uruguayense
Gymnocalycium valnicekianum → Gymnocalycium mostii
Gymnocalycium vatteri → Gymnocalycium ochoterenae ssp. vatteri
Gymnocalycium weissianum → Gymnocalycium hossei
Gymnocalycium zegarrae → Gymnocalycium pflanzii
Gymnocereus altissimus → Browningia altissima
Gymnocereus amstutziae → Browningia amstutziae
Gymnocereus microspermus → Browningia microsperma

Haageocereus acanthocladus → Haageocereus pseudomelanostele
Haageocereus achaetus → Haageocereus acranthus ssp. acranthus
Haageocereus acranthus
Haageocereus acranthus ssp. **acranthus**
Haageocereus acranthus ssp. **olowinskianus**
Haageocereus akersii → Haageocereus pseudomelanostele
Haageocereus albisetatus → ×Haagespostoa albisetata
Haageocereus albispinus
Haageocereus ambiguus → Haageocereus decumbens
Haageocereus aureispinus → Haageocereus pseudomelanostele ssp. aureispinus
Haageocereus australis
Haageocereus cephalomacrostibas → Echinopsis cephalomacrostibas
Haageocereus chalaensis
Haageocereus chosicensis → Haageocereus pseudomelanostele
Haageocereus chrysacanthus → Haageocereus pseudomelanostele
Haageocereus chryseus
Haageocereus churinensis → Weberbauerocereus churinensis
Haageocereus clavatus → Haageocereus pseudomelanostele
Haageocereus clavispinus → Haageocereus acranthus ssp. acranthus
Haageocereus climaxanthus → ×Haagespostoa climaxantha
Haageocereus comosus → Haageocereus sp.
Haageocereus crassiareolatus → Haageocereus pseudomelanostele
Haageocereus cuzcoensis → Weberbauerocereus cuzcoensis
Haageocereus decumbens
Haageocereus deflexispinus → Haageocereus acranthus ssp. acranthus
Haageocereus dichromus → Haageocereus pseudomelanostele
Haageocereus divaricatispinus → Haageocereus pseudomelanostele
Haageocereus fascicularis
Haageocereus fulvus
Haageocereus horrens → Haageocereus pacalaensis
Haageocereus icensis
Haageocereus icosagonoides
Haageocereus johnsonii → Weberbauerocereus johnsonii
Haageocereus lachayensis → Haageocereus acranthus ssp. acranthus
Haageocereus lanugispinus
Haageocereus laredensis → Haageocereus pacalaensis
©*Haageocereus limensis* → ?Haageocereus acranthus ssp. acranthus
Haageocereus litoralis → Haageocereus decumbens
Haageocereus longiareolatus → Haageocereus pseudomelanostele
Haageocereus longicomus → Weberbauerocereus longicomus
Haageocereus mamillatus → Haageocereus decumbens
Haageocereus multangularis → ?Haageocereus pseudomelanostele
Haageocereus multicolorispinus → Haageocereus australis
Haageocereus olowinskianus → Haageocereus acranthus ssp. olowinskianus
Haageocereus pacalaensis
Haageocereus pachystele → Haageocereus pseudomelanostele
Haageocereus paradoxus → Cleistocactus acanthurus ssp. acanthurus
Haageocereus peculiaris → Cleistocactus peculiaris
Haageocereus peniculatus → Haageocereus albispinus
Haageocereus piliger → Haageocereus pseudomelanostele
Haageocereus platinospinus
Haageocereus pluriflorus
Haageocereus pseudoacranthus → Haageocereus acranthus ssp. acranthus
Haageocereus pseudomelanostele
Haageocereus pseudomelanostele ssp. **aureispinus**
Haageocereus pseudomelanostele ssp. **carminiflorus**
Haageocereus pseudomelanostele ssp. **pseudomelanostele**
Haageocereus pseudomelanostele ssp. **turbidus**
Haageocereus pseudoversicolor
Haageocereus rauhii → Weberbauerocereus rauhii

Haageocereus repens → Haageocereus pacalaensis
Haageocereus rubrospinus → Haageocereus sp.
Haageocereus salmonoideus → Haageocereus sp.
Haageocereus seticeps → Haageocereus sp.
Haageocereus setosus → Haageocereus pseudomelanostele
Haageocereus smaragdiflorus → Haageocereus sp.
Haageocereus subtilispinus
Haageocereus symmetros → Haageocereus pseudomelanostele
Haageocereus tenuis
Haageocereus tenuispinus → Haageocereus pacalaensis
Haageocereus torataensis → Weberbauerocereus torataensis
Haageocereus turbidus → Haageocereus pseudomelanostele ssp. turbidus
Haageocereus versicolor
Haageocereus viridiflorus → Haageocereus pseudomelanostele
Haageocereus vulpes
Haageocereus weberbaueri → Weberbauerocereus weberbaueri
Haageocereus winterianus → Weberbauerocereus winterianus
Haageocereus zangalensis
Haageocereus zehnderi → Haageocereus pseudomelanostele
Haageocereus zonatus → Haageocereus acranthus ssp. acranthus

×***Haagespostoa albisetata*** (Espostoa melanostele × Haageocereus pseudomelanostele)
×***Haagespostoa climaxantha*** (Espostoa melanostele × Haageocereus albispinus)

Hamatocactus crassihamatus → Sclerocactus uncinatus ssp. crassihamatus
Hamatocactus hamatacanthus → Ferocactus hamatacanthus
Hamatocactus setispinus → Thelocactus setispinus
Hamatocactus sinuatus → Ferocactus hamatacanthus ssp. sinuatus
Hamatocactus uncinatus → Sclerocactus uncinatus ssp. uncinatus

Harrisia aboriginum
Harrisia adscendens
Harrisia balansae
®*Harrisia bonplandii* → Harrisia pomanensis
Harrisia brookii
Harrisia deeringii → Harrisia simpsonii
Harrisia divaricata
Harrisia donae-antoniae → Harrisia gracilis
Harrisia earlei
Harrisia eriophora
Harrisia fernowii
Harrisia fragrans
Harrisia gracilis
Harrisia guelichii → Harrisia balansae
Harrisia hahniana → Echinopsis hahniana
Harrisia hurstii
Harrisia jusbertii → Harrisia 'Jusbertii'
Harrisia martinii
Harrisia nashii
Harrisia pomanensis
®*Harrisia pomanensis* ssp. *bonplandii* → Harrisia pomanensis
Harrisia pomanensis ssp. ***regelii***
Harrisia pomanensis ssp. *tarijensis* → Harrisia pomanensis
Harrisia portoricensis
Harrisia regelii → Harrisia pomanensis ssp. regelii
Harrisia serruliflora → Harrisia divaricata
Harrisia simpsonii
Harrisia taetra
Harrisia taylori
Harrisia tetracantha
Harrisia tortuosa

Haseltonia columna-trajani → Cephalocereus apicicephalium

Hatiora bambusoides → Hatiora salicornioides
Hatiora cylindrica → Hatiora salicornioides
Hatiora epiphylloides
Hatiora epiphylloides ssp. **epiphylloides**
Hatiora epiphylloides ssp. **bradei**
Hatiora gaertneri
Hatiora ×***graeseri*** (H. gaertneri × H. rosea)
Hatiora herminiae
Hatiora rosea
Hatiora salicornioides

Heliabravoa chende → Polaskia chende

Helianthocereus antezanae → Echinopsis antezanae
Helianthocereus atacamensis → Echinopsis atacamensis
Helianthocereus bertramianus → Echinopsis bertramiana
Helianthocereus conaconensis → Echinopsis conaconensis
Ⓤ*Helianthocereus crassicaulis* → Echinopsis crassicaulis
Helianthocereus escayachensis → Echinopsis escayachensis
Helianthocereus grandiflorus → Echinopsis huascha
Helianthocereus herzogianus → Echinopsis tarijensis ssp. herzogianus
Helianthocereus huascha → Echinopsis huascha
Helianthocereus hyalacanthus → Echinopsis huascha
Helianthocereus narvaecensis → ?Echinopsis tarijensis
Helianthocereus orurensis → ?Echinopsis ferox
Helianthocereus pasacana → Echinopsis atacamensis ssp. pasacana
Helianthocereus pecheretianus → Echinopsis huascha
Helianthocereus poco → Echinopsis tarijensis
Ⓤ*Helianthocereus pseudocandicans* → Echinopsis candicans
Helianthocereus randallii → Echinopsis formosa
Ⓤ*Helianthocereus tarijensis* → Echinopsis tarijensis

Heliocereus aurantiacus → Disocactus aurantiacus
Heliocereus cinnabarinus → Disocactus cinnabarinus
Heliocereus elegantissimus → Disocactus schrankii
Heliocereus heterodoxus → Disocactus cinnabarinus
Heliocereus luzmariae → Disocactus schrankii
Heliocereus schrankii → Disocactus schrankii
Heliocereus speciosissimus → Disocactus speciosus
Heliocereus speciosus → Disocactus speciosus

Hertrichocereus beneckei → Stenocereus beneckei

Hildewintera aureispina → Cleistocactus winteri

Homalocephala texensis → Echinocactus texensis

Horridocactus aconcaguensis → Eriosyce curvispina
Horridocactus andicola → Eriosyce curvispina
Horridocactus armatus → Eriosyce curvispina
Horridocactus atroviridis → Eriosyce crispa ssp. atroviridis
Horridocactus carrizalensis → Eriosyce crispa ssp. atroviridis
Horridocactus choapensis → Eriosyce curvispina
Horridocactus crispus → Eriosyce crispa ssp. crispa
Horridocactus curvispinus → Eriosyce curvispina
Horridocactus echinus → Eriosyce taltalensis ssp. echinus
Horridocactus engleri → Eriosyce engleri
Horridocactus froehlichianus → ?Eriosyce curvispina
Horridocactus garaventae → Eriosyce garaventae

Horridocactus geissei → ?Eriosyce kunzei
Horridocactus grandiflorus → Eriosyce curvispina
Horridocactus heinrichianus → Eriosyce heinrichiana
Horridocactus kesselringianus → ?Eriosyce curvispina
Horridocactus lissocarpus → Eriosyce marksiana
Horridocactus marksianus → Eriosyce marksiana
Horridocactus nigricans → ?Eriosyce limariensis
Ⓤ*Horridocactus robustus* → Eriosyce curvispina
Horridocactus tuberisulcatus → Eriosyce curvispina
Horridocactus vallenarensis → Eriosyce kunzei

Hylocereus antiguensis → Hylocereus trigonus
Hylocereus calcaratus
Hylocereus compressus → Hylocereus triangularis
Hylocereus costaricensis
Hylocereus cubensis → Hylocereus triangularis
Hylocereus escuintlensis
Hylocereus estebanensis
Ⓜ*Hylocereus extensus* → ?Hylocereus lemairei
Hylocereus guatemalensis
Hylocereus lemairei
Hylocereus microcladus
Hylocereus minutiflorus
Hylocereus monacanthus
Hylocereus napoleonis → Hylocereus trigonus
Hylocereus ocamponis
Hylocereus peruvianus
Hylocereus plumieri → Hylocereus trigonus
Hylocereus polyrhizus
Hylocereus purpusii
Hylocereus scandens
Hylocereus schomburgkii → Hylocereus sp.
Hylocereus stenopterus
Hylocereus triangularis
Hylocereus trigonus
Hylocereus trinitatensis → Hylocereus lemairei
Hylocereus undatus
Hylocereus venezuelensis → Hylocereus lemairei

Hymenorebutia aurea → Echinopsis aurea
Hymenorebutia chlorogona → Echinopsis densispina
Hymenorebutia chrysantha → Echinopsis chrysantha
Hymenorebutia cintiensis → Echinopsis lateritia
Hymenorebutia drijveriana → Echinopsis haematantha
Hymenorebutia kuehnrichii → Echinopsis haematantha
Hymenorebutia pusilla → Echinopsis tiegeliana
Hymenorebutia quinesensis → Echinopsis aurea
Hymenorebutia tiegeliana → Echinopsis tiegeliana
Hymenorebutia torataensis → Echinopsis lateritia
Hymenorebutia torreana → Echinopsis lateritia

Islaya bicolor → Eriosyce islayensis
Islaya brevicylindrica → Eriosyce islayensis
Islaya copiapoides → Eriosyce islayensis
Islaya divaricatiflora → Eriosyce islayensis
Islaya flavida → Eriosyce islayensis
Islaya grandiflorens → Eriosyce islayensis
Islaya grandis → Eriosyce islayensis
Islaya islayensis → Eriosyce islayensis
Islaya krainziana → Eriosyce islayensis

Islaya maritima → Eriosyce islayensis
Islaya minor → Eriosyce islayensis
Islaya minuscula → Eriosyce islayensis
Islaya mol(l)endensis → Eriosyce islayensis
Islaya mollendensis → Eriosyce islayensis
Islaya omasensis → Eriosyce omasensis
Islaya paucispina → Eriosyce islayensis
Islaya paucispinosa → Eriosyce islayensis
Islaya unguispina → Eriosyce islayensis

Isolatocereus dumortieri → Stenocereus dumortieri

Jasminocereus galapagensis → Jasminocereus thouarsii
Jasminocereus howellii → Jasminocereus thouarsii
Jasminocereus sclerocarpus → Jasminocereus thouarsii
Jasminocereus thouarsii

Kadenicarpus pseudomacrochele → Turbinicarpus pseudomacrochele ssp. pseudomacrochele

Krainzia guelzowiana → Mammillaria guelzowiana
Krainzia longiflora → Mammillaria longiflora ssp. longiflora

Lasiocereus fulvus
Lasiocereus rupicola

Lemaireocereus aragonii → Stenocereus aragonii
Lemaireocereus beneckei → Stenocereus beneckei
Lemaireocereus cartwrightianus → Armatocereus cartwrightianus
Lemaireocereus chende → Polaskia chende
Lemaireocereus chichipe → Polaskia chichipe
Lemaireocereus deficiens → Stenocereus griseus
Lemaireocereus dumortieri → Stenocereus dumortieri
Lemaireocereus eichlamii → Stenocereus eichlamii
Lemaireocereus godingianus → Armatocereus godingianus
Lemaireocereus griseus → Stenocereus griseus
Lemaireocereus hollianus → Pachycereus hollianus
Lemaireocereus humilis → Armatocereus humilis
®*Lemaireocereus hystrix* → Stenocereus fimbriatus
Lemaireocereus laetus → Armatocereus laetus
Lemaireocereus longispinus → Stenocereus eichlamii
Lemaireocereus martinezii → Stenocereus martinezii
Lemaireocereus mixtecensis → Polaskia chichipe
Lemaireocereus montanus → Stenocereus montanus
Lemaireocereus pruinosus → Stenocereus pruinosus
Lemaireocereus queretaroensis → Stenocereus queretaroensis
Lemaireocereus quevedonis → Stenocereus quevedonis
Lemaireocereus stellatus → Stenocereus stellatus
Lemaireocereus thurberi → Stenocereus thurberi
Lemaireocereus treleasei → Stenocereus treleasei
Lemaireocereus weberi → Pachycereus weberi

Leocereus bahiensis
Leocereus bahiensis ssp. *barreirensis* → Leocereus bahiensis
Leocereus bahiensis ssp. *exiguospinus* → Leocereus bahiensis
Leocereus bahiensis ssp. *robustispinus* → Leocereus bahiensis
Leocereus bahiensis ssp. *urandianus* → Leocereus bahiensis
Leocereus estevesii → Leocereus bahiensis
Leocereus glaziovii → Arthrocereus glaziovii
Leocereus melanurus → Arthrocereus melanurus

Leocereus paulensis → Coleocephalocereus fluminensis ssp. fluminensis
Leocereus urandianus → Leocereus bahiensis

Lepidocoryphantha macromeris → Coryphantha macromeris
Lepidocoryphantha runyonii → Coryphantha macromeris ssp. runyonii

Lepismium aculeatum
Lepismium bolivianum
Lepismium brevispinum
Lepismium chrysanthum → ?Rhipsalis dissimilis
Lepismium crenatum
Lepismium cruciforme
Lepismium dissimile → Rhipsalis dissimilis
Lepismium epiphyllanthoides → Rhipsalis dissimilis
Lepismium erectum → Lepismium ianthothele
Lepismium floccosum → Rhipsalis floccosa ssp. floccosa
Lepismium gibberulum → Rhipsalis floccosa ssp. pulvinigera
Lepismium grandiflorum → Rhipsalis grandiflora
Lepismium houlletianum
Lepismium ianthothele
Lepismium incachacanum
Lepismium lineare → Lepismium warmingianum
Lepismium lorentzianum
Lepismium lumbricoides
Lepismium mataralense → Lepismium ianthothele
Lepismium megalanthum → Rhipsalis neves-armondii
Lepismium micranthum
Lepismium miyagawae
Lepismium monacanthum
Lepismium neves-armondii → Rhipsalis neves-armondii
Lepismium pacheco-leonii → Rhipsalis pacheco-leonis
Lepismium paradoxum → Rhipsalis paradoxa ssp. paradoxa
Lepismium paranganiense
Lepismium pittieri → Rhipsalis floccosa ssp. pittieri
Lepismium pulvinigerum → Rhipsalis floccosa ssp. pulvinigera
Lepismium puniceodiscus → Rhipsalis puniceodiscus
Lepismium saxatile → Rhipsalis dissimilis
Lepismium trigonum → Rhipsalis trigona
Lepismium tucumanense → Rhipsalis floccosa ssp. tucumanensis
Lepismium warmingianum

Leptocereus arboreus
Leptocereus assurgens
Leptocereus carinatus
Leptocereus ekmanii
Leptocereus grantianus
Leptocereus leonii
Leptocereus maxonii
Leptocereus paniculatus
Leptocereus prostratus
Leptocereus quadricostatus
Leptocereus santamarinae
Leptocereus scopulophilus
Leptocereus sylvestris
Leptocereus weingartianus
Leptocereus wrightii

Leptocladodia microhelia → Mammillaria microhelia
Leptocladodia microheliopsis → Mammillaria microhelia

Leuchtenbergia principis

Leucostele rivierei → Echinopsis atacamensis ssp. pasacana

Lobeira macdougallii♦ → Disocactus macdougallii♦

®*Lobivia acanthoplegma* → Echinopsis cinnabarina
Lobivia aculeata → Echinopsis pentlandii
Lobivia adpressispina → Echinopsis pugionacantha
Lobivia aguilarii → Echinopsis obrepanda ssp. calorubra
Lobivia akersii → Echinopsis tegeleriana
Lobivia albipectinata → Rebutia albopectinata
Lobivia allegraiana → Echinopsis hertrichiana
Lobivia amblayensis → Echinopsis haematantha
Lobivia andalgalensis → Echinopsis huascha
Lobivia arachnacantha → Echinopsis ancistrophora ssp. arachnacantha
Lobivia argentea → Echinopsis pentlandii
Lobivia atrovirens → ?Rebutia pygmaea
Lobivia auranitida → Rebutia einsteinii
®*Lobivia aurantiaca* → Echinopsis pentlandii
Lobivia aurea → Echinopsis aurea
Lobivia aureolilacina → Echinopsis ferox
Lobivia aureosenilis → Echinopsis pampana
Lobivia backebergiana → Echinopsis ferox
Lobivia backebergii → Echinopsis backebergii
Lobivia backebergii ssp. *hertrichiana* → Echinopsis hertrichiana
Lobivia backebergii ssp. *schieliana* → Echinopsis schieliana
Lobivia backebergii ssp. *wrightiana* → Echinopsis backebergii
Lobivia backebergii ssp. *zecheri* → Echinopsis backebergii
Lobivia binghamiana → Echinopsis hertrichiana
Lobivia boliviensis → Echinopsis pentlandii
Lobivia brachyantha → Rebutia steinmannii
Lobivia breviflora → Echinopsis sanguiniflora
Lobivia bruchii → Echinopsis bruchii
Lobivia brunneo-rosea → Echinopsis pentlandii
Lobivia buiningiana → Echinopsis marsoneri
Lobivia cabradae → Echinopsis haematantha
Lobivia cachensis → Echinopsis saltensis
Lobivia caespitosa → Echinopsis maximiliana
Lobivia caineana → Echinopsis caineana
Lobivia calorubra → Echinopsis obrepanda ssp. calorubra
Lobivia camataquiensis → Echinopsis lateritia
Lobivia campicola → Echinopsis pugionacantha
Lobivia cardenasiana → Echinopsis ancistrophora ssp. cardenasiana
Lobivia cariquinensis → Echinopsis maximiliana
Lobivia carminantha → Echinopsis lateritia
Lobivia cerasiflora → Echinopsis haematantha
Lobivia charazanensis → Echinopsis maximiliana
Lobivia charcasina → Echinopsis cinnabarina
Lobivia chionantha → Echinopsis thionantha
Lobivia chorrillosensis → Echinopsis haematantha
Lobivia chrysantha → Echinopsis chrysantha
Lobivia chrysantha ssp. *jajoiana* → Echinopsis marsoneri
Lobivia chrysantha ssp. *marsoneri* → Echinopsis marsoneri
Lobivia chrysochete → Echinopsis chrysochete
Lobivia cinnabarina → Echinopsis cinnabarina
®*Lobivia cinnabarina* ssp. *acanthoplegma* → Echinopsis cinnabarina
Lobivia cinnabarina ssp. *prestoana* → Echinopsis cinnabarina
Lobivia cinnabarina ssp. *taratensis* → Echinopsis cinnabarina

Lobivia cintiensis → Echinopsis lateritia
Lobivia claeysiana → Echinopsis ferox
Lobivia corbula → Echinopsis maximiliana
Lobivia coriquinensis → Echinopsis maximiliana
Lobivia cornuta → Echinopsis pugionacantha
Lobivia crassicaulis → Echinopsis crassicaulis
Lobivia cruciaureispina → Echinopsis maximiliana
Lobivia culpinensis → Echinopsis pugionacantha
Lobivia cumingii → Rebutia neocumingii ssp. neocumingii
Lobivia cylindracea → Echinopsis aurea
Lobivia cylindrica → Echinopsis aurea
Lobivia draxleriana → Echinopsis cinnabarina
Lobivia drijveriana → Echinopsis haematantha
Lobivia duursmaiana → Echinopsis sanguiniflora
Lobivia echinata → Echinopsis hertrichiana
Lobivia einsteinii → Rebutia einsteinii
Lobivia elongata → Echinopsis haematantha
Lobivia emmae → Echinopsis saltensis
®*Lobivia euanthema* → Rebutia aureiflora
Lobivia eucaliptana → Rebutia steinmannii
Lobivia fallax ssp. *aurea* → Echinopsis aurea
Lobivia famatimensis → Echinopsis famatinensis
Lobivia ferox → Echinopsis ferox
Lobivia formosa → Echinopsis formosa
Lobivia formosa ssp. *bruchii* → Echinopsis bruchii
Lobivia formosa ssp. *grandis* → Echinopsis bruchii
Lobivia formosa ssp. *tarijensis* → Echinopsis tarijensis
Lobivia fricii → Echinopsis tiegeliana
Lobivia glauca → Echinopsis marsoneri
Lobivia glaucescens → Echinopsis pampana
Lobivia grandiflora → Echinopsis huascha
Lobivia grandis → Echinopsis bruchii
Lobivia haageana → Echinopsis marsoneri
Lobivia haagei → Rebutia pygmaea
Lobivia haematantha → Echinopsis haematantha
Lobivia haematantha ssp. *chorrillosensis* → Echinopsis haematantha
Lobivia haematantha ssp. *densispina* → Echinopsis densispina
Lobivia haematantha ssp. *kuehnrichii* → Echinopsis haematantha
Lobivia hastifera → Echinopsis ferox
Lobivia hermanniana → Echinopsis maximiliana
Lobivia hertrichiana → Echinopsis hertrichiana
Lobivia higginsiana → Echinopsis pentlandii
®*Lobivia hoffmanniana* → Rebutia steinbachii
Lobivia horrida → Echinopsis ferox
Lobivia hualfinensis → Echinopsis haematantha
Lobivia huascha → Echinopsis huascha
Lobivia huilcanota → Echinopsis hertrichiana
Lobivia hystrix → Echinopsis chrysochete
Lobivia incaica → Echinopsis hertrichiana
Lobivia incuiensis → Echinopsis tegeleriana
Lobivia intermedia → Echinopsis maximiliana
Lobivia iridescens → Echinopsis marsoneri
Lobivia jajoiana → Echinopsis marsoneri
Lobivia johnsoniana → Echinopsis pentlandii
Lobivia jujuiensis → Echinopsis chrysantha
Lobivia kieslingii → Echinopsis formosa
Lobivia kuehnrichii → Echinopsis haematantha
Lobivia kupperiana → Echinopsis lateritia
®*Lobivia larabei* → Echinopsis hertrichiana

Lobivia larae → Echinopsis pentlandii
Lobivia lateritia → Echinopsis lateritia
Lobivia laui → Echinopsis hertrichiana
Lobivia lauramarca → Echinopsis pentlandii
Lobivia leptacantha → Echinopsis schieliana
Lobivia leucorhodon → Echinopsis pentlandii
Lobivia leucoviolacea → Echinopsis pentlandii
Lobivia longispina → Echinopsis ferox
Lobivia markusii → Echinopsis chrysochete
Lobivia marsoneri → Echinopsis marsoneri
Lobivia maximiliana → Echinopsis maximiliana
Lobivia maximiliana ssp. *caespitosa* → Echinopsis maximiliana
Lobivia maximiliana ssp. *quiabayensis* → Echinopsis schieliana
Lobivia maximiliana ssp. *westii* → Echinopsis maximiliana
Lobivia miniatiflora → Echinopsis maximiliana
Lobivia minuta → Echinopsis hertrichiana
Lobivia mirabunda → Echinopsis haematantha
Lobivia mistiensis → Echinopsis pampana
Lobivia mizquensis → Echinopsis obrepanda ssp. calorubra
®*Lobivia muhriae* → Echinopsis marsoneri
®*Lobivia multicostata* → Echinopsis pentlandii
Lobivia napina → Echinopsis densispina
Lobivia nealeana → Echinopsis saltensis
®*Lobivia neocinnabarina* → Echinopsis cinnabarina
Lobivia neohaageana → Rebutia pygmaea
Lobivia nigricans → Rebutia nigricans
Lobivia nigrostoma → Echinopsis marsoneri
Lobivia oligotricha → Echinopsis cinnabarina
Lobivia omasuyana → Echinopsis pentlandii
Lobivia orurensis → Rebutia pygmaea
Lobivia oxyalabastra → Echinopsis backebergii
Lobivia pachyacantha → Echinopsis ferox
Lobivia pampana → Echinopsis pampana
Lobivia peclardiana → Echinopsis tiegeliana
Lobivia pectinata → Rebutia pygmaea
Lobivia pentlandii → Echinopsis pentlandii
Lobivia pictiflora → Echinopsis ferox
Lobivia planiceps → Echinopsis hertrichiana
Lobivia pojoensis → Echinopsis ancistrophora ssp. pojoensis
Lobivia polaskiana → Echinopsis chrysantha
Lobivia polycephala → Echinopsis sanguiniflora
Lobivia potosina → Echinopsis ferox
Lobivia prestoana → Echinopsis cinnabarina
Lobivia pseudocachensis → Echinopsis saltensis
Lobivia pseudocariquinensis → Echinopsis maximiliana
®*Lobivia pseudocinnabarina* → Echinopsis cinnabarina
Lobivia pugionacantha → Echinopsis pugionacantha
Lobivia purpureominiata → Echinopsis huascha
Lobivia pusilla → Echinopsis tiegeliana
Lobivia quiabayensis → Echinopsis schieliana
Lobivia raphidacantha → Echinopsis pentlandii
Lobivia rauschii → Echinopsis yuquina
Lobivia rebutioides → Echinopsis densispina
Lobivia ritteri → Rebutia ritteri
Lobivia rosarioana → Echinopsis formosa
Lobivia rossii → Echinopsis pugionacantha ssp. rossii
Lobivia rubescens → Echinopsis marsoneri
Lobivia salitrensis → Echinopsis pugionacantha
Lobivia saltensis → Echinopsis saltensis

Lobivia sanguiniflora → Echinopsis sanguiniflora
Lobivia scheeri → Echinopsis pentlandii
Lobivia schieliana → Echinopsis schieliana
Lobivia schneideriana → Echinopsis pentlandii
Lobivia schreiteri → Echinopsis schreiteri
Lobivia scoparia → Echinopsis densispina
Lobivia scopulina → Echinopsis lateritia
Lobivia shaferi → Echinopsis aurea
Lobivia shaferi ssp. *fallax* → Echinopsis aurea
Lobivia shaferi ssp. *leucomalla* → Echinopsis aurea
Lobivia shaferi ssp. *rubriflora* → Echinopsis aurea
Lobivia sicuaniensis → Echinopsis maximiliana
Lobivia silvestrii → Echinopsis chamaecereus
Lobivia simplex → Echinopsis hertrichiana
Lobivia steinmannii ssp. *melanocentra* → Rebutia steinmannii
Lobivia stilowiana → Echinopsis schreiteri
Lobivia taratensis → Echinopsis cinnabarina
Lobivia thionantha → Echinopsis thionantha
Lobivia tiegeliana → Echinopsis tiegeliana
Lobivia titicacensis → Echinopsis pentlandii
Lobivia tuberculosa → Echinopsis marsoneri
Lobivia uitewaaleana → Echinopsis marsoneri
Lobivia varians → Echinopsis pentlandii
Lobivia varispina → Echinopsis ferox
Lobivia versicolor → Echinopsis pugionacantha
Lobivia vilcabambae → Echinopsis hertrichiana
Lobivia walteri → Echinopsis walteri
Lobivia walterspielii → Echinopsis cinnabarina
Lobivia wegheiana → Echinopsis pentlandii
Lobivia westii → Echinopsis maximiliana
®*Lobivia wilkeae* → Echinopsis ferox
Lobivia winteriana → Echinopsis backebergii
Lobivia wrightiana → Echinopsis backebergii
Lobivia zecheri → Echinopsis backebergii
Lobivia zudanensis → Echinopsis cinnabarina

Lophocereus gatesii → Pachycereus gatesii
Lophocereus mieckleyanus → Pachycereus schottii
Lophocereus sargentianus → Pachycereus schottii
Lophocereus schottii → Pachycereus schottii

Lophophora diffusa
Lophophora diffusa ssp. *fricii* → Lophophora williamsii
Lophophora diffusa ssp. *viridescens* → Lophophora williamsii
Lophophora echinata → Lophophora williamsii
Lophophora fricii → Lophophora williamsii
Lophophora jourdaniana → Lophophora williamsii
Lophophora lutea → Lophophora diffusa
Lophophora williamsii

Loxanthocereus acanthurus → Cleistocactus acanthurus ssp. acanthurus
Loxanthocereus aticensis → Cleistocactus sextonianus
Loxanthocereus bicolor → Cleistocactus acanthurus ssp. acanthurus
Loxanthocereus brevispinus → Cleistocactus peculiaris
Loxanthocereus camanaensis → Cleistocactus sextonianus
Loxanthocereus canetensis → Cleistocactus acanthurus ssp. acanthurus
Loxanthocereus cantaensis → Cleistocactus peculiaris
Loxanthocereus convergens → Cleistocactus acanthurus ssp. acanthurus
Loxanthocereus crassiserpens → Cleistocactus crassiserpens

Loxanthocereus cullmannianus → Cleistocactus acanthurus ssp. acanthurus
Loxanthocereus deserticola → ?Cleistocactus clavispinus
Loxanthocereus erectispinus → ?Cleistocactus × Espostoa hybr.
Loxanthocereus eremiticus → Cleistocactus acanthurus ssp. acanthurus
Loxanthocereus erigens → Cleistocactus acanthurus ssp. acanthurus
Loxanthocereus eriotrichus → Cleistocactus acanthurus ssp. acanthurus
Loxanthocereus eulalianus → Cleistocactus acanthurus ssp. acanthurus
Loxanthocereus faustianus → Cleistocactus acanthurus ssp. faustianus
Loxanthocereus ferrugineus → ?Cleistocactus clavispinus
Loxanthocereus formosus → Matucana formosa
Loxanthocereus gracilis → Cleistocactus sextonianus
Loxanthocereus gracilispinus → Cleistocactus acanthurus ssp. acanthurus
Loxanthocereus granditessellatus → Cleistocactus serpens
Loxanthocereus hoffmannii → ?Cleistocactus × Espostoa hybr.
Loxanthocereus hystrix → Cleistocactus hystrix
Loxanthocereus jajoianus → Cleistocactus sepium
Loxanthocereus keller-badensis → Cleistocactus acanthurus ssp. acanthurus
Loxanthocereus madisoniorum → Matucana madisoniorum
Loxanthocereus montanus → Cleistocactus hystrix
Loxanthocereus multifloccosus → Cleistocactus acanthurus ssp. acanthurus
Loxanthocereus nanus → Cleistocactus sextonianus
Loxanthocereus neglectus → Cleistocactus acanthurus ssp. acanthurus
Loxanthocereus otuscensis → Cleistocactus serpens
Loxanthocereus pacaranensis → Cleistocactus acanthurus ssp. acanthurus
Loxanthocereus pachycladus → Cleistocactus pachycladus
Loxanthocereus parvitesselatus → Cleistocactus serpens
Loxanthocereus peculiaris → Cleistocactus peculiaris
Loxanthocereus piscoensis → Cleistocactus pachycladus
Loxanthocereus pullatus → Cleistocactus acanthurus ssp. pullatus
Loxanthocereus puquiensis → Cleistocactus sextonianus
Loxanthocereus riomajensis → Cleistocactus sextonianus
Loxanthocereus sextonianus → Cleistocactus sextonianus
Loxanthocereus splendens → Cleistocactus sextonianus
Loxanthocereus sulcifer → Cleistocactus serpens
Loxanthocereus trujilloensis → Cleistocactus chotaensis
Loxanthocereus variabilis → Cleistocactus sextonianus
Loxanthocereus xylorhizus → Cleistocactus xylorhizus
Loxanthocereus yauyosensis → Cleistocactus pachycladus

Lymanbensonia micrantha → Lepismium micranthum

Machaerocereus eruca → Stenocereus eruca
Machaerocereus gummosus → Stenocereus gummosus

Maihuenia albolanata → Maihuenia patagonica
Maihuenia brachydelphys → Maihuenia patagonica
Maihuenia cumulata → Maihuenia patagonica
Maihuenia latispina → Maihuenia patagonica
Maihuenia patagonica
Maihuenia philippii → Maihuenia poeppigii
Maihuenia poeppigii
Maihuenia valentinii → Maihuenia patagonica

Maihueniopsis albomarginata → Opuntia darwinii
Maihueniopsis archiconoidea → Opuntia archiconoidea
⊗*Maihueniopsis atacamensis* → Opuntia sp.
Maihueniopsis boliviana → Opuntia boliviana
Maihueniopsis camachoi → Opuntia camachoi
Maihueniopsis colorea → Opuntia colorea

®*Maihueniopsis conoidea* → Opuntia conoidea
Maihueniopsis crassispina → Opuntia crassipina
Maihueniopsis darwinii → Opuntia darwinii
Maihueniopsis domeykoensis → Opuntia domeykoensis
Maihueniopsis glomerata → Opuntia glomerata
Maihueniopsis grandiflora → Opuntia llanos-de-huanta
Maihueniopsis hypogaea → Opuntia glomerata
Maihueniopsis leoncito → Opuntia glomerata
Maihueniopsis leptoclada → Opuntia glomerata
Maihueniopsis mandragora → Opuntia minuta
Maihueniopsis minuta → Opuntia minuta
Maihueniopsis molfinoi → Opuntia glomerata
Maihueniopsis molinensis → Opuntia molinensis
Maihueniopsis neuquensis → Opuntia darwinii
Maihueniopsis nigrispina → Opuntia nigrispina
Maihueniopsis ovallei → Opuntia glomerata
Maihueniopsis ovata → Opuntia ovata
Maihueniopsis pentlandii → Opuntia pentlandii
Maihueniopsis rahmeri → Opuntia rahmeri
Maihueniopsis tarapacana → Opuntia tarapacana
Maihueniopsis wagenknechtii → Opuntia wagenknechtii

Mamillopsis diguetii → Mammillaria senilis
Mamillopsis senilis → Mammillaria senilis

Mammillaria acultzingensis → Mammillaria haageana ssp. acultzingensis
Mammillaria alamensis → Mammillaria sheldonii
Mammillaria albata
Mammillaria albescens → Mammillaria decipiens ssp. albescens
Mammillaria albiarmata → Mammillaria coahuilensis ssp. albiarmata
Mammillaria albicans
Mammillaria albicans ssp. **albicans**
Mammillaria albicans ssp. **fraileana**
Mammillaria albicoma
®*Mammillaria albidula* → Mammillaria haageana ssp. conspicua
Mammillaria albiflora
Mammillaria albilanata
Mammillaria albilanata ssp. **albilanata**
Mammillaria albilanata ssp. **oaxacana**
Mammillaria albilanata ssp. **reppenhagenii**
Mammillaria albilanata ssp. **tegelbergiana**
Mammillaria albrechtiana → Mammillaria rekoi
Mammillaria amajacensis
Mammillaria angelensis → Mammillaria dioica ssp. angelensis
Mammillaria anniana
Mammillaria antesbergeriana → Mammillaria wagneriana
Mammillaria apamensis → Mammillaria wiesingeri ssp. apamensis
Mammillaria apozolensis → Mammillaria petterssonii
Mammillaria applanata → Mammillaria heyderi ssp. hemisphaerica
Mammillaria arida → Mammillaria petrophila ssp. arida
Mammillaria armatissima → Mammillaria gigantea
Mammillaria armillata
Mammillaria armillata ssp. **armillata**
Mammillaria armillata ssp. **cerralboa**
Mammillaria arroyensis → Mammillaria formosa
Mammillaria ascensionis → Mammillaria glassii ssp. ascensionis
®*Mammillaria atroflorens* → Mammillaria mystax
Mammillaria aureiceps → Mammillaria rhodantha ssp. aureiceps
Mammillaria aureilanata
Mammillaria aureispina → Mammillaria rekoi ssp. aureispina

Mammillaria aureoviridis → Mammillaria crinita ssp. leucantha
Mammillaria auriareolis → Mammillaria parkinsonii
Mammillaria auricantha → Mammillaria standleyi
Mammillaria auricoma → Mammillaria spinosissima ssp. tepoxtlana
Mammillaria aurihamata → Mammillaria crinita ssp. leucantha
Ⓤ*Mammillaria aurisaeta* → Mammillaria picta ssp. picta
Mammillaria auritricha → Mammillaria standleyi
Ⓤ*Mammillaria avila-camachoi* → Mammillaria perbella
Mammillaria backebergiana
Mammillaria backebergiana ssp. **backebergiana**
Mammillaria backebergiana ssp. **ernestii**
Mammillaria balsasoides → Mammillaria beneckei
Mammillaria bambusiphila → Mammillaria xaltianguensis ssp. bambusiphila
Mammillaria barbata
Ⓤ*Mammillaria barkeri* → Mammillaria beneckei
Mammillaria baumii
Mammillaria baxteriana → Mammillaria petrophila ssp. baxteriana
Mammillaria beiselii → Mammillaria karwinskiana ssp. beiselii
Mammillaria bella → Mammillaria nunezii ssp. bella
Mammillaria bellacantha → Mammillaria canelensis
Mammillaria bellisiana → Mammillaria sonorensis
Mammillaria beneckei
Mammillaria berkiana
Mammillaria bernalensis → Mammillaria compressa ssp. compressa
Mammillaria blossfeldiana
Mammillaria bocasana
Mammillaria bocasana ssp. **bocasana**
Mammillaria bocasana ssp. **eschauzieri**
Mammillaria bocensis
Mammillaria boelderliana
Mammillaria bombycina
Mammillaria bombycina ssp. **bombycina**
Mammillaria bombycina ssp. **perezdelarosae**
Mammillaria bonavitii → Mammillaria rhodantha
Mammillaria boolii
Mammillaria brachytrichion → Mammillaria mercadensis
Mammillaria brandegeei
Mammillaria brandegeei ssp. **brandegeei**
Mammillaria brandegeei ssp. **gabbii**
Mammillaria brandegeei ssp. **glareosa**
Mammillaria brandegeei ssp. **lewisiana**
Mammillaria brauneana → Mammillaria klissingiana
Mammillaria bravoae → Mammillaria hahniana ssp. bravoae
Mammillaria brevicrinita → Mammillaria crinita ssp. leucantha
Mammillaria bucareliensis → Mammillaria magnimamma
Mammillaria buchenaui → Mammillaria crucigera ssp. crucigera
Mammillaria bullardiana → Mammillaria hutchisoniana ssp. hutchisoniana
Mammillaria buxbaumiana → Mammillaria densispina
Mammillaria cadereytana → Mammillaria crinita ssp. crinita
Mammillaria cadereytensis → Mammillaria perbella
Mammillaria caerulea → Mammillaria formosa ssp. chionocephala
Mammillaria calacantha → Mammillaria rhodantha
Mammillaria calleana → Mammillaria crinita ssp. wildii
Mammillaria camptotricha → Mammillaria decipiens ssp. camptotricha
Mammillaria candida → Mammilloydia candida
Mammillaria canelensis
Mammillaria capensis
Mammillaria carmenae
Mammillaria carnea
Mammillaria carretii

Mammillaria casoi → Mammillaria mystax
Mammillaria celsiana → Mammillaria sp.
®*Mammillaria celsiana* → Mammillaria muehlenpfordtii
Mammillaria centralifera → Mammillaria compressa ssp. centralifera
Mammillaria centraliplumosa → Mammillaria spinosissima ssp. spinosissima
Mammillaria centricirrha → Mammillaria magnimamma
Mammillaria cerralboa → Mammillaria armillata ssp. cerralboa
®*Mammillaria chavezei* → Mammillaria barbata
Mammillaria chiapensis → Mammillaria columbiana ssp. yucatanensis
Mammillaria chica → Mammillaria stella-de-tacubaya
Mammillaria chionocephala → Mammillaria formosa ssp. chionocephala
Mammillaria claviformis → Mammillaria duoformis
Mammillaria coahuilensis
Mammillaria coahuilensis ssp. **albiarmata**
Mammillaria coahuilensis ssp. **coahuilensis**
Mammillaria collina → Mammillaria haageana ssp. elegans
Mammillaria collinsii → Mammillaria karwinskiana ssp. collinsii
Mammillaria colonensis → Mammillaria beneckei
Mammillaria columbiana
Mammillaria columbiana ssp. **columbiana**
Mammillaria columbiana ssp. **yucatanensis**
Mammillaria compacticaulis → Mammillaria matudae
Mammillaria compressa
Mammillaria compressa ssp. **centralifera**
Mammillaria compressa ssp. **compressa**
Mammillaria confusa → Mammillaria karwinskiana ssp. karwinskiana
Mammillaria conspicua → Mammillaria haageana ssp. conspicua
Mammillaria cowperae → Mammillaria moelleriana
Mammillaria craigii
Mammillaria crassa → Mammillaria wagneriana
Mammillaria crassimammillis → Mammillaria winterae ssp. aramberri
Mammillaria crassior → Mammillaria spinosissima ssp. tepoxtlana
Mammillaria criniformis → Mammillaria crinita
Mammillaria crinita
Mammillaria crinita ssp. **crinita**
Mammillaria crinita ssp. **leucantha**
Mammillaria crinita ssp. *scheinvariana* → Mammillaria scheinvariana
Mammillaria crinita ssp. **wildii**
Mammillaria crispiseta → Mammillaria mystax
Mammillaria crucigera
Mammillaria crucigera ssp. **crucigera**
Mammillaria crucigera ssp. **tlalocii**
Mammillaria dasyacantha → Escobaria dasyacantha
®*Mammillaria dasyacantha* → Mammillaria laui ssp. dasyacantha
Mammillaria dawsonii → Mammillaria brandegeei ssp. glareosa
Mammillaria dealbata → Mammillaria haageana
Mammillaria decipiens
Mammillaria decipiens ssp. **albescens**
Mammillaria decipiens ssp. **camptotricha**
Mammillaria decipiens ssp. **decipiens**
Mammillaria deherdtiana
Mammillaria deherdtiana ssp. **deherdtiana**
Mammillaria deherdtiana ssp. **dodsonii**
Mammillaria densispina
Mammillaria denudata → Mammillaria lasiacantha ssp. lasiacantha
Mammillaria diguetii → Mammillaria senilis
Mammillaria dioica
Mammillaria dioica ssp. **angelensis**
Mammillaria dioica ssp. **dioica**
Mammillaria dioica ssp. **estebanensis**

Mammillaria discolor
Mammillaria discolor ssp. **discolor**
Mammillaria discolor ssp. **esperanzaensis**
Mammillaria dixanthocentron
Mammillaria dodsonii → Mammillaria deherdtiana ssp. dodsonii
Mammillaria donatii → Mammillaria haageana
Ⓜ*Mammillaria droegeana* → Mammillaria microhelia
Mammillaria dumetorum → Mammillaria schiedeana ssp. dumetorum
Mammillaria duoformis
Mammillaria durangicola → Mammillaria grusonii
Mammillaria durispina → Mammillaria polythele ssp. durispina
Mammillaria duwei
Mammillaria dyckiana → Mammillaria haageana
Ⓤ*Mammillaria ebenacantha* → Mammillaria karwinskiana
Mammillaria echinaria → Mammillaria elongata ssp. echinaria
Mammillaria egregia → Mammillaria lasiacantha ssp. egregia
Mammillaria eichlamii → Mammillaria voburnensis ssp. eichlamii
Mammillaria ekmanii
¶*Mammillaria elegans* → ?Mammillaria geminispina
Ⓜ*Mammillaria elegans* → Mammillaria haageana ssp. elegans
Mammillaria elongata
Mammillaria elongata ssp. **echinaria**
Mammillaria elongata ssp. **elongata**
Ⓜ*Mammillaria erectacantha* → Mammillaria wiesingeri ssp. apamensis
Mammillaria erectohamata → Mammillaria crinita ssp. leucantha
Mammillaria eriacantha
Mammillaria ernestii → Mammillaria backebergiana ssp. ernestii
Mammillaria erythra → Mammillaria mystax
Mammillaria erythrocalyx → Mammillaria duoformis
Mammillaria erythrosperma
Mammillaria eschauzieri → Mammillaria bocasana ssp. eschauzieri
Mammillaria esperanzaensis → Mammillaria discolor ssp. esperanzaensis
Mammillaria esseriana → Mammillaria compressa
Mammillaria estanzuelensis → Mammilloydia candida
Mammillaria estebanensis → Mammillaria dioica ssp. estebanensis
Mammillaria evermanniana
Ⓜ*Mammillaria fasciculata* → Mammillaria thornberi ssp. thornberi
Mammillaria felicis → Mammillaria voburnensis ssp. voburnensis
Mammillaria felipensis → Mammillaria nana
Mammillaria fera-rubra → Mammillaria rhodantha ssp. fera-rubra
Mammillaria fittkaui
Mammillaria fittkaui ssp. **fittkaui**
Mammillaria fittkaui ssp. **limonensis**
Mammillaria fittkaui ssp. *mathildae* → Mammillaria mathildae
Mammillaria flavescens → ?Mammillaria nivosa
Mammillaria flavicentra
Ⓤ*Mammillaria flavihamata* → Mammillaria rettigiana
Mammillaria floresii → Mammillaria standleyi
Mammillaria formosa
Mammillaria formosa ssp. **chionocephala**
Mammillaria formosa ssp. **formosa**
Mammillaria formosa ssp. **microthele**
Mammillaria formosa ssp. **pseudocrucigera**
Mammillaria fraileana → Mammillaria albicans ssp. fraileana
Mammillaria freudenbergeri → Mammillaria winterae ssp. winterae
Mammillaria fuauxiana → Mammillaria albilanata ssp. albilanata
Mammillaria fuscohamata → Mammillaria jaliscana
Mammillaria gabbii → Mammillaria brandegeei ssp. gabbii
Mammillaria garessii → Mammillaria barbata
Mammillaria gasseriana → Mammillaria stella-de-tacubaya

Mammillaria gasterantha → Mammillaria spinosissima
Mammillaria gatesii → Mammillaria petrophila ssp. petrophila
Mammillaria gaumeri → Mammillaria heyderi ssp. gaumeri
Mammillaria geminispina
Mammillaria geminispina ssp. **geminispina**
Mammillaria geminispina ssp. **leucocentra**
Mammillaria gigantea
Mammillaria gilensis → Mammillaria rettigiana
Mammillaria giselae → Mammillaria schiedeana ssp. giselae
Mammillaria glareosa → Mammillaria brandegeei ssp. glareosa
Mammillaria glassii
Mammillaria glassii ssp. **ascensionis**
Mammillaria glassii ssp. **glassii**
Mammillaria glochidiata
Mammillaria glomerata → Mammillaria prolifera
Mammillaria goldii → Mammillaria saboae ssp. goldii
Mammillaria goodridgii
Mammillaria gracilis → Mammillaria vetula ssp. gracilis
Mammillaria grahamii
Mammillaria grusonii
Mammillaria gueldemanniana → Mammillaria sheldonii
Mammillaria guelzowiana
Mammillaria guerreronis
Mammillaria guiengolensis → Mammillaria beneckei
Mammillaria guillauminiana
Mammillaria guirocobensis → Mammillaria sheldonii
Mammillaria gummifera → Mammillaria heyderi ssp. gummifera
Mammillaria haageana
Mammillaria haageana ssp. **acultzingensis**
Mammillaria haageana ssp. **conspicua**
Mammillaria haageana ssp. **elegans**
Mammillaria haageana ssp. **haageana**
Mammillaria haageana ssp. **san-angelensis**
Mammillaria haageana ssp. **schmollii**
Mammillaria haasii → Mammillaria spinosissima ssp. spinosissima
Mammillaria hahniana
Mammillaria hahniana ssp. **bravoae**
Mammillaria hahniana ssp. **hahniana**
Mammillaria hahniana ssp. **mendeliana**
Mammillaria hahniana ssp. **woodsii**
Mammillaria halbingeri
Mammillaria halei
Mammillaria hamata
Mammillaria hamiltonhoytea → Mammillaria gigantea
Mammillaria hastifera → Mammillaria gigantea
Mammillaria haudeana → Mammillaria saboae ssp. haudeana
Mammillaria heidiae
Mammillaria hemisphaerica → Mammillaria heyderi ssp. hemisphaerica
Mammillaria hennisii → Mammillaria columbiana ssp. columbiana
Mammillaria hernandezii
Mammillaria herrerae
Mammillaria hertrichiana
Mammillaria heyderi
Mammillaria heyderi ssp. *coahuilensis* → Mammillaria coahuilensis ssp. coahuilensis
Mammillaria heyderi ssp. **gaumeri**
Mammillaria heyderi ssp. **gummifera**
Mammillaria heyderi ssp. **hemisphaerica**
Mammillaria heyderi ssp. **heyderi**
Mammillaria heyderi ssp. **macdougalii**
Mammillaria heyderi ssp. **meiacantha**

Mammillaria hidalgensis → Mammillaria polythele ssp. polythele
Mammillaria hirsuta → Mammillaria bocasana ssp. eschauzieri
Mammillaria hoffmanniana → Mammillaria polythele ssp. polythele
Mammillaria huajuapensis → Mammillaria mystax
Mammillaria hubertmulleri → Mammillaria nunezii ssp. nunezii
Mammillaria huiguerensis → Mammillaria petterssonii
Mammillaria huitzilopochtli
Mammillaria humboldtii
Mammillaria hutchisoniana
Mammillaria hutchisoniana ssp. **hutchisoniana**
Mammillaria hutchisoniana ssp. **louisae**
Mammillaria ignota → Mammillaria albilanata ssp. oaxacana
Mammillaria igualensis → Mammillaria albilanata ssp. albilanata
Mammillaria inae ['inaiae'] → Mammillaria sheldonii
Mammillaria infernillensis → Mammillaria perbella
Mammillaria ingens → Mammillaria polythele ssp. obconella
Mammillaria insularis
Mammillaria isotensis → Mammillaria backebergiana ssp. ernestii
Mammillaria jaliscana
Mammillaria jaliscana ssp. **jaliscana**
Mammillaria jaliscana ssp. **zacatecasensis**
Mammillaria johnstonii
Mammillaria jozef-bergeri → Mammillaria karwinskiana
Mammillaria karwinskiana
Mammillaria karwinskiana ssp. **beiselii**
Mammillaria karwinskiana ssp. **collinsii**
Mammillaria karwinskiana ssp. **karwinskiana**
Mammillaria karwinskiana ssp. **nejapensis**
Mammillaria kelleriana → Mammillaria polythele ssp. durispina
Mammillaria kewensis → Mammillaria polythele ssp. polythele
Mammillaria kleiniorum → Mammillaria jaliscana
Mammillaria klissingiana
Mammillaria knebeliana → Mammillaria bocasana ssp. eschauzieri
Mammillaria knippeliana
Mammillaria kraehenbuehlii
Mammillaria krasuckae → Mammillaria rekoi
®*Mammillaria kuentziana* → Mammillaria vetula ssp. vetula
Mammillaria kunthii → Mammillaria haageana
Mammillaria kunzeana → Mammillaria bocasana ssp. eschauzieri
Mammillaria lanata → Mammillaria supertexta
Mammillaria lanigera → Mammillaria albilanata ssp. oaxacana
Mammillaria lanisumma → Mammillaria standleyi
Mammillaria lasiacantha
Mammillaria lasiacantha ssp. **egregia**
Mammillaria lasiacantha ssp. **hyalina**
Mammillaria lasiacantha ssp. **lasiacantha**
Mammillaria lasiacantha ssp. **magallanii**
Mammillaria laui
Mammillaria laui ssp. **dasyacantha**
Mammillaria laui ssp. **laui**
Mammillaria laui ssp. **subducta**
Mammillaria lengdobleriana
Mammillaria lenta
Mammillaria leona → Mammillaria pottsii
Mammillaria leptacantha → Mammillaria rekoi ssp. leptacantha
Mammillaria leucantha → Mammillaria crinita ssp. leucantha
Mammillaria leucocentra → Mammillaria geminispina ssp. leucocentra
Mammillaria lewisiana → Mammillaria brandegeei ssp. lewisiana
Mammillaria limonensis → Mammillaria fittkaui ssp. limonensis
Mammillaria linaresensis → Mammillaria melanocentra ssp. linaresensis

Mammillaria lindsayi
Mammillaria lloydii
Mammillaria longicoma → Mammillaria bocasana ssp. eschauzieri
Mammillaria longiflora
Mammillaria longiflora ssp. **longiflora**
Mammillaria longiflora ssp. **stampferi**
Mammillaria longimamma
Mammillaria louisae → Mammillaria hutchisoniana ssp. louisae
Mammillaria luethyi
Mammillaria macdougalii → Mammillaria heyderi ssp. macdougalii
Mammillaria macracantha → Mammillaria magnimamma
Mammillaria magallanii → Mammillaria lasiacantha ssp. magallanii
Mammillaria magneticola → Mammillaria vetula ssp. vetula
Mammillaria magnifica
Mammillaria magnimamma
Mammillaria mainiae
Mammillaria mammillaris
Mammillaria marcosii
Mammillaria maritima → Mammillaria pondii ssp. maritima
Mammillaria marksiana
Mammillaria marnieriana → Mammillaria sheldonii
Mammillaria marshalliana → Mammillaria petrophila ssp. baxteriana
Mammillaria martinezii → Mammillaria supertexta
Mammillaria mathildae
Mammillaria matudae
Mammillaria mayensis → Mammillaria standleyi
Mammillaria mazatlanensis
Mammillaria mazatlanensis ssp. **mazatlanensis**
Mammillaria mazatlanensis ssp. **patonii**
Mammillaria meiacantha → Mammillaria heyderi ssp. meiacantha
Mammillaria meissneri → Mammillaria haageana ssp. schmollii
Mammillaria melaleuca
Mammillaria melanocentra
Mammillaria melanocentra ssp. **linaresensis**
Mammillaria melanocentra ssp. **melanocentra**
Mammillaria melanocentra ssp. **rubrograndis**
Mammillaria melispina → Mammillaria sp.
Mammillaria mendeliana → Mammillaria hahniana ssp. mendeliana
Mammillaria mercadensis
Mammillaria meridiorosei → Mammillaria wrightii ssp. wilcoxii
Mammillaria mexicensis → Mammillaria sp.
Mammillaria meyranii
®*Mammillaria microcarpa* → Mammillaria grahamii
®*Mammillaria microcarpa* ssp. *grahamii* → Mammillaria grahamii
Mammillaria microhelia
Mammillaria microheliopsis → Mammillaria microhelia
Mammillaria micromeris ssp. *unguispina* → Epithelantha micromeris ssp. unguispina
Mammillaria microthele → Mammillaria formosa ssp. microthele
Mammillaria miegiana
Mammillaria mieheana
Mammillaria milleri → Mammillaria grahamii
Mammillaria mitlensis → Mammillaria rekoi
Mammillaria mixtecensis → Mammillaria mystax
Mammillaria moeller-valdeziana → Mammillaria crinita ssp. leucantha
Mammillaria moelleriana
Mammillaria mollendorffiana → Mammillaria rhodantha ssp. mollendorffiana
Mammillaria mollihamata → Mammillaria crinita ssp. crinita
®*Mammillaria monancistracantha* → Mammillaria nana
Mammillaria montensis → Mammillaria standleyi
Mammillaria monticola → Mammillaria albilanata ssp. oaxacana

Mammillaria morganiana
Mammillaria morricalii → Mammillaria barbata
Mammillaria movensis → Mammillaria sonorensis
Mammillaria muehlenpfordtii
Mammillaria multiceps → Mammillaria prolifera ssp. texana
Mammillaria multidigitata
Mammillaria multiformis → Mammillaria erythrosperma
Mammillaria multihamata → ?Mammillaria marcosii
Mammillaria multiseta → Mammillaria karwinskiana
Mammillaria mundtii → Mammillaria wiesingeri ssp. apamensis
Mammillaria mystax
Mammillaria nagliana → Mammillaria karwinskiana ssp. collinsii
Mammillaria nana
Mammillaria napina
Mammillaria nazasensis → Mammillaria pennispinosa ssp. nazasensis
Mammillaria nejapensis → Mammillaria karwinskiana ssp. nejapensis
Mammillaria neobertrandiana → Mammillaria lasiacantha ssp. magallanii
Mammillaria neocrucigera → Mammillaria parkinsonii
Mammillaria neomystax → Mammillaria karwinskiana
Mammillaria neopalmeri
Mammillaria neophaeacantha → Mammillaria polythele ssp. polythele
Mammillaria neopotosina → Mammillaria muehlenpfordtii
Mammillaria neoschwarzeana → Mammillaria bocensis
Mammillaria nivosa
Mammillaria noureddineana → Mammillaria albilanata ssp. oaxacana
Mammillaria nunezii
Mammillaria nunezii ssp. **bella**
Mammillaria nunezii ssp. **nunezii**
Mammillaria obconella → Mammillaria polythele ssp. obconella
Mammillaria obscura → Mammillaria petterssonii
Mammillaria occidentalis → Mammillaria mazatlanensis ssp. mazatlanensis
Mammillaria ochoterenae → Mammillaria discolor
Mammillaria ocotillensis → Mammillaria gigantea
Mammillaria oliviae → Mammillaria grahamii
¶*Mammillaria orcuttii* → Mammillaria carnea
Mammillaria orestera → Mammillaria barbata
Mammillaria ortegae → Mammillaria sp.
Mammillaria ortizrubiana → Mammilloydia candida
Mammillaria oteroi
Mammillaria pachycylindrica → Mammillaria grusonii
Mammillaria pachyrhiza → Mammillaria discolor
Mammillaria pacifica → Mammillaria petrophila ssp. baxteriana
Mammillaria painteri
Mammillaria papasquiarensis → Mammillaria grusonii
Mammillaria parensis → Mammillaria rhodantha ssp. pringlei
Mammillaria parkinsonii
Mammillaria parrasensis → Mammillaria heyderi
Mammillaria patonii → Mammillaria mazatlanensis ssp. patonii
Mammillaria pectinifera♦
Mammillaria peninsularis
Mammillaria pennispinosa
Mammillaria pennispinosa ssp. **nazasensis**
Mammillaria pennispinosa ssp. **pennispinosa**
Mammillaria perbella
Mammillaria perezdelarosae → Mammillaria bombycina ssp. perezdelarosae
Mammillaria petrophila
Mammillaria petrophila ssp. **arida**
Mammillaria petrophila ssp. **baxteriana**
Mammillaria petrophila ssp. **petrophila**
Mammillaria petterssonii

Mammillaria phitauiana
Mammillaria picta
Mammillaria picta ssp. **picta**
Mammillaria picta ssp. **viereckii**
Mammillaria pilcayensis → Mammillaria spinosissima ssp. pilcayensis
Mammillaria pilensis → Mammillaria petterssonii
Mammillaria pilispina
®*Mammillaria pitcayensis* → Mammillaria spinosissima ssp. pilcayensis
Mammillaria plumosa
Mammillaria polyedra
Mammillaria polythele
Mammillaria polythele ssp. **durispina**
Mammillaria polythele ssp. **obconella**
Mammillaria polythele ssp. **polythele**
Mammillaria pondii
Mammillaria pondii ssp. **maritima**
Mammillaria pondii ssp. **pondii**
Mammillaria pondii ssp. **setispina**
Mammillaria poselgeri
Mammillaria posseltiana → Mammillaria rettigiana
Mammillaria pottsii
Mammillaria praelii → Mammillaria karwinskiana
Mammillaria priessnitzii → Mammillaria magnimamma
Mammillaria pringlei → Mammillaria rhodantha ssp. pringlei
Mammillaria prolifera
Mammillaria prolifera ssp. **arachnoidea**
Mammillaria prolifera ssp. **haitiensis**
Mammillaria prolifera ssp. **prolifera**
Mammillaria prolifera ssp. **texana**
Mammillaria prolifera ssp. **zublerae**
Mammillaria pseudocrucigera → Mammillaria formosa ssp. pseudocrucigera
Mammillaria pseudoperbella → Mammillaria sp.
Mammillaria pseudorekoi → Mammillaria rekoi
Mammillaria pseudoscrippsiana → Mammillaria scrippsiana
®*Mammillaria pseudosimplex* → Mammillaria mammillaris
Mammillaria puberula → Mammillaria crinita ssp. leucantha
Mammillaria pubispina → Mammillaria crinita
Mammillaria pullihamata → Mammillaria rekoi
Mammillaria pygmaea → Mammillaria crinita ssp. crinita
Mammillaria queretarica → Mammillaria perbella
Mammillaria radiaissima → Mammillaria baumii
Mammillaria rayonensis → Mammillaria pilispina
Mammillaria rectispina → Mammillaria dioica
Mammillaria rekoi
Mammillaria rekoi ssp. **aureispina**
Mammillaria rekoi ssp. **leptacantha**
Mammillaria rekoi ssp. **rekoi**
Mammillaria rekoiana → Mammillaria rekoi ssp. rekoi
Mammillaria reppenhagenii → Mammillaria albilanata ssp. reppenhagenii
Mammillaria rettigiana
Mammillaria rhodantha
Mammillaria rhodantha ssp. **aureiceps**
Mammillaria rhodantha ssp. **fera-rubra**
Mammillaria rhodantha ssp. **mccartenii**
Mammillaria rhodantha ssp. **mollendorffiana**
Mammillaria rhodantha ssp. **pringlei**
Mammillaria rhodantha ssp. **rhodantha**
Mammillaria ritteriana → Mammillaria formosa ssp. chionocephala
Mammillaria rosensis → Mammillaria parkinsonii
Mammillaria roseoalba

Mammillaria roseocentra → Mammillaria sp.
Mammillaria rossiana → ?Mammillaria duoformis
Mammillaria rubida → Mammillaria bocensis
Mammillaria rubrograndis → Mammillaria melanocentra ssp. rubrograndis
Mammillaria ruestii → Mammillaria columbiana ssp. yucatanensis
Mammillaria saboae
Mammillaria saboae ssp. **goldii**
Mammillaria saboae ssp. **haudeana**
Mammillaria saboae ssp. **saboae**
Mammillaria saetigera → Mammillaria sp.
Mammillaria saffordii → Mammillaria carretii
Mammillaria saint-pieana → Mammillaria gigantea
Mammillaria san-angelensis → Mammillaria haageana ssp. san-angelensis
Mammillaria sanchez-mejoradae
Mammillaria sanjuanensis → Mammillaria rekoi
Mammillaria sanluisensis → Mammillaria pilispina
Mammillaria santaclarensis → Mammillaria barbata
Mammillaria sartorii
Mammillaria saxicola → Mammillaria magnimamma
Mammillaria scheinvariana
Mammillaria schelhasii → Mammillaria crinita
Mammillaria schiedeana
Mammillaria schiedeana ssp. **dumetorum**
Mammillaria schiedeana ssp. **giselae**
Mammillaria schiedeana ssp. **schiedeana**
Mammillaria schieliana → Mammillaria picta ssp. picta
Mammillaria schmollii → Mammillaria discolor
Mammillaria schumannii
®*Mammillaria schwartzii* → Mammillaria coahuilensis ssp. coahuilensis
Mammillaria schwarzii
Mammillaria scrippsiana
Mammillaria seideliana → Mammillaria sp.
Mammillaria seitziana → Mammillaria sp.
Mammillaria sempervivi
Mammillaria senilis
Mammillaria setispina → Mammillaria pondii ssp. setispina
Mammillaria sheldonii
Mammillaria shurliana → Mammillaria blossfeldiana
Mammillaria silvatica → Mammillaria nunezii
Mammillaria simplex → Mammillaria mammillaris
Mammillaria sinistrohamata
Mammillaria slevinii
Mammillaria soehlemannii → Mammillaria columbiana ssp. columbiana
Mammillaria solisii → Mammillaria nunezii ssp. nunezii
Mammillaria solisioides♦
Mammillaria sonorensis
Mammillaria sororia → Mammillaria sp.
Mammillaria sphacelata
Mammillaria sphacelata ssp. **sphacelata**
Mammillaria sphacelata ssp. **viperina**
Mammillaria sphaerica
Mammillaria spinosissima
Mammillaria spinosissima ssp. **pilcayensis**
Mammillaria spinosissima ssp. **spinosissima**
Mammillaria spinosissima ssp. **tepoxtlana**
Mammillaria stampferi → Mammillaria longiflora ssp. stampferi
Mammillaria standleyi
Mammillaria stella-de-tacubaya
Mammillaria strobilina → Mammillaria karwinskiana
Mammillaria subducta → Mammillaria laui ssp. subducta

Mammillaria subdurispina → Mammillaria polythele ssp. durispina
Mammillaria subtilis → Mammillaria pilispina
Mammillaria supertexta
Mammillaria supraflumen → Mammillaria nunezii
Mammillaria surculosa
Mammillaria swinglei → Mammillaria sheldonii
Mammillaria tayloriorum
Mammillaria tegelbergiana → Mammillaria albilanata ssp. tegelbergiana
Mammillaria tenampensis → Mammillaria sartorii
Mammillaria tepexicensis
Mammillaria tesopacensis → Mammillaria sonorensis
Mammillaria tetracantha → Mammillaria polythele ssp. polythele
Mammillaria tetrancistra
Mammillaria tezontle → Mammillaria crinita ssp. leucantha
Mammillaria theresae
Mammillaria thornberi
Mammillaria thornberi ssp. **thornberi**
Mammillaria thornberi ssp. **yaquensis**
Mammillaria tlalocii → Mammillaria crucigera ssp. tlalocii
Mammillaria tolimensis → Mammillaria compressa ssp. compressa
Mammillaria tonalensis
Mammillaria trichacantha → Mammillaria sp.
Mammillaria tropica → Mammillaria karwinskiana ssp. collinsii
Mammillaria uberiformis → Mammillaria longimamma
Mammillaria umbrina → Mammillaria sp.
Mammillaria uncinata
Mammillaria unihamata → Mammillaria weingartiana
Mammillaria vagaspina → Mammillaria magnimamma
Mammillaria vallensis → Mammillaria magnimamma
Mammillaria vari(e)aculeata
Mammillaria variabilis → Mammillaria fittkaui
Mammillaria vaupelii → Mammillaria haageana
Mammillaria verhaertiana → ?Mammillaria phitauiana
Mammillaria verticealba → Mammillaria rhodantha ssp. mccartenii
Mammillaria vetula
Mammillaria vetula ssp. **gracilis**
Mammillaria vetula ssp. **vetula**
Mammillaria viereckii → Mammillaria picta ssp. viereckii
Mammillaria viescensis → Mammillaria stella-de-tacubaya
Mammillaria viperina → Mammillaria sphacelata ssp. viperina
Mammillaria virginis → Mammillaria spinosissima ssp. spinosissima
Mammillaria viridiflora → Mammillaria barbata
Mammillaria voburnensis
Mammillaria voburnensis ssp. **eichlamii**
Mammillaria voburnensis ssp. **voburnensis**
Mammillaria vonwyssiana → Mammillaria sp.
Mammillaria wagneriana
Mammillaria weingartiana
Mammillaria wiesingeri
Mammillaria wiesingeri ssp. **apamensis**
Mammillaria wiesingeri ssp. **wiesingeri**
Mammillaria wilcoxii → Mammillaria wrightii ssp. wilcoxii
Mammillaria wildii → Mammillaria crinita ssp. wildii
Mammillaria winterae
Mammillaria winterae ssp. **aramberri**
Mammillaria winterae ssp. **winterae**
®*Mammillaria woburnensis* → Mammillaria voburnensis
Mammillaria wohlschlageri → Mammillaria lasiacantha ssp. hyalina
Mammillaria woodsii → Mammillaria hahniana ssp. woodsii
Mammillaria wrightii

Mammillaria wrightii ssp. **wilcoxii**
Mammillaria wrightii ssp. **wrightii**
Mammillaria wuthenauiana → Mammillaria nunezii ssp. nunezii
Mammillaria xaltianguensis
Mammillaria xaltianguensis ssp. **bambusiphila**
Mammillaria xaltianguensis ssp. **xaltianguensis**
Mammillaria xanthina → Mammillaria standleyi
Mammillaria xochipilli → Mammillaria polythele ssp. polythele
Mammillaria yaquensis → Mammillaria thornberi ssp. yaquensis
Mammillaria yucatanensis → Mammillaria columbiana ssp. yucatanensis
Mammillaria zacatecasensis → Mammillaria jaliscana ssp. zacatecasensis
Mammillaria zahniana → Mammillaria winterae ssp. winterae
Mammillaria zeilmanniana
Mammillaria zephyranthoides
Mammillaria zeyeriana → Mammillaria grusonii
Mammillaria zopilotensis → Mammillaria guerreronis
Mammillaria zublerae → Mammillaria prolifera ssp. zublerae
Mammillaria zuccariniana → Mammillaria magnimamma

Mammilloydia candida
Mammilloydia candida ssp. *ortizrubiana* → Mammilloydia candida

Marenopuntia marenae → Opuntia marenae

Marginatocereus marginatus → Pachycereus marginatus

Marniera chrysocardium → Selenicereus chrysocardium

Marshallocereus aragonii → Stenocereus aragonii
Marshallocereus thurberi → Stenocereus thurberi

Matucana aurantiaca
Matucana aurantiaca ssp. **aurantiaca**
Matucana aurantiaca ssp. **currundayensis**
Matucana aureiflora
Matucana blancii → Matucana haynei ssp. herzogiana
Matucana breviflora → Matucana haynei ssp. hystrix
Matucana calliantha → Matucana krahnii
Matucana calocephala → Matucana haynei ssp. myriacantha
Matucana calvescens → Matucana aurantiaca ssp. aurantiaca
Matucana celendinensis
Matucana cereoides → Matucana haynei ssp. haynei
Matucana comacephala
Matucana crinifera → Matucana haynei ssp. herzogiana
Matucana currundayensis → Matucana aurantiaca ssp. currundayensis
Matucana elongata → Matucana haynei ssp. haynei
Matucana formosa
Matucana fruticosa
Matucana hastifera → Matucana aurantiaca ssp. currundayensis
Matucana haynei
Matucana haynei ssp. **haynei**
Matucana haynei ssp. **herzogiana**
Matucana haynei ssp. **hystrix**
Matucana haynei ssp. **myriacantha**
Matucana herzogiana → Matucana haynei ssp. herzogiana
Matucana huagalensis
Matucana humboldtii → Cleistocactus icosagonus
Matucana hystrix → Matucana haynei ssp. hystrix
Matucana intertexta
Matucana krahnii

Matucana madisoniorum
Matucana megalantha → Matucana haynei ssp. herzogiana
Matucana mirabilis → Cleistocana mirabilis
Matucana multicolor → Matucana haynei ssp. hystrix
Matucana myriacantha → Matucana haynei ssp. myriacantha
Matucana oreodoxa
Matucana pallarensis → Matucana aurantiaca ssp. aurantiaca
Matucana paucicostata
Matucana polzii
Matucana pujupatii
Matucana purpureoalba → Matucana haynei ssp. myriacantha
Matucana ritteri
Matucana supertexta → Matucana haynei ssp. haynei
Matucana tuberculata
Matucana tuberculosa → Matucana tuberculata
Matucana variabilis → Matucana haynei ssp. haynei
Matucana weberbaueri
Matucana winteri → Matucana haynei ssp. myriacantha
Matucana yanganucensis → Matucana haynei ssp. herzogiana

Ⓜ*Mediocactus coccineus* → Selenicereus setaceus
Mediocactus hahnianus → Echinopsis hahniana
Mediocactus hassleri → Selenicereus setaceus
Mediocactus lindmanii → Selenicereus setaceus
Mediocactus megalanthus → Selenicereus megalanthus
Mediocactus pomifer → Hylocereus trigonus

Mediolobivia auranitida → Rebutia einsteinii
Mediolobivia aureiflora → Rebutia aureiflora
Mediolobivia brachyantha → Rebutia steinmannii
Mediolobivia brunescens → Rebutia brunescens
Mediolobivia conoidea → Rebutia einsteinii
Mediolobivia costata → Rebutia steinmannii
Mediolobivia elegans → Rebutia aureiflora
Mediolobivia eos → Rebutia pygmaea
Ⓤ*Mediolobivia euanthema* → Rebutia aureiflora
Mediolobivia eucaliptana → Rebutia steinmannii
Mediolobivia haefneriana → Rebutia pygmaea
Mediolobivia hirsutissima → Echinopsis tiegeliana
Mediolobivia ithyacantha → Rebutia fiebrigii
Mediolobivia neopygmaea → ?Rebutia einsteinii
Mediolobivia nigricans → Rebutia nigricans
Mediolobivia pectinata → Rebutia pygmaea
Mediolobivia pygmaea → Rebutia pygmaea
Mediolobivia ritteri → Rebutia ritteri
Mediolobivia schmiedcheniana → Rebutia einsteinii

Melocactus acispinosus → Melocactus bahiensis ssp. bahiensis
Melocactus actinacanthus → Melocactus matanzanus
Melocactus acunae → Melocactus harlowii
Melocactus acunae ssp. *lagunaensis* → Melocactus harlowii
Melocactus ×albicephalus (M. ernestii × M. glaucescens)
Melocactus amethystinus → Melocactus bahiensis ssp. amethystinus
Melocactus ammotrophus → Melocactus bahiensis ssp. amethystinus
Ⓜ*Melocactus amoenus* → Melocactus curvispinus ssp. caesius
Melocactus amstutziae → Melocactus peruvianus
Melocactus andinus
Melocactus arcuatispinus → Melocactus zehntneri
Melocactus axiniphorus → Melocactus concinnus
Melocactus azulensis → Melocactus ernestii ssp. ernestii

Melocactus azureus
Melocactus azureus ssp. **azureus**
Melocactus azureus ssp. **ferreophilus**
Melocactus bahiensis
Melocactus bahiensis ssp. **amethystinus**
Melocactus bahiensis ssp. **bahiensis**
Melocactus barbarae → Melocactus macracanthos
Melocactus bellavistensis
Melocactus bellavistensis ssp. **bellavistensis**
Melocactus bellavistensis ssp. **onychacanthus**
Melocactus borhidii → Melocactus harlowii
Melocactus bozsingianus → Melocactus macracanthos
Melocactus brederooianus → Melocactus bahiensis ssp. bahiensis
Melocactus broadwayi
Melocactus caesius → Melocactus curvispinus ssp. caesius
Melocactus canescens → Melocactus zehntneri
Melocactus caroli-linnaei
Melocactus communis → Melocactus intortus
Melocactus concinnus
Melocactus conoideus♦
®*Melocactus coronatus* → Melocactus caroli-linnaei
¶*Melocactus coronatus* → Melocactus intortus
Melocactus cremnophilus → Melocactus oreas ssp. cremnophilus
Melocactus curvicornis → Melocactus zehntneri
Melocactus curvispinus
Melocactus curvispinus ssp. **caesius**
Melocactus curvispinus ssp. **curvispinus**
Melocactus curvispinus ssp. **dawsonii**
Melocactus dawsonii → Melocactus curvispinus ssp. dawsonii
Melocactus deinacanthus♦
Melocactus deinacanthus ssp. *florschuetzianus* → Melocactus ernestii ssp. longicarpus
Melocactus deinacanthus ssp. *longicarpus* → Melocactus ernestii ssp. longicarpus
Melocactus delessertianus → Melocactus curvispinus ssp. curvispinus
Melocactus depressus → Melocactus violaceus ssp. violaceus
Melocactus diersianus → Melocactus levitestatus
Melocactus douradaensis → Melocactus zehntneri
Melocactus ellemeetii → Melocactus violaceus ssp. margaritaceus
Melocactus ernestii
Melocactus ernestii ssp. **ernestii**
Melocactus ernestii ssp. **longicarpus**
Melocactus erythracanthus → Melocactus ernestii ssp. ernestii
®*Melocactus erythranthus* → Melocactus ernestii ssp. ernestii
Melocactus estevesii
Melocactus evae → Melocactus harlowii
Melocactus ferreophilus → Melocactus azureus ssp. ferreophilus
Melocactus florschuetzianus → Melocactus ernestii ssp. longicarpus
Melocactus fortalezensis → Melocactus peruvianus
Melocactus giganteus → Melocactus zehntneri
Melocactus glaucescens♦
Melocactus glauxianus → Melocactus bahiensis ssp. amethystinus
Melocactus griseoloviridis → Melocactus bahiensis ssp. amethystinus
Melocactus guaricensis → Melocactus neryi
Melocactus guitartii → Melocactus curvispinus
Melocactus harlowii
Melocactus helvolilanatus → Melocactus zehntneri
Melocactus hispaniolicus → ?Melocactus lemairei
Melocactus holguinensis → Melocactus curvispinus
Melocactus* ×*horridus (M. ernestii × M. zehntneri)
Melocactus huallanc(a)ensis → Melocactus peruvianus
Melocactus inclinatus → Melocactus macracanthos

¶*Melocactus inconcinnus* → Melocactus bahiensis ssp. bahiensis
Ⓜ*Melocactus inconcinnus* → Melocactus salvadorensis
Melocactus interpositus → Melocactus ernestii ssp. ernestii
Melocactus intortus
Melocactus intortus ssp. **domingensis**
Melocactus intortus ssp. **intortus**
Melocactus jakusii → Melocactus curvispinus
Melocactus jansenianus → Melocactus peruvianus
Melocactus krainzianus → Melocactus azureus ssp. azureus
Melocactus lanssensianus
Melocactus laui → Melocactus macracanthos
Melocactus lemairei
Melocactus lensselinkianus → Melocactus bahiensis ssp. amethystinus
Melocactus levitestatus
Melocactus lobelii → Melocactus curvispinus ssp. caesius
Melocactus loboguerreroi → Melocactus curvispinus ssp. curvispinus
Melocactus longicarpus → Melocactus ernestii ssp. longicarpus
Melocactus longispinus → Melocactus ernestii ssp. ernestii
Melocactus macracanthos
Melocactus macrodiscus → Melocactus zehntneri
Ⓤ*Melocactus margaritaceus* → Melocactus violaceus ssp. margaritaceus
Melocactus matanzanus
Melocactus maxonii → Melocactus curvispinus ssp. curvispinus
Melocactus mazelianus
Melocactus melocactoides → ?Melocactus violaceus
Melocactus montanus → Melocactus ernestii ssp. longicarpus
Melocactus mulequensis → Melocactus ernestii ssp. longicarpus
Melocactus nagyi → Melocactus harlowii
Melocactus neomontanus → Melocactus ernestii ssp. longicarpus
Melocactus neryi
Melocactus nitidus → Melocactus ernestii ssp. ernestii
Melocactus oaxacensis → Melocactus curvispinus ssp. curvispinus
Melocactus obtusipetalus → ?Melocactus curvispinus ssp. curvispinus
Ⓤ*Melocactus ocujalii* → Melocactus harlowii
Melocactus onychacanthus → Melocactus bellavistensis ssp. onychacanthus
Melocactus oreas
Ⓜ*Melocactus oreas* ssp. *bahiensis* → Melocactus oreas ssp. oreas
Melocactus oreas ssp. **cremnophilus**
Melocactus oreas ssp. *ernestii* → Melocactus ernestii ssp. ernestii
Melocactus oreas ssp. **oreas**
Melocactus oreas ssp. *rubrisaetosus* → Melocactus oreas ssp. oreas
Melocactus pachyacanthus
Melocactus pachyacanthus ssp. **pachyacanthus**
Melocactus pachyacanthus ssp. **viridis**
Melocactus paucispinus♦
Melocactus pedernalensis → Melocactus intortus ssp. domingensis
Melocactus perezassoi
Melocactus peruvianus
Melocactus pruinosus → ?Melocactus concinnus
Melocactus radoczii → Melocactus harlowii
Melocactus robustispinus → Melocactus concinnus
Melocactus roraimensis → Melocactus smithii
Melocactus rubrisaetosus → Melocactus oreas ssp. oreas
Melocactus rubrispinus → Melocactus levitestatus
Melocactus ruestii → Melocactus curvispinus ssp. curvispinus
Melocactus ruestii ssp. *cintalapensis* → Melocactus curvispinus ssp. curvispinus
Melocactus ruestii ssp. *maxonii* → Melocactus curvispinus ssp. curvispinus
Melocactus ruestii ssp. *oaxacensis* → Melocactus curvispinus ssp. curvispinus
Melocactus ruestii ssp. *sanctae-rosae* → Melocactus curvispinus ssp. curvispinus
Melocactus salvador → Melocactus curvispinus ssp. curvispinus

Melocactus salvadorensis
Melocactus saxicola → Melocactus zehntneri
Melocactus schatzlii
Melocactus schulzianus → Melocactus neryi
Melocactus securituberculatus → Melocactus levitestatus
Melocactus smithii
Melocactus trujilloensis → Melocactus peruvianus
Melocactus uebelmannii → Melocactus levitestatus
Melocactus unguispinus → Melocactus peruvianus
Melocactus violaceus
Melocactus violaceus ssp. **margaritaceus**
Melocactus violaceus ssp. *natalensis* → Melocactus violaceus ssp. violaceus
Melocactus violaceus ssp. **ritteri**
Melocactus violaceus ssp. **violaceus**
Melocactus warasii → Melocactus levitestatus
Melocactus zehntneri
Melocactus zehntneri ssp. *canescens* → Melocactus zehntneri
Melocactus zehntneri ssp. *robustispinus* → Melocactus concinnus

Meyerocactus horizonthalonius → Echinocactus horizonthalonius

Micranthocereus albicephalus
Micranthocereus aureispinus → Micranthocereus albicephalus
Micranthocereus auriazureus
Micranthocereus densiflorus → Micranthocereus flaviflorus
Micranthocereus dolichospermaticus
Micranthocereus estevesii
Micranthocereus flaviflorus
Micranthocereus flaviflorus ssp. *densiflorus* → Micranthocereus flaviflorus
Micranthocereus haematocarpus → Micranthocereus purpureus
Micranthocereus lehmannianus → Micranthocereus purpureus
Micranthocereus monteazulensis → Micranthocereus albicephalus
Micranthocereus polyanthus
Micranthocereus purpureus
Micranthocereus ruficeps → Micranthocereus purpureus
Micranthocereus streckeri
Micranthocereus uilianus → Micranthocereus flaviflorus
Micranthocereus violaciflorus

Micropuntia barkleyana → Opuntia pulchella
Micropuntia brachyrhopalica → Opuntia pulchella
Micropuntia gracilicylindrica → Opuntia pulchella
Micropuntia pulchella → Opuntia pulchella
Micropuntia pygmaea → Opuntia pulchella
Micropuntia tuberculosirhopalica → Opuntia pulchella
Micropuntia wiegandii → Opuntia pulchella

Mila albisaetacens → Mila caespitosa
Mila alboareolata → Mila nealeana
Mila breviseta → Mila nealeana
Mila caespitosa
Mila caespitosa ssp. *nealeana* → Mila nealeana
Mila cereoides → Mila caespitosa
Mila colorea
Mila densiseta → Mila nealeana
Mila fortalezensis → Mila caespitosa
Mila kubeana → Mila nealeana
Mila lurinensis → Mila nealeana
Mila nealeana
Mila pugionifera
Mila sublanata → Mila caespitosa

Miqueliopuntia miquelii → Opuntia miquelii

Mirabella albicaulis → Cereus albicaulis
Mirabella minensis → Cereus mirabella

Mitrocereus fulviceps → Pachycereus fulviceps
Mitrocereus militaris♦ → Pachycereus militaris♦
Mitrocereus ruficeps → Neobuxbaumia macrocephala

Monvillea adelmarii → Cereus adelmarii
Monvillea albicaulis → Cereus albicaulis
Monvillea alticostata → Praecereus euchlorus ssp. euchlorus
Monvillea amazonica → Praecereus euchlorus ssp. amazonicus
Monvillea apoloensis → Praecereus euchlorus ssp. amazonicus
Monvillea ballivianii → Praecereus euchlorus ssp. amazonicus
Monvillea campinensis → Praecereus euchlorus ssp. euchlorus
¶*Monvillea cavendishii* → Acanthocereus sp.
⊗*Monvillea cavendishii* → ?Praecereus euchlorus ssp. euchlorus
Monvillea chacoana → Praecereus saxicola
Monvillea diffusa → Praecereus euchlorus ssp. diffusus
Monvillea ebenacantha → Cereus spegazzinii
Monvillea euchlora → Praecereus euchlorus ssp. euchlorus
Monvillea haageana → Cereus haageanus
Monvillea insularis → Cereus insularis
Monvillea jaenensis → Praecereus euchlorus ssp. jaenensis
⊗*Monvillea kroenleinii* → Cereus kroenleinii
Monvillea lauterbachii → Praecereus euchlorus ssp. euchlorus
Monvillea leucantha → Praecereus euchlorus ssp. euchlorus
Monvillea lindenzweigiana → Cereus spegazzinii
Monvillea maritima → Praecereus euchlorus ssp. diffusus
Monvillea minensis → Cereus mirabella
Monvillea parapetiensis → Praecereus saxicola
⊗*Monvillea paxtoniana* → ?Praecereus euchlorus ssp. euchlorus
Monvillea phatnosperma → Cereus phatnospermus
Monvillea piedadensis → Praecereus euchlorus ssp. euchlorus
Monvillea pugionifera → Praecereus euchlorus ssp. diffusus
Monvillea rhodoleucantha → Praecereus saxicola
Monvillea saddiana → Cereus saddianus
Monvillea saxicola → Praecereus saxicola
Monvillea smithiana → Praecereus euchlorus ssp. smithianus
Monvillea spegazzinii → Cereus spegazzinii

Morangaya pensilis → Echinocereus pensilis

Morawetzia doelziana → Oreocereus doelzianus
Morawetzia sericata → Oreocereus doelzianus
Morawetzia varicolor → Oreocereus varicolor

×***Myrtgerocactus lindsayi*** (Bergerocactus emoryi × Myrtillocactus cochal)

Myrtillocactus chende → Polaskia chende
Myrtillocactus chichipe → Polaskia chichipe
Myrtillocactus chiotilla → Escontria chiotilla
Myrtillocactus cochal
Myrtillocactus eichlamii
Myrtillocactus geometrizans
Myrtillocactus grandiareolatus → Myrtillocactus geometrizans
Myrtillocactus schenckii

®*Navajoa fickeisenii*♦ → Pediocactus peeblesianus♦
Navajoa peeblesiana♦ → Pediocactus peeblesianus♦
Navajoa peeblesiana ssp. *fickeisenii*♦ → Pediocactus peeblesianus♦

Neoabbottia grantiana → Leptocereus grantianus
Neoabbottia paniculata → Leptocereus paniculatus

Neobesseya asperispina → Escobaria missouriensis ssp. asperispina
Neobesseya macdougallii → Ortegocactus macdougallii
Neobesseya missouriensis → Escobaria missouriensis ssp. missouriensis
Neobesseya notesteinii → Escobaria missouriensis ssp. missouriensis
Neobesseya rosiflora → Escobaria missouriensis ssp. missouriensis
Neobesseya similis → Escobaria missouriensis ssp. missouriensis
Neobesseya wissmannii → Escobaria missouriensis ssp. missouriensis

Neobinghamia climaxantha → ×Haagespostoa climaxantha
Neobinghamia mirabilis → ×Espostocactus mirabilis
Neobinghamia multiareolata → ×Haagespostoa albisetata
Neobinghamia villigera → ×Haagespostoa albisetata

Neobuxbaumia euphorbioides
Neobuxbaumia laui
Neobuxbaumia macrocephala
Neobuxbaumia mezcalaensis
Neobuxbaumia multiareolata
Neobuxbaumia polylopha
Neobuxbaumia sanchezmejoradae → Neobuxbaumia laui
Neobuxbaumia scoparia
Neobuxbaumia squamulosa
®*Neobuxbaumia tetazo* → Neobuxbaumia tetetzo
Neobuxbaumia tetetzo

Neocardenasia herzogiana → Neoraimondia herzogiana

Neochilenia aerocarpa → Eriosyce aerocarpa
®*Neochilenia andreaeana* → Eriosyce andreaeana
Neochilenia aricensis → Eriosyce recondita ssp. iquiquensis
Neochilenia aspillagae → Eriosyce aspillagae
®*Neochilenia atra* → ?Eriosyce odieri
Neochilenia calderana → Eriosyce taltalensis ssp. taltalensis
®*Neochilenia carneoflora* → Eriosyce odieri ssp. glabrescens
Neochilenia chilensis → Eriosyce chilensis
Neochilenia chorosensis → Eriosyce heinrichiana
Neochilenia confinis → Eriosyce confinis
Neochilenia deherdtiana → Eriosyce heinrichiana
Neochilenia dimorpha → Eriosyce heinrichiana ssp. intermedia
Neochilenia duripulpa → Eriosyce napina ssp. lembckei
Neochilenia eriocephala → Eriosyce taltalensis ssp. echinus
Neochilenia eriosyzoides → Eriosyce kunzei
Neochilenia esmeraldana → Eriosyce esmeraldana
Neochilenia floccosa → Eriosyce taltalensis ssp. echinus
Neochilenia fobeana → ?Eriosyce taltalensis ssp. paucicostata
Neochilenia fusca → ?Eriosyce taltalensis ssp. paucicostata
Neochilenia glabrescens → Eriosyce odieri ssp. glabrescens
Neochilenia glaucescens → Eriosyce taltalensis ssp. echinus
Neochilenia gracilis → Eriosyce taltalensis ssp. taltalensis
Neochilenia hankeana → ?Eriosyce taltalensis ssp. paucicostata
Neochilenia huascensis → Eriosyce crispa ssp. atroviridis
®*Neochilenia imitans* → Eriosyce napina ssp. lembckei
Neochilenia intermedia → Eriosyce taltalensis ssp. taltalensis

Neochilenia iquiquensis → Eriosyce recondita ssp. iquiquensis
Neochilenia jussieui → ?Eriosyce heinrichiana
Neochilenia krausii → Eriosyce krausii
Neochilenia kunzei → Eriosyce kunzei
Ⓤ*Neochilenia lembckei* → Eriosyce napina ssp. lembckei
Neochilenia malleolata → Eriosyce krausii
Neochilenia mitis → Eriosyce napina
Ⓤ*Neochilenia monte-amargensis* → Eriosyce odieri ssp. odieri
Neochilenia napina → Eriosyce napina
Ⓤ*Neochilenia neofusca* → ?Eriosyce taltalensis
Ⓤ*Neochilenia neoreichei* → Eriosyce napina ssp. lembckei
Neochilenia nigriscoparia → Eriosyce crispa ssp. crispa
Neochilenia occulta → ?Eriosyce heinrichiana
Neochilenia odieri → Eriosyce odieri ssp. odieri
Neochilenia odoriflora → Eriosyce curvispina
Neochilenia paucicostata → Eriosyce taltalensis ssp. paucicostata
Neochilenia pilispina → Eriosyce taltalensis ssp. pilispina
Ⓤ*Neochilenia pseudoreichei* → ?Eriosyce odieri
Neochilenia pulchella → Eriosyce taltalensis ssp. taltalensis
Neochilenia pygmaea → Eriosyce taltalensis ssp. taltalensis
Neochilenia recondita → Eriosyce recondita
Neochilenia reichei → ?Eriosyce odieri
Neochilenia residua → Eriosyce recondita ssp. iquiquensis
Neochilenia robusta → Eriosyce curvispina
Neochilenia rupicola → Eriosyce taltalensis ssp. taltalensis
Neochilenia scoparia → Eriosyce taltalensis ssp. taltalensis
Neochilenia setosiflora → Eriosyce heinrichiana ssp. intermedia
Neochilenia simulans → Eriosyce heinrichiana ssp. simulans
Neochilenia taltalensis → Eriosyce taltalensis
Neochilenia tenebrica → Eriosyce tenebrica
Neochilenia totoralensis → Eriosyce crispa ssp. atroviridis
Neochilenia transitensis → Eriosyce kunzei
Ⓤ*Neochilenia trapichensis* → Eriosyce heinrichiana
Neochilenia wagenknechtii → Eriosyce heinrichiana ssp. intermedia

Neodawsonia apicicephalium → Cephalocereus nizandensis
Neodawsonia nizandensis → Cephalocereus nizandensis
Neodawsonia totolapensis → Cephalocereus totolapensis

Neoevansia haackiana → Peniocereus sp.
Neoevansia lazaro-cardenasii → Peniocereus lazaro-cardenasii
Neoevansia striata → Peniocereus striatus
Neoevansia zopilotensis → Peniocereus zopilotensis

Neogomesia agavoides♦ → Ariocarpus agavoides♦

Ⓤ*Neolloydia beguinii* (1922) → Sclerocactus erectocentrus♦
ⓂⓊ*Neolloydia beguinii* (1923) → Turbinicarpus beguinii♦
Neolloydia ceratites → Neolloydia conoidea
Neolloydia clavata → Coryphantha clavata
Neolloydia conoidea
Neolloydia durangensis → Sclerocactus unguispinus
Neolloydia erectocentra♦ → Sclerocactus erectocentrus♦
Neolloydia gautii → ?Sclerocactus warnockii
Neolloydia gielsdorfiana♦ → Turbinicarpus gielsdorfianus♦
Neolloydia grandiflora → Neolloydia conoidea
Ⓤ*Neolloydia hoferi* → Turbinicarpus hoferi♦
Neolloydia horripila♦ → Turbinicarpus horripilus♦
Neolloydia intertexta → Sclerocactus intertextus

Neolloydia johnsonii → Sclerocactus johnsonii
Neolloydia knuthiana♦ → Turbinicarpus knuthianus♦
Ⓤ*Neolloydia krainziana* → Turbinicarpus pseudomacrochele♦
Neolloydia laui♦ → Turbinicarpus laui♦
Neolloydia lophophoroides♦ → Turbinicarpus lophophoroides♦
Neolloydia macdowellii → Thelocactus macdowellii
Neolloydia mandragora♦ → Turbinicarpus mandragora♦
Neolloydia mariposensis♦ → Sclerocactus mariposensis♦
Neolloydia matehualensis
Neolloydia odorata → Coryphantha odorata
¶*Neolloydia pilispina* → Mammillaria pilispina
Ⓜ*Neolloydia pilispina*♦ → Turbinicarpus laui♦
Neolloydia pseudomacrochele♦
→ Turbinicarpus pseudomacrochele ssp. pseudomacrochele♦
Neolloydia pseudopectinata♦ → Turbinicarpus pseudopectinatus♦
Neolloydia pulleineana → Coryphantha pulleineana
Ⓤ*Neolloydia roseiflora* → Turbinicarpus roseiflorus♦
Neolloydia saueri♦ → Turbinicarpus saueri♦
Neolloydia schmiedickeana♦ → Turbinicarpus schmiedickeanus ssp. schmiedickeanus♦
¶*Neolloydia smithii* → ?Thelocactus conothelos ssp. conothelos♦
Ⓜ*Neolloydia smithii*♦ → Turbinicarpus beguinii♦
Neolloydia subterranea♦ → Turbinicarpus subterraneus♦
Neolloydia texensis → Neolloydia conoidea
Neolloydia unguispina → Sclerocactus unguispinus
Neolloydia valdeziana♦ → Turbinicarpus valdezianus♦
Neolloydia viereckii♦ → Turbinicarpus viereckii♦
Neolloydia warnockii → Sclerocactus warnockii

Neolobivia echinata → Echinopsis hertrichiana
Neolobivia hertrichiana → Echinopsis hertrichiana
Neolobivia incaica → Echinopsis hertrichiana
Neolobivia minuta → Echinopsis hertrichiana
Neolobivia vilcabambae → Echinopsis hertrichiana
Neolobivia winteriana → Echinopsis backebergii

Ⓤ*Neoporteria andreaeana* → Eriosyce andreaeana
Neoporteria aricensis → Eriosyce recondita ssp. iquiquensis
Neoporteria armata → Eriosyce curvispina
Neoporteria aspillagae → Eriosyce aspillagae
Neoporteria atrispinosa → Eriosyce villosa
Neoporteria atroviridis → Eriosyce crispa ssp. atroviridis
Neoporteria backebergii → Eriosyce strausiana
Neoporteria bicolor → Eriosyce islayensis
Neoporteria bulbocalyx → Eriosyce bulbocalyx
Neoporteria calderana → Eriosyce taltalensis ssp. taltalensis
Ⓤ*Neoporteria carrizalensis* → Eriosyce crispa ssp. atroviridis
Neoporteria castanea → Eriosyce subgibbosa ssp. subgibbosa
Neoporteria castaneoides → Eriosyce subgibbosa
Neoporteria catamarcensis → ?Echinopsis sp.
Neoporteria cephalophora → Eriosyce villosa
Neoporteria chilensis → Eriosyce chilensis
Neoporteria choapensis → Eriosyce curvispina
Neoporteria chorosensis → Eriosyce heinrichiana
Neoporteria clavata → Eriosyce subgibbosa ssp. clavata
Neoporteria coimasensis → Eriosyce senilis ssp. coimasensis
Neoporteria confinis → Eriosyce confinis
Neoporteria crispa → Eriosyce crispa ssp. crispa
Neoporteria curvispina → Eriosyce curvispina
Neoporteria deherdtiana → Eriosyce heinrichiana

Neoporteria dimorpha → Eriosyce heinrichiana ssp. intermedia
Neoporteria dubia → Eriosyce bulbocalyx
Neoporteria echinus → Eriosyce taltalensis ssp. echinus
Neoporteria engleri → Eriosyce engleri
Neoporteria eriocephala → Eriosyce taltalensis ssp. echinus
Neoporteria eriosyzoides → Eriosyce kunzei
Neoporteria esmeraldana → Eriosyce esmeraldana
Neoporteria floccosa → Eriosyce taltalensis ssp. echinus
Neoporteria fusca → ?Eriosyce taltalensis ssp. paucicostata
Neoporteria garaventae → Eriosyce garaventae
ⓤ*Neoporteria gerocephala* → Eriosyce senilis ssp. senilis
Neoporteria hankeana → ?Eriosyce taltalensis ssp. paucicostata
Neoporteria heinrichiana → Eriosyce heinrichiana
Neoporteria heteracantha → Eriosyce subgibbosa ssp. subgibbosa
Neoporteria horrida → Eriosyce curvispina
Neoporteria huascensis → Eriosyce crispa ssp. crispa
Neoporteria intermedia → Eriosyce taltalensis ssp. taltalensis
Neoporteria iquiquensis → Eriosyce recondita ssp. iquiquensis
Neoporteria islayensis → Eriosyce islayensis
Neoporteria jussieui → ?Eriosyce heinrichiana
Neoporteria krainziana → Eriosyce islayensis
Neoporteria kunzei → Eriosyce kunzei
Neoporteria laniceps → Eriosyce villosa
Neoporteria limariensis → Eriosyce limariensis
Neoporteria litoralis → Eriosyce subgibbosa ssp. subgibbosa
Neoporteria marksiana → Eriosyce marksiana
Neoporteria megliolii → Eriosyce bulbocalyx
ⓤ*Neoporteria melanacantha* → Eriosyce villicumensis
Neoporteria microsperma → Eriosyce subgibbosa ssp. clavata
ⓤ*Neoporteria monte-amargensis* → Eriosyce odieri ssp. odieri
Neoporteria multicolor → Eriosyce senilis ssp. senilis
Neoporteria napina → Eriosyce napina
Neoporteria nidus → ?Eriosyce kunzei
Neoporteria nigricans → ?Eriosyce limariensis
Neoporteria nigrihorrida → Eriosyce subgibbosa ssp. clavata
Neoporteria occulta → ?Eriosyce heinrichiana
Neoporteria odieri → Eriosyce odieri ssp. odieri
ⓤ*Neoporteria omasensis* → Eriosyce omasensis
Neoporteria paucicostata → Eriosyce taltalensis ssp. paucicostata
Neoporteria pilispina → Eriosyce taltalensis ssp. pilispina
Neoporteria polyraphis → Eriosyce villosa
Neoporteria pulchella → Eriosyce taltalensis ssp. taltalensis
Neoporteria recondita → Eriosyce recondita
Neoporteria reichii → ?Eriosyce odieri
Neoporteria residua → Eriosyce recondita ssp. iquiquensis
Neoporteria ritteri → Eriosyce heinrichiana ssp. intermedia
Neoporteria robusta → Eriosyce senilis ssp. coimasensis
Neoporteria rupicola → Eriosyce taltalensis ssp. taltalensis
Neoporteria sanjuanensis → Eriosyce strausiana
Neoporteria scoparia → Eriosyce taltalensis ssp. taltalensis
Neoporteria setiflora → Eriosyce strausiana
Neoporteria setosiflora → Eriosyce heinrichiana ssp. intermedia
Neoporteria simulans → Eriosyce heinrichiana ssp. simulans
Neoporteria sociabilis → Eriosyce sociabilis
Neoporteria strausiana → Eriosyce strausiana
Neoporteria subaiana → Eriosyce garaventae
Neoporteria subcylindrica → Eriosyce subgibbosa ssp. subgibbosa
Neoporteria subgibbosa → Eriosyce subgibbosa
Neoporteria taltalensis → Eriosyce taltalensis

Neoporteria totoralensis → Eriosyce crispa ssp. atroviridis
Neoporteria transiens → Eriosyce taltalensis ssp. taltalensis
Neoporteria transitensis → Eriosyce kunzei
Neoporteria tuberisulcata → Eriosyce curvispina
Neoporteria umadeave → Eriosyce umadeave
Neoporteria vallenarensis → Eriosyce subgibbosa ssp. clavata
Neoporteria vallenarensis → Eriosyce kunzei
Neoporteria villicumensis → Eriosyce villicumensis
Neoporteria villosa → Eriosyce villosa
Neoporteria volliana → Eriosyce strausiana
Neoporteria wagenknechtii → Eriosyce subgibbosa ssp. clavata
®*Neoporteria woutersiana* → Eriosyce taltalensis ssp. paucicostata

Neoraimondia arequipensis
Neoraimondia aticensis → Neoraimondia arequipensis
®*Neoraimondia gigantea* → Neoraimondia arequipensis
Neoraimondia herzogiana
Neoraimondia macrostibas → Neoraimondia arequipensis
®*Neoraimondia peruviana* → Neoraimondia arequipensis
Neoraimondia roseiflora → Neoraimondia arequipensis

Neowerdermannia chilensis
Neowerdermannia chilensis ssp. **chilensis**
Neowerdermannia chilensis ssp. **peruviana**
Neowerdermannia peruviana → Neowerdermannia chilensis ssp. peruviana
Neowerdermannia vorwerkii

Nopalea auberi → Opuntia auberi
Nopalea cochenillifera → Opuntia cochenillifera
Nopalea dejecta → Opuntia dejecta
Nopalea escuintlensis → Opuntia inaperta
Nopalea gaumeri → Opuntia inaperta
Nopalea guatemalensis → Opuntia lutea
Nopalea inaperta → Opuntia inaperta
Nopalea karwinskiana → Opuntia karwinskiana
Nopalea lutea → Opuntia lutea
Nopalea nuda → Opuntia nuda

Nopalxochia ackermannii → Disocactus ackermannii
Nopalxochia conzattiana → Disocactus ackermannii
Nopalxochia horichii → Disocactus kimnachii
Nopalxochia macdougallii♦ → Disocactus macdougallii♦
Nopalxochia phyllanthoides → Disocactus phyllanthoides

Normanbokea pseudopectinata♦ → Turbinicarpus pseudopectinatus♦
Normanbokea valdeziana♦ → Turbinicarpus valdezianus♦

Notocactus acuatus → Parodia erinacea
Notocactus acutus → Parodia ottonis
Notocactus agnetae → Parodia concinna ssp. agnetae
Notocactus alacriportanus → Parodia alacriportana ssp. alacriportana
Notocactus allosiphon → Parodia allosiphon
Notocactus ampliocostatus → Parodia schumanniana
Notocactus apricus → Parodia concinna
Notocactus arachnites → Parodia werneri
Notocactus arechavaletae → Parodia ottonis
Notocactus arnostianus → Parodia arnostiana
Notocactus blaauwianus → Parodia concinna ssp. blauuwiana
Notocactus bommeljei → Parodia tabularis ssp. bommeljei

Notocactus brederooianus → Parodia tabularis ssp. bommeljei
Notocactus brevihamatus → Parodia alacriportana ssp. brevihamata
Notocactus buenekeri → Parodia alacriportana ssp. buenekeri
Notocactus buiningii → Parodia buiningii
Notocactus caespitosus → Parodia concinna
Notocactus calvescens → Parodia turbinata
Notocactus campestrensis → Parodia oxycostata
Notocactus carambeiensis → Parodia carambeiensis
Notocactus catarinensis → Parodia alacriportana ssp. buenekeri
Notocactus claviceps → Parodia schumanniana ssp. claviceps
Notocactus concinnioides → Parodia concinna
Notocactus concinnus → Parodia concinna ssp. concinna
Notocactus corynodes → Parodia sellowii
Notocactus crassigibbus → Parodia crassigibba
Notocactus cristatoides → Parodia mammulosa
Notocactus curvispinus → Parodia curvispina
Notocactus eremiticus → Parodia concinna
Notocactus erinaceus → Parodia erinacea
Notocactus erubescens → Parodia erubescens
Notocactus erythracanthus → Parodia mammulosa ssp. erythracantha
Notocactus eugeniae → Parodia mammulosa ssp. eugeniae
Ⓤ*Notocactus eurypleurus* → Parodia oxycostata ssp. gracilis
Notocactus ferrugineus → Parodia werdermanniana
Notocactus floricomus → Parodia mammulosa
Notocactus fricii → Parodia sellowii
Notocactus fuscus → Parodia fusca
Notocactus gibberulus → Parodia concinna
Notocactus glaucinus → Parodia oxycostata
Notocactus glomeratus → Parodia rudibuenekeri ssp. glomerata
Notocactus gracilis → Parodia oxycostata ssp. gracilis
Notocactus graessneri → Parodia haselbergii ssp. graessneri
Ⓤ*Notocactus grandiensis* → Parodia ottonis
Notocactus gutierrezii → Parodia mueller-melchersii ssp. gutierrezii
Notocactus harmonianus → Parodia oxycostata ssp. gracilis
Notocactus haselbergii → Parodia haselbergii ssp. haselbergii
Notocactus herteri → Parodia herteri
Notocactus horstii → Parodia horstii
Notocactus hypocrateriformis → Parodia mammulosa
Notocactus ibicuiensis → Parodia oxycostata ssp. gracilis
Notocactus incomptus → Parodia oxycostata
Notocactus joadii → Parodia concinna
Notocactus laetivirens → Parodia muricata
Notocactus langsdorfii → Parodia langsdorfii
Notocactus leninghausii → Parodia leninghausii
Notocactus leprosorum → Parodia langsdorfii
Notocactus leucocarpus → Parodia sellowii
Notocactus linkii → Parodia linkii
Notocactus longispinus → Parodia langsdorfii
Notocactus macambarensis → Parodia mammulosa
Notocactus macracanthus → Parodia sellowii
Notocactus macrogonus → Parodia sellowii
Notocactus magnificus → Parodia magnifica
Notocactus maldonadensis → Parodia neoarechavaletae
Notocactus mammulosus → Parodia mammulosa ssp. mammulosa
Notocactus megalanthus → ?Parodia mammulosa
Notocactus megapotamicus → Parodia linkii
Notocactus memorialis → ?Parodia werdermanniana
Notocactus meonacanthus → Parodia meonacantha
Notocactus miniatispinus → Parodia oxycostata

Notocactus minimus → Parodia tenuicylindrica
Notocactus minusculus → Parodia nothominuscula
Notocactus mueller-melchersii → Parodia mueller-melchersii ssp. mueller-melchersii
Notocactus mueller-moelleri → Parodia mammulosa
Notocactus multicostatus → Parodia concinna ssp. blauuwiana
Notocactus muricatus → Parodia muricata
Notocactus neoarechavaletae → Parodia neoarechavaletae
Notocactus neobuenekeri → Parodia scopa ssp. neobuenekeri
Notocactus neohorstii → Parodia neohorstii
Notocactus nigrispinus → Parodia nigrispina
Notocactus obscurus → Parodia sp.
Notocactus olimarensis → Parodia concinna
Notocactus orthacanthus → ?Parodia mammulosa
Notocactus ottonis → Parodia ottonis ssp. ottonis
Notocactus ottonis ssp. *horstii* → Parodia ottonis ssp. horstii
Notocactus oxycostatus → Parodia oxycostata
Notocactus pampeanus → Parodia mammulosa ssp. submammulosa
Notocactus pauciareolatus → Parodia sellowii
Notocactus paulus → Parodia mammulosa
Notocactus permutatus → Parodia permutata
Notocactus polyacanthus → Parodia langsdorfii
Notocactus prolifer → Parodia langsdorfii
Notocactus pseudoherteri → Parodia herteri
Notocactus pulvinatus → Parodia langsdorfii
Notocactus purpureus → Parodia horstii
Notocactus rauschii → Parodia nothorauschii
Notocactus rechensis → Parodia rechensis
Notocactus ritterianus → Parodia mammulosa
Notocactus roseiflorus → Parodia rutilans
Notocactus roseoluteus → Parodia mammulosa
Notocactus rubricostatus → Parodia sellowii
Notocactus rubriflorus → Parodia herteri
Notocactus rubrigemmatus → Parodia concinna
Notocactus rubropedatus → Parodia curvispina
Notocactus rudibuenekeri → Parodia rudibuenekeri
Notocactus ruoffii → ?Parodia ottonis
Notocactus rutilans → Parodia rutilans ssp. rutilans
Notocactus schaeferianus → Parodia turbinata
Notocactus schlosseri → Parodia erubescens
Notocactus schumannianus → Parodia schumanniana ssp. schumanniana
Notocactus schumannianus ssp. *nigrispinus* → Parodia nigrispina
Notocactus scopa → Parodia scopa ssp. scopa
Notocactus scopa ssp. *marchesii* → Parodia scopa ssp. marchesii
Notocactus securituberculatus → Parodia oxycostata
Notocactus sellowii → Parodia sellowii
Notocactus sessiliflorus → Parodia sellowii
Notocactus soldtianus → Parodia scopa ssp. scopa
Notocactus spinibarbis → Parodia nothorauschii
Notocactus stegmannii → Parodia sellowii
Notocactus stockingeri → Parodia stockingeri
Notocactus submammulosus → Parodia mammulosa ssp. submammulosa
Notocactus succineus → Parodia scopa ssp. succinea
Notocactus tabularis → Parodia tabularis ssp. tabularis
Notocactus tenuicylindricus → Parodia tenuicylindrica
Notocactus tenuispinus → Parodia ottonis
Notocactus tephracanthus → Parodia sellowii
Notocactus tetracanthus → Parodia erinacea
Notocactus turbinatus → Parodia turbinata
Notocactus uebelmannianus → Parodia werneri ssp. werneri

Notocactus uebelmannianus ssp. *pleiocephalus* → Parodia werneri ssp. pleiocephala
Notocactus uruguayus → Parodia ottonis
Notocactus vanvlietii → Parodia werdermanniana
Notocactus veenianus → Parodia rutilans ssp. veeniana
Notocactus villa-velhensis → Parodia carambeiensis
Notocactus vorwerkianus → Parodia sellowii
Notocactus warasii → Parodia warasii
Notocactus werdermannianus → Parodia werdermanniana
Notocactus winkleri → Parodia mueller-melchersii ssp. winkleri

Nyctocereus castellanosii → Peniocereus serpentinus
Nyctocereus chontalensis → Selenicereus chontalensis
Nyctocereus guatemalensis → Peniocereus hirschtianus
Nyctocereus hirschtianus → Peniocereus hirschtianus
Nyctocereus neumannii → Peniocereus hirschtianus
Nyctocereus oaxacensis → Peniocereus oaxacensis
Nyctocereus serpentinus → Peniocereus serpentinus

Obregonia denegrii♦

Oehmea beneckei → Mammillaria beneckei

Opuntia abjecta → Opuntia triacantha
Opuntia abyssi
Opuntia acanthocarpa
Opuntia acanthocarpa ssp. ***ganderi***
Opuntia acaulis
Opuntia aciculata
Opuntia ×aequatorialis (?O. pubescens × O. soederstromiana)
Opuntia affinis → Opuntia velutina
Opuntia aggeria
Opuntia agglomerata
Opuntia albicarpa → Opuntia sp.
Opuntia albisaetacens
Opuntia alcahes
Opuntia alcerrecensis
Opuntia alexanderi
Opuntia alko-tuna
Opuntia allairei → Opuntia humifusa
Opuntia ammophila
Opuntia amyclaea
Opuntia anacantha
Opuntia anahuacensis → Opuntia dillenii
Opuntia angustata → Opuntia phaeacantha
Opuntia anteojoensis
Opuntia antillana
Opuntia aoracantha
Opuntia apurimacensis
Opuntia arbuscula
Opuntia arcei
Opuntia archiconoidea
Opuntia arechavaletae → Opuntia monacantha
Opuntia arenaria → Opuntia polyacantha
Opuntia argentina → Opuntia brasiliensis
Opuntia armata
Opuntia arrastradillo → Opuntia stenopetala
Opuntia articulata
Opuntia asplundii → Opuntia boliviana
Opuntia assumptionis
Opuntia atacamensis

Opuntia atrispina
Opuntia atrocapensis → Opuntia dillenii
Opuntia atroglobosa → Opuntia nigrispina
Opuntia atropes
Opuntia atrovirens
Opuntia atroviridis → Opuntia floccosa
Opuntia auberi
Opuntia aulacothele → ?Opuntia weberi
Opuntia aurantiaca
Opuntia aurea
Opuntia aureispina
Opuntia austrina
Opuntia azurea
Opuntia backebergii
Opuntia bahamana → Opuntia stricta
Opuntia bahiensis
Opuntia ×bakeri → Opuntia ×aequatorialis
Opuntia ballii → Opuntia macrorhiza
Opuntia basilaris
Opuntia bella
Opuntia bensonii
Opuntia bergeriana → Opuntia elatior
Opuntia ×bernichiana (O. humifusa × O. stricta)
Ⓜ*Opuntia berteri* → Opuntia sphaerica
Opuntia bigelovii
Opuntia bisetosa
Ⓤ*Opuntia bispinosa* → Opuntia anacantha
Opuntia blancii
Opuntia boldinghii
Opuntia boliviana
Opuntia boliviensis → Opuntia soehrensii
Opuntia bonaerensis → Opuntia paraguayensis
Opuntia bonplandii
Opuntia borinquensis
Opuntia brachyacantha → Opuntia sulphurea ssp. brachyacantha
Opuntia brachyarthra
Opuntia brachyclada
Opuntia bradleyi
Opuntia bradtiana
Opuntia brasiliensis
Opuntia brasiliensis ssp. *bahiensis*
Opuntia brasiliensis ssp. *subacarpa*
Opuntia bravoana
Opuntia brevispina → Opuntia alcahes
Opuntia bruchii → Opuntia alexanderi
Opuntia brunneogemmia → Opuntia monacantha
Opuntia brunnescens → Opuntia sulphurea
Opuntia bulbispina
Opuntia burrageana
Opuntia calcicola → Opuntia humifusa
Opuntia californica
Opuntia camachoi
Opuntia camanchica
Opuntia campestris → Opuntia sphaerica
Opuntia candelabriformis → Opuntia spinulifera
Opuntia canina → Opuntia anacantha
Opuntia cantabrigiensis → Opuntia engelmannii
Opuntia canterae
Opuntia caracassana
Opuntia cardenche

Opuntia cardiosperma
Opuntia caribaea
Opuntia catingicola → Opuntia ×quipa
Opuntia cedergreniana → Opuntia soehrensii
Opuntia chaffeyi
Opuntia chakensis
Opuntia chavena
Opuntia chichensis
Opuntia chihuahuensis
Opuntia chisosensis
Opuntia chlorotica
Opuntia cholla
Opuntia chuquisacana → Opuntia vestita
Opuntia cineracea
Opuntia ciribe → Opuntia bigelovii
Opuntia clavarioides
Opuntia clavata
Opuntia cochabambensis
Opuntia cochenillifera
Opuntia cognata
Opuntia colorea
Opuntia colubrina
Opuntia ×columbiana (O. fragilis × O. polyacantha)
Opuntia comonduensis → Opuntia tapona
®*Opuntia compressa* → Opuntia humifusa
Opuntia ×congesta (O. acanthocarpa × O. whipplei)
Opuntia conjungens
®***Opuntia conoidea***
Opuntia cordobensis → Opuntia ficus-indica
Opuntia corotilla
Opuntia corrugata
Opuntia covillei → Opuntia vaseyi
Opuntia crassa
Opuntia crassicylindrica
Opuntia crassipina
Opuntia cretochaeta → Opuntia hyptiacantha
Opuntia crispicrinita → Opuntia floccosa
Opuntia crystalenia
Opuntia ×cubensis (O. dillenii × O. militaris)
Opuntia cuija → Opuntia engelmannii
Opuntia cumulicola → Opuntia humifusa
Opuntia curassavica
Opuntia ×curvispina (O. chlorotica × O. phaeacantha)
Opuntia cylindrarticulata → Opuntia dactylifera
Opuntia cylindrica
Opuntia cylindrolanata → Opuntia floccosa
Opuntia cymochila
Opuntia dactylifera
Opuntia darrahiana
Opuntia darwinii
Opuntia davisii
Opuntia deamii
Opuntia decumbens
Opuntia dejecta
Opuntia delaetiana
Opuntia delicata → Opuntia macrorhiza
Opuntia demissa → Opuntia sp.
Opuntia densispina
Opuntia depauperata
Opuntia depressa

Opuntia dillenii
Opuntia dimorpha → Opuntia sphaerica
Opuntia discata → Opuntia engelmannii
Opuntia discolor
Opuntia distans → Opuntia quimilo
Opuntia dobbieana → Opuntia soederstromiana
Opuntia domeykoensis
Opuntia domingensis → Opuntia antillana
Opuntia drummondii → Opuntia pusilla
Opuntia dumetorum
Opuntia durangensis
Opuntia eburnispina → Opuntia sp.
Opuntia echinacea → Opuntia boliviana
Opuntia echinocarpa
Opuntia echios
Opuntia edwardsii
Opuntia eichlamii
Opuntia ekmanii
Opuntia elata
Opuntia elatior
Opuntia elizondoana
Opuntia ellisiana
Opuntia emoryi
Opuntia engelmannii
Opuntia erectoclada
Opuntia erinacea
Opuntia estevesii
Opuntia exaltata → Opuntia subulata
Opuntia excelsa
Opuntia falcata
Opuntia feracantha
Opuntia ferocior → Opuntia chichensis
Opuntia ficus-indica
Opuntia flexospina → Opuntia engelmannii
Opuntia flexuosa
Opuntia floccosa
Opuntia ×fosbergii (O. bigelovii × O. echinocarpa or O. acanthocarpa ssp. ganderi)
Opuntia fragilis
Opuntia fragilis ssp. *brachyarthra* → Opuntia brachyarthra
Opuntia frigida
Opuntia fulgida
Opuntia fuliginosa
Opuntia fulvicoma
Opuntia fuscoatra → Opuntia humifusa
Opuntia galapageia
Opuntia galerasensis
Opuntia ganderi → Opuntia acanthocarpa ssp. ganderi
Opuntia geometrica → Opuntia alexanderi
Opuntia glaucescens → Opuntia stenopetala
Opuntia glomerata
⊗*Opuntia glomerata* → Opuntia articulata
Opuntia gosseliniana
Opuntia grahamii
Opuntia grandiflora → Opuntia sp.
Opuntia grandis → Opuntia stenopetala
Opuntia grosseana
Opuntia guatemalensis
Opuntia guatinensis
Opuntia guerrana → Opuntia robusta
Opuntia guilanchi

Opuntia haitiensis → Opuntia moniliformis
Opuntia halophila
Opuntia hamiltonii → Opuntia rosarica
Opuntia heacockiae
Opuntia heliabravoana → Opuntia spinulifera
Opuntia heliae → Opuntia puberula
Opuntia helleri
Opuntia hernandezii → Opuntia tomentosa
Opuntia herrfeldtii → Opuntia rufida
Opuntia heteromorpha
Opuntia hickenii → Opuntia darwinii
Opuntia hirschii
Opuntia hitchcockii
Opuntia hoffmannii → Opuntia pubescens
Opuntia hondurensis
Opuntia howeyi
Opuntia huajuapensis
Opuntia humifusa
Opuntia hyptiacantha
Opuntia hystricina → Opuntia polyacantha
Opuntia ianthinantha
Opuntia ignescens
Opuntia ignota → Opuntia corotilla
Opuntia imbricata
Opuntia impedata → Opuntia humifusa
Opuntia inaequilateralis
Opuntia inamoena
Opuntia inaperta
Opuntia infesta
Opuntia insularis
Opuntia invicta
Opuntia italica → Opuntia humifusa
Opuntia jaliscana
Opuntia jamaicensis
Opuntia joconostle
Opuntia juniperina → Opuntia sphaerocarpa
Opuntia karwinskiana
Opuntia ×kelvinensis (O. fulgida × O. spinosior)
Opuntia keyensis → Opuntia stricta
Opuntia kiska-loro → Opuntia anacantha
Opuntia kleiniae
Opuntia kuehnrichiana → Opuntia sphaerica
Opuntia kunzei
Opuntia laetevirens → Opuntia sp.
Opuntia laevis
Opuntia lagopus
Opuntia lagunae
Opuntia larreyi
Opuntia lasiacantha
Opuntia lata → Opuntia ammophila
Opuntia leoncito → Opuntia glomerata
Opuntia leptocaulis
Opuntia leucophaea
Opuntia leucotricha
Opuntia lilae
Opuntia limitata
Opuntia lindheimeri → Opuntia engelmannii
Opuntia lindsayi
Opuntia linguiformis
Opuntia littoralis

Opuntia llanos-de-huanta
Opuntia lloydii
Opuntia longiareolata
Opuntia longispina
Opuntia lubrica → Opuntia rufida
Opuntia ×lucayana (O. dillenii × O. nashii)
Opuntia lutea
Opuntia macateei → Opuntia pusilla
Opuntia macbridei → Opuntia quitensis
Opuntia macdougaliana → Opuntia tomentosa
Opuntia mackensenii → Opuntia cymochila
Opuntia macracantha
Opuntia macrarthra → Opuntia stricta
Opuntia macrocalyx → Opuntia microdasys
Opuntia macrocentra
Opuntia macrorhiza
Opuntia magnifica → Opuntia stricta
Opuntia maldonadensis → Opuntia sp.
Opuntia mamillata
Opuntia mandragora → Opuntia minuta
Opuntia marenae
Opuntia marnieriana → Opuntia stenopetala
Opuntia martiniana
Opuntia matudae → Opuntia hyptiacantha
Opuntia maxonii → Opuntia puberula
Opuntia megacantha
Opuntia megapotamica
Opuntia megarhiza
Opuntia megasperma
Opuntia melanosperma → Opuntia dillenii
Opuntia microcarpa → Opuntia engelmannii
Opuntia microdasys
Opuntia microdisca
Opuntia mieckleyi
Opuntia militaris → Opuntia triacantha
Opuntia millspaughii
Opuntia minuscula
Opuntia minuta
Opuntia miquelii
Opuntia mira → Opuntia sphaerica
Opuntia mistiensis
Opuntia moelleri
Opuntia mojavensis
Opuntia molesta
Opuntia molinensis
Opuntia monacantha
Opuntia monacantha ssp. *brunneogemmia*
Opuntia moniliformis
Opuntia montevideensis
®*Opuntia multiareolata* → Opuntia soehrensii
Opuntia ×multigeniculata (O. echinocarpa × O. whipplei)
Opuntia munzii
Opuntia nashii
Opuntia nejapensis
Opuntia nemoralis → Opuntia humifusa
Opuntia ×neoarbuscula (O. arbuscula × O. spinosior)
Opuntia neoargentina → Opuntia brasiliensis
Opuntia neochrysacantha
Opuntia neuquensis → Opuntia darwinii
Opuntia nicholii → Opuntia polyacantha

Opuntia nigrispina
Opuntia nitens → Opuntia dillenii
Opuntia noodtiae → Opuntia dactylifera
Opuntia nuda
®*Opuntia obliqua* → Opuntia soehrensii
Opuntia ×occidentalis (O. littoralis × O. engelmannii × O. phaeacantha)
Opuntia ochrocentra
Opuntia orbiculata
Opuntia oricola
Opuntia orurensis
Opuntia ovata
Opuntia p(a)ediophila → Opuntia aoracantha
Opuntia pachona
Opuntia pachypus
Opuntia pailana
Opuntia pallida → Opuntia rosea
Opuntia palmadora
Opuntia palmadora ssp. *catingicola* → Opuntia ×quipa
Opuntia palmeri → Opuntia chlorotica
Opuntia papyracantha → Opuntia articulata
Opuntia paraguayensis
Opuntia parishii
Opuntia parmentieri → Opuntia sp.
Opuntia parryi
Opuntia parviclada
Opuntia pascoensis → Opuntia pubescens
Opuntia patagonica → Maihuenia patagonica
Opuntia penicilligera
Opuntia pennellii
Opuntia pentlandii
Opuntia pes-corvi → Opuntia pusilla
Opuntia pestifer → Opuntia pubescens
Opuntia phaeacantha
Opuntia picardae → Opuntia moniliformis
Opuntia picardoi
Opuntia pilifera
Opuntia pisciformis → Opuntia sp.
Opuntia pittieri
Opuntia pituitosa
Opuntia platyacantha → Opuntia darwinii
Opuntia plumbea
®*Opuntia poecilacantha* → Opuntia microdisca
Opuntia pollardii → Opuntia austrina
Opuntia polyacantha
Opuntia polycarpa → Opuntia austrina
Opuntia posnanskyana → Opuntia verschaffeltii
Opuntia pottsii → Opuntia macrorhiza
Opuntia prasina
Opuntia procumbens → Opuntia engelmannii
Opuntia prolifera
Opuntia pseudo-udonis → Opuntia floccosa
Opuntia puberula
Opuntia pubescens
Opuntia puelchana → Opuntia tunicata
Opuntia pulchella
Opuntia pumila
Opuntia punta-caillan
Opuntia purpurea → Opuntia nigrispina
Opuntia pusilla
Opuntia pycnantha

Opuntia pyriformis
Opuntia pyrrhacantha
Opuntia pyrrhantha
Opuntia quimilo
Opuntia ×quipa (O. inamoena × O. palmadora)
Opuntia quitensis
Opuntia rafinesquei → Opuntia humifusa
Opuntia rahmeri
Opuntia ramosissima
Opuntia rastrera
Opuntia rauhii → Opuntia floccosa
Opuntia rauppiana → Opuntia sphaerica
Opuntia reflexispina
Opuntia reicheana → Opuntia glomerata
Opuntia repens
Opuntia retrorsa → Opuntia anacantha
Opuntia rhodantha → Opuntia polyacantha
Opuntia rileyi
Opuntia ritteri
®*Opuntia riviereana* → Opuntia stenopetala
Opuntia robinsonii
Opuntia roborensis
Opuntia robusta
Opuntia rosarica
Opuntia rosea
Opuntia rossiana
Opuntia rubescens
Opuntia rubiflora → Opuntia humifusa
Opuntia rubrifolia → Opuntia sp.
Opuntia rufida
Opuntia russellii → Opuntia ovata
Opuntia rutila
Opuntia rzedowskii → Opuntia lasiacantha
Opuntia salagria
Opuntia salmiana
Opuntia salvadorensis
Opuntia sanctae-barbarae
Opuntia sanguinea
Opuntia santa-rita
Opuntia santamaria
Opuntia sarca → Opuntia tomentosa
Opuntia saxatilis
Opuntia saxicola
Opuntia scheeri
Opuntia scheinvariana → Opuntia puberula
Opuntia schickendantzii
Opuntia schottii
Opuntia schulzii → Opuntia brasiliensis
Opuntia schumannii
Opuntia securigera
Opuntia serpentina → Opuntia californica
Opuntia setispina → Opuntia macrorhiza
Opuntia shaferi
Opuntia shreveana → Opuntia santa-rita
Opuntia silvestris
Opuntia skottsbergii → Pterocactus hickenii
Opuntia soederstromiana
Opuntia soehrensii
Opuntia sphaerica
Opuntia sphaerocarpa

Opuntia ×spinosibacca (O. aureispina × O. phaeacantha)
Opuntia spinosior
Opuntia spinosissima
Opuntia spinulifera
Opuntia spraguei
Opuntia stanlyi → Opuntia emoryi
Opuntia steiniana → Opuntia verschaffeltii
Opuntia stenarthra
Opuntia stenopetala
Opuntia streptacantha
Opuntia stricta
Opuntia strigil
Opuntia subarmata → Opuntia engelmannii
Opuntia subsphaerocarpa
Opuntia subterranea
Opuntia subulata
Opuntia sulphurea
Opuntia sulphurea ssp. ***brachyacantha***
Opuntia sulphurea ssp. ***spinibarbis***
Opuntia superbospina
Opuntia tapona
Opuntia tarapacana
Opuntia tardospina → Opuntia engelmannii
Opuntia tayapayensis → Opuntia pubescens
Opuntia taylorii
Opuntia tehuacana
Opuntia tehuantepecana
Opuntia tenuiflora
Opuntia tenuispina → Opuntia macrorhiza
Opuntia tephrocactoides → Opuntia floccosa
Opuntia tesajo
Opuntia testudinis-crus → Opuntia moniliformis
Opuntia ×tetracantha (O. acanthocarpa × O. leptocaulis)
Opuntia thornberi → Opuntia acanthocarpa
Opuntia thurberi
Opuntia ticnamarensis
Opuntia tilcarensis → Opuntia soehrensii
Opuntia tomentella
Opuntia tomentosa
Opuntia tracyi → Opuntia pusilla
Opuntia treleasei
Opuntia triacantha
Opuntia trichophora
Opuntia tricolor → Opuntia engelmannii
Opuntia tumida
Opuntia tuna
Opuntia tuna-blanca → Opuntia ficus-indica
Opuntia tunicata
Opuntia turbinata
Opuntia turgida → Opuntia ammophila
Opuntia udonis → Opuntia floccosa
Opuntia undulata
Opuntia unguispina
Opuntia urbaniana
Opuntia utahensis → Opuntia sp.
Opuntia utkilio → Opuntia anacantha
Opuntia vaginata
Opuntia ×vaseyi (O. littoralis × O. phaeacantha)
Opuntia velutina
Opuntia verschaffeltii

Opuntia versicolor
Opuntia verticosa → Opuntia floccosa
Opuntia vestita
Opuntia vilis
®*Opuntia violacea* → Opuntia macrocentra
Opuntia* ×*viridiflora (O. imbricata × O. whipplei)
Opuntia viridirubra
Opuntia viridirubra ssp. ***rubrogemmia***
Opuntia vitelliniflora
Opuntia vitelliniflora ssp. ***interjecta***
Opuntia* ×*vivipara (O. arbuscula × O. versicolor)
¶*Opuntia vulgaris* → Opuntia ficus-indica
⊗*Opuntia vulgaris* Miller → Opuntia humifusa
⊗*Opuntia vulgaris* auctt → Opuntia monacantha
Opuntia vulpina → Opuntia sulphurea
Opuntia wagenknechtii
Opuntia weberi
Opuntia weingartiana → Opuntia shaferi
Opuntia wentiana → Opuntia caracassana
Opuntia werneri
Opuntia wetmorei
Opuntia whipplei
Opuntia whitneyana → Opuntia basilaris
Opuntia wigginsii
Opuntia wilcoxii
Opuntia wolfei
Opuntia woodsii → Opuntia phaeacantha
Opuntia wootonii
Opuntia wrightiana → Opuntia kunzei
Opuntia yanganucensis
Opuntia zacana → Opuntia echios
Opuntia zebrina → Opuntia dillenii
Opuntia zehnderi

Oreocereus australis → Oreocereus hempelianus
Oreocereus celsianus
Oreocereus crassiniveus → Oreocereus trollii
Oreocereus doelzianus
⊗*Oreocereus fossulatus* → Oreocereus pseudofossulatus
Oreocereus hempelianus
Oreocereus hendriksenianus → Oreocereus leucotrichus
Oreocereus leucotrichus
Oreocereus maximus → Oreocereus celsianus
Oreocereus neocelsianus → Oreocereus celsianus
Oreocereus piscoensis → Cleistocactus pachycladus
Oreocereus pseudofossulatus
Oreocereus rettigii → Oreocereus hempelianus
Oreocereus ritteri
Oreocereus tacnaensis
Oreocereus trollii
Oreocereus varicolor

Oroya baumannii → Oroya peruviana
Oroya borchersii
Oroya gibbosa → Oroya peruviana
Oroya laxiareolata → Oroya peruviana
Oroya neoperuviana → Oroya peruviana
Oroya peruviana
Oroya subocculta → Oroya peruviana

Ortegocactus macdougallii

×***Pacherocactus orcuttii*** (Bergerocactus emoryi × Pachycereus pringlei)

Pachycereus aragonii → Stenocereus aragonii
Pachycereus calvus → Pachycereus pringlei
Pachycereus chrysomallus♦ → Pachycereus militaris♦
Pachycereus columna-trajani → Cephalocereus apicicephalium
Pachycereus foetidus → Pachycereus gaumeri
Pachycereus fulviceps
Pachycereus gatesii
Pachycereus gaumeri
Pachycereus gigas → Pachycereus weberi
Pachycereus grandis
Pachycereus hollianus
Pachycereus lepidanthus
Pachycereus marginatus
Pachycereus militaris♦
Pachycereus orcuttii → Pacherocactus orcuttii
Pachycereus pecten-aboriginum
Pachycereus pecten-aboriginum ssp. *tehuantepecanus* → Pachycereus pecten-aboriginum
Pachycereus pringlei
Pachycereus ruficeps → Neobuxbaumia macrocephala
Pachycereus schottii
Pachycereus tehuantepecanus → Pachycereus pecten-aboriginum
Pachycereus weberi

Parodia aconquijaensis → Parodia microsperma
Parodia agasta → Parodia ritteri
Parodia aglaisma → Parodia ritteri
Parodia alacriportana
Parodia alacriportana ssp. **alacriportana**
Parodia alacriportana ssp. **brevihamata**
Parodia alacriportana ssp. **buenekeri**
Parodia alacriportana ssp. **catarinensis**
Parodia albofuscata → Parodia microsperma
Parodia allosiphon
®*Parodia amblayensis* → Parodia microsperma
Parodia ampliocostata → Parodia schumanniana
Parodia andreae → Parodia procera
Parodia andreaeoides → Parodia procera
Parodia applanata → Parodia schwebsiana
Parodia argerichiana → Parodia microsperma
Parodia arnostiana
Parodia atroviridis → Parodia microsperma ssp. horrida
Parodia augustinii → Parodia ocampoi
Parodia aureicentra
Parodia aureispina → Parodia microsperma
Parodia ayopayana
Parodia backebergiana → Parodia tuberculata
Parodia belenensis → Parodia microsperma
Parodia bellavistana → Parodia formosa
Parodia belliata → Parodia ritteri
Parodia bermejoensis → Parodia maassii
Parodia betaniana → Parodia microsperma
Parodia bilbaoensis → Parodia taratensis
Parodia borealis → Parodia ayopayana
U %*Parodia brasiliensis* → XXXX
Parodia brevihamata → Parodia alacriportana ssp. brevihamata
Parodia buenekeri → Parodia alacriportana ssp. buenekeri
Parodia buiningii
Parodia buxbaumiana → Parodia ayopayana

Parodia cabracorralensis → Parodia microsperma
Parodia cachiana → Parodia microsperma ssp. horrida
Parodia caespitosa → Parodia concinna
Parodia caineana → Parodia taratensis
Parodia camargensis → Parodia ritteri
®*Parodia camblayana* → Parodia ritteri
Parodia campestris → Parodia microsperma ssp. microsperma
Parodia candidata → Parodia tuberculata
Parodia capillitaensis → Parodia microsperma
Parodia carambeiensis
Parodia carapariana → Parodia formosa
Parodia cardenasii → Parodia formosa
Parodia carminata → Parodia tilcarensis
Parodia carrerana → Parodia ritteri
Parodia castanea → Parodia ritteri
Parodia catamarcensis → Parodia microsperma
Parodia catarinensis → Parodia alacriportana ssp. buenekeri
Parodia cebilarensis → Parodia microsperma
Parodia chaetocarpa → Parodia formosa
Parodia challamarcana → Parodia procera
Parodia chirimoyarana → Parodia formosa
Parodia chlorocarpa → Parodia microsperma
Parodia chrysacanthion
Parodia cintiensis → Parodia ritteri
Parodia claviceps → Parodia schumanniana ssp. claviceps
Parodia columnaris
Parodia comarapana
Parodia commutans
Parodia comosa → Parodia ayopayana
Parodia compressa → Parodia ocampoi
Parodia concinna
Parodia concinna ssp. **agnetae**
Parodia concinna ssp. **blauuwiana**
Parodia concinna ssp. **concinna**
Parodia copavilquensis → Parodia ocampoi
Parodia cotacajensis → Parodia ayopayana
Parodia crassigibba
Parodia culpinensis → Parodia subterranea
Parodia curvispina
Parodia dextrohamata → Parodia microsperma ssp. horrida
Parodia dichroacantha → Parodia microsperma ssp. horrida
Parodia echinopsoides → Parodia procera
Parodia echinus → Parodia ayopayana
Parodia elachisantha → Parodia haselbergii ssp. graessneri
Parodia elachista → Parodia ocampoi
Parodia elata → Parodia ayopayana
Parodia elegans → Parodia microsperma
Parodia erinacea
Parodia erubescens
Parodia erythrantha → Parodia microsperma
Parodia escayachensis → Parodia maassii
Parodia exquisita → Parodia ocampoi
Parodia faustiana → Parodia nivosa
Parodia fechseri → Parodia microsperma
Parodia firmisissima → Parodia tuberculata
Parodia formosa
Parodia friciana → Parodia tilcarensis
Parodia fulvispina → Parodia ritteri
Parodia fusca
Parodia fuscato-viridis → Parodia microsperma

Parodia gibbulosa → Parodia ocampoi
Parodia gibbulosoides → Parodia ocampoi
Parodia glischrocarpa → Parodia microsperma
Parodia gracilis → Parodia procera
Parodia graessneri → Parodia haselbergii ssp. graessneri
Parodia grandiflora → Parodia microsperma
Parodia grossei → Parodia schumanniana
Parodia guachipasana → Parodia microsperma
Parodia gummifera♦
Parodia gutekunstiana → Parodia stuemeri
Parodia haageana → Parodia maassii
Parodia haselbergii
Parodia haselbergii ssp. **graessneri**
Parodia haselbergii ssp. **haselbergii**
Parodia hausteiniana
Parodia herteri
Parodia herzogii → Parodia microsperma ssp. microsperma
Parodia heteracantha → Parodia microsperma ssp. horrida
Parodia heyeriana → Parodia microsperma
Parodia horrida → Parodia microsperma ssp. horrida
Parodia horstii
Parodia hummeliana → Parodia microsperma
Parodia idiosa → Parodia tuberculata
Parodia ignorata → Parodia tuberculata
Parodia jujuyana → Parodia tilcarensis
Parodia kilianana → Parodia microsperma ssp. horrida
Parodia koehresiana → Parodia maassii
Parodia krahnii → Parodia taratensis
Parodia krasuckana → Parodia tuberculata
Parodia lamprospina → Parodia maassii
Parodia langsdorfii
Parodia laui → Parodia hausteiniana
Parodia legitima → Parodia columnaris
Parodia lembckei → Parodia microsperma
Parodia leninghausii
Parodia liliputana → Blossfeldia liliputana
Parodia linkii
Parodia lohaniana → Parodia microsperma ssp. horrida
Parodia lychnosa → Parodia procera
Parodia maassii
Parodia macednosa → Parodia ayopayana
Parodia macrancistra → Parodia microsperma ssp. microsperma
Parodia magnifica
Parodia mairanana → Parodia comarapana
Parodia malyana → Parodia microsperma
®*Parodia malyana* ssp. *igneiflora* → Parodia microsperma
Parodia mammulosa
Parodia mammulosa ssp. **brasiliensis**
Parodia mammulosa ssp. **erythracantha**
Parodia mammulosa ssp. **eugeniae**
Parodia mammulosa ssp. **mammulosa**
Parodia mammulosa ssp. **submammulosa**
Parodia maxima → Parodia commutans
Parodia mendeziana → Parodia maassii
Parodia meonacantha
Parodia mercedesiana → Parodia microsperma
Parodia mesembrina → Parodia microsperma
Parodia microsperma
Parodia microsperma ssp. **horrida**
Parodia microsperma ssp. **microsperma**

Parodia microthele → Parodia microsperma
Parodia miguillensis → Parodia ayopayana
Parodia minima → Parodia schwebsiana
Parodia minuscula → Parodia microsperma
Parodia minuta → Parodia ocampoi
Parodia miranda → Parodia subterranea
Parodia mueller-melchersii
Parodia mueller-melchersii ssp. **gutierrezii**
Parodia mueller-melchersii ssp. **mueller-melchersii**
Parodia mueller-melchersii ssp. **winkleri**
Parodia muhrii → Parodia aureicentra
Parodia multicostata → Parodia tuberculata
Parodia muricata
Parodia mutabilis → Parodia microsperma
Parodia nana → Parodia microsperma
Parodia neglecta → Parodia comarapana
Parodia neglectoides → Parodia comarapana
Parodia neoarechavaletae
Parodia neohorstii
Parodia nigresca → Parodia subterranea
Parodia nigrispina
Parodia nivosa
Parodia nothominuscula
Parodia nothorauschii
Parodia obtusa → Parodia commutans
Ⓤ*Parodia obtusa* ssp. *atochana* → Parodia maassii
Parodia ocampoi
Parodia occulta → Parodia subterranea
Parodia otaviana → Parodia maassii
Parodia ottonis
Parodia ottonis ssp. **horstii**
Parodia ottonis ssp. **ottonis**
Parodia otuyensis → Parodia tuberculata
Parodia oxycostata
Parodia oxycostata ssp. **gracilis**
Parodia oxycostata ssp. **oxycostata**
Parodia pachysa → Parodia formosa
Parodia papagayana → Parodia microsperma
Parodia paraguayensis → Parodia ottonis
Parodia parvula → Parodia formosa
Parodia penicillata
Parodia permutata
Parodia perplexa → Parodia procera
Parodia piltziorum → Parodia microsperma ssp. horrida
Parodia pluricentralis → Parodia microsperma ssp. horrida
Parodia prestoensis → Parodia procera
Parodia procera
Parodia pseudoayopayana → Parodia ayopayana
Parodia pseudoprocera → Parodia procera
Ⓤ*Parodia pseudoprocera* ssp. *aurantiaciflora* → Parodia procera
Parodia pseudostuemeri → Parodia tilcarensis
Parodia pseudosubterranea → Parodia subterranea
Parodia punae → Parodia ocampoi
Parodia purpureo-aurea → Parodia formosa
Parodia pusilla → Parodia formosa
Parodia quechua → Parodia tuberculata
Parodia rauschii → Parodia aureicentra
Parodia rechensis
Parodia rigida → Parodia microsperma ssp. horrida
Parodia rigidispina → Parodia microsperma

Parodia riograndensis → Parodia procera
Parodia ritteri
Parodia roseoalba → Parodia ritteri
Parodia rostrum-sperma → Parodia ritteri
Parodia rubellihamata → Parodia microsperma
Parodia rubida → Parodia ritteri
Parodia rubricentra → Parodia stuemeri
Parodia rubriflora → Parodia microsperma
Parodia rubrispina → Parodia stuemeri
Parodia rubristaminea → Parodia microsperma
Parodia rudibuenekeri
Parodia rudibuenekeri ssp. **glomerata**
Parodia rudibuenekeri ssp. **rudibunekeri**
Parodia rutilans
Parodia rutilans ssp. **rutilans**
Parodia rutilans ssp. **veeniana**
Parodia saint-pieana
Parodia salitrensis → Parodia subterranea
Parodia salmonea → Parodia schwebsiana
Parodia sanguiniflora → Parodia microsperma
Parodia schuetziana → Parodia tilcarensis
Parodia schumanniana
Parodia schumanniana ssp. **claviceps**
Parodia schumanniana ssp. **schumanniana**
Parodia schwebsiana
Parodia scopa
Parodia scopa ssp. **marchesii**
Parodia scopa ssp. **neobuenekeri**
Parodia scopa ssp. **scopa**
Parodia scopa ssp. **succinea**
Parodia scopaoides → Parodia microsperma
Parodia scoparia → Parodia tilcarensis
Parodia sellowii
Parodia separata → Parodia procera
Parodia setifera → Parodia microsperma
Parodia setispina → Parodia formosa
®*Parodia setosa* → Parodia tilcarensis
Parodia sotomayorensis → Parodia tuberculata
Parodia spanisa → Parodia microsperma
Parodia spegazziniana → Parodia microsperma
Parodia splendens → Parodia ritteri
Parodia stereospina → Parodia tuberculata
Parodia stockingeri
Parodia stuemeri
Parodia submammulosa → Parodia mammulosa ssp. submammulosa
Parodia submammulosa ssp. *minor* → Parodia mammulosa ssp. submammulosa
Parodia subterranea
Parodia subtilihamata → Parodia procera
Parodia succinea → Parodia scopa ssp. succinea
Parodia sucrensis → Parodia tuberculata
Parodia superba → Parodia microsperma ssp. horrida
Parodia suprema → Parodia maassii
Parodia tabularis
Parodia tabularis ssp. **bommeljei**
Parodia tabularis ssp. **tabularis**
Parodia tafiensis → Parodia microsperma
Parodia talaensis → Parodia microsperma ssp. microsperma
Parodia tarabucina → Parodia tuberculata
Parodia taratensis
Parodia tenuicylindrica

Parodia thieleana → Parodia maassii
Parodia thionantha → Parodia microsperma ssp. microsperma
Parodia tilcarensis
Parodia tillii → Parodia formosa
Parodia tojoensis → Parodia ritteri
Parodia tolombona → Parodia microsperma ssp. horrida
Parodia tredecimcostata → Parodia procera
Parodia tuberculata
Parodia tuberculosi-costata → Parodia microsperma
Parodia tucumanensis → Parodia microsperma ssp. microsperma
Parodia tumbayana → Parodia tilcarensis
Parodia turbinata
Parodia turecekiana
Parodia uebelmanniana → Parodia microsperma
Parodia uhligiana → Parodia nivosa
Parodia varicolor → Parodia aureicentra
Parodia wagneriana → Parodia microsperma
Parodia warasii
Parodia weberiana → Parodia microsperma ssp. microsperma
Parodia werdermanniana
Parodia werneri
Parodia werneri ssp. **pleiocephala**
Parodia werneri ssp. **werneri**
Parodia weskampiana → Parodia microsperma
Parodia winbergii → Parodia formosa
Parodia yamparaezi → Parodia tuberculata
Parodia zaletaewana → Parodia subterranea
Parodia zecheri → Parodia ocampoi
®*Parodia zecheri* ssp. *elachista* → Parodia ocampoi

Pediocactus alonsoi♦ → Turbinicarpus alonsoi♦
Pediocactus bonatzii♦ → Turbinicarpus bonatzii♦
Pediocactus bradyi♦
Pediocactus bradyi ssp. ***despainii***♦
Pediocactus bradyi ssp. ***winkleri***♦
Pediocactus brevihamatus → Sclerocactus brevihamatus ssp. brevihamatus
Pediocactus brevihamatus ssp. *tobuschii* → Sclerocactus brevihamatus ssp. tobuschii
Pediocactus cloverae → Sclerocactus parviflorus
Pediocactus cloverae ssp. *brackii* → Sclerocactus parviflorus
Pediocactus conoideus → Neolloydia conoidea
Pediocactus despainii♦ → Pediocactus bradyi ssp. despainii♦
Pediocactus erectocentrus♦ → Sclerocactus erectocentrus♦
Pediocactus gautii → ?Sclerocactus warnockii
Pediocactus gielsdorfianus♦ → Turbinicarpus gielsdorfianus♦
Pediocactus glaucus♦ → Sclerocactus glaucus♦
Pediocactus hermannii
Pediocactus hoferi♦ → Turbinicarpus hoferi♦
Pediocactus horripilus♦ → Turbinicarpus horripilus♦
Pediocactus intertextus → Sclerocactus intertextus
Pediocactus johnsonii → Sclerocactus johnsonii
Pediocactus knowltonii♦
Pediocactus knuthianus♦ → Turbinicarpus knuthianus♦
Pediocactus laui♦ → Turbinicarpus laui♦
Pediocactus lophophoroides♦ → Turbinicarpus lophophoroides♦
Pediocactus mandragora♦ → Turbinicarpus mandragora♦
Pediocactus mariposensis♦ → Sclerocactus mariposensis♦
Pediocactus mesae-verdae♦ → Sclerocactus mesae-verdae♦
Pediocactus nigrispinus
Pediocactus nigrispinus ssp. ***beastonii***
Pediocactus nigrispinus ssp. ***puebloensis***

Pediocactus nyensis → Sclerocactus nyensis
Pediocactus papyracanthus♦ → Sclerocactus papyracanthus♦
Pediocactus paradinei♦
Pediocactus parviflorus → Sclerocactus parviflorus ssp. parviflorus
Pediocactus parviflorus ssp. *havasupaiensis* → Sclerocactus parviflorus ssp. havasupaiensis
Pediocactus parviflorus ssp. *intermedius* → Sclerocactus parviflorus ssp. intermedius
Pediocactus parviflorus ssp. *terrae-canyonae*
→ Sclerocactus parviflorus ssp. terrae-canyonae
Pediocactus peeblesianus♦
Pediocactus polyancistrus → Sclerocactus polyancistrus
Pediocactus pseudomacrochele♦
→ Turbinicarpus pseudomacrochele ssp. pseudomacrochele♦
Pediocactus pseudopectinatus♦ → Turbinicarpus pseudopectinatus♦
Pediocactus pubispinus♦ → Sclerocactus pubispinus♦
Pediocactus pubispinus ssp. *sileri* → Sclerocactus sileri
Pediocactus rioverdensis♦ → Turbinicarpus rioverdensis♦
Pediocactus robustior → Pediocactus simpsonii ssp. robustior
Pediocactus saueri♦ → Turbinicarpus saueri♦
Pediocactus scheeri → Sclerocactus scheeri
Pediocactus schmiedickeanus♦ → Turbinicarpus schmiedickeanus♦
Pediocactus sileri♦
Pediocactus simpsonii
Pediocactus simpsonii ssp. ***bensonii***
Pediocactus simpsonii ssp. *bradyi*♦ → Pediocactus bradyi♦
Pediocactus simpsonii ssp. ***idahoensis***
Pediocactus simpsonii ssp. ***indranus***
Pediocactus simpsonii ssp. ***robustior***
¶*Pediocactus smithii*♦ → ?Thelocactus conothelos ssp. conothelos♦
Pediocactus spinosior → Sclerocactus spinosior
Pediocactus spinosior ssp. *blainei* → Sclerocactus spinosior ssp. blainei
Pediocactus subterraneus♦ → Turbinicarpus subterraneus♦
Pediocactus swobodae♦ → Turbinicarpus swobodae♦
Pediocactus uncinatus → Sclerocactus uncinatus ssp. uncinatus
Pediocactus unguispinus → Sclerocactus unguispinus
Pediocactus valdezianus♦ → Turbinicarpus valdezianus♦
Pediocactus viereckii♦ → Turbinicarpus viereckii♦
Pediocactus warnockii → Sclerocactus warnockii
Pediocactus wetlandicus♦ → Sclerocactus glaucus♦
Pediocactus whipplei → Sclerocactus whipplei
Pediocactus whipplei ssp. *busekii* → Sclerocactus whipplei ssp. busekii
Pediocactus winkleri♦ → Pediocactus bradyi ssp. winkleri♦
Pediocactus wrightiae♦ → Sclerocactus wrightiae♦
Pediocactus ysabelae♦ → Turbinicarpus ysabelae♦

Pelecyphora aselliformis♦
Pelecyphora pseudopectinata♦ → Turbinicarpus pseudopectinatus♦
Pelecyphora pulcherrima♦ → Turbinicarpus pseudopectinatus♦
Pelecyphora strobiliformis♦

Peniocereus castellae
Peniocereus cuixmalensis
Peniocereus diguetii → Peniocereus striatus
Peniocereus fosterianus
Peniocereus greggii
Peniocereus hirschtianus
Peniocereus johnstonii
Peniocereus lazaro-cardenasii
Peniocereus macdougallii
Peniocereus maculatus
Peniocereus marianus

Peniocereus oaxacensis
Peniocereus occidentalis
®*Peniocereus papillosus* → Peniocereus viperinus
Peniocereus rosei
Peniocereus serpentinus
Peniocereus striatus
Peniocereus tepalcatepecanus
Peniocereus viperinus
Peniocereus zopilotensis

Pereskia aculeata
Pereskia amapola → Pereskia nemorosa
Pereskia aureiflora
Pereskia autumnalis → Pereskia lychnidiflora
Pereskia bahiensis
Pereskia bleo
®*Pereskia bleo* → Pereskia grandifolia ssp. grandifolia
Pereskia colombiana → Pereskia guamacho
Pereskia conzattii → Pereskia lychnidiflora
Pereskia cubensis → Pereskia zinniiflora
Pereskia diaz-romeroana
Pereskia grandifolia
Pereskia grandifolia ssp. **grandifolia**
Pereskia grandifolia ssp. **violacea**
Pereskia guamacho
Pereskia higuerana → Pereskiopsis spathulata
Pereskia horrida
Pereskia horrida ssp. **horrida**
Pereskia horrida ssp. **rauhii**
Pereskia humboldtii → Pereskia horrida ssp. horrida
Pereskia lychnidiflora
Pereskia marcanoi
Pereskia moorei → Pereskia sacharosa
Pereskia nemorosa
Pereskia pereskia → Pereskia aculeata
Pereskia pflanzii → Quiabentia verticillata
Pereskia pititache → Pereskia lychnidiflora
Pereskia portulacifolia
Pereskia quisqueyana
Pereskia sacharosa
Pereskia saipinensis → Pereskia sacharosa
Pereskia sparsiflora → Pereskia sacharosa
Pereskia stenantha
Pereskia tampicana → Pereskia grandifolia ssp. grandifolia
Pereskia vargasii → Pereskia horrida ssp. horrida
Pereskia verticillata → Quiabentia verticillata
Pereskia weberiana
Pereskia zehntneri → Quiabentia zehntneri
Pereskia zinniiflora

Pereskiopsis aquosa
Pereskiopsis blakeana
Pereskiopsis chapistle → Pereskiopsis rotundifolia
Pereskiopsis diguetii
Pereskiopsis gatesii → ?Pereskiopsis porteri
Pereskiopsis kellermanii
Pereskiopsis opuntiiflora → Pereskia lychnidiflora
Pereskiopsis pititache → Pereskia lychnidiflora
Pereskiopsis porteri
Pereskiopsis rotundifolia

Pereskiopsis scandens → Pereskiopsis kellermanii
Pereskiopsis spathulata
Pereskiopsis velutina → Pereskiopsis diguetii

®*Pfeiffera brevispina* → Lepismium brevispinum
Pfeiffera crenata → Lepismium crenatum
Pfeiffera erecta → Lepismium ianthothele
Pfeiffera gracilis → Lepismium ianthothele
Pfeiffera ianthothele → Lepismium ianthothele
Pfeiffera incahuasina → Lepismium monacanthum
Pfeiffera mataralensis → Lepismium ianthothele
Pfeiffera micrantha → Lepismium micranthum
Pfeiffera miyagawae → Lepismium miyagawae
Pfeiffera monacantha → Lepismium monacanthum
Pfeiffera multigona → Lepismium ianthothele
Pfeiffera paranganiensis → Lepismium paranganiense

Phellosperma tetrancistra → Mammillaria tetrancistra

Philippicereus castaneus → Eulychnia castanea

Pierrebraunia bahiensis → Arrojadoa bahiensis

Pilocopiapoa solaris → Copiapoa solaris

Pilosocereus albisummus
Pilosocereus alensis
Pilosocereus arenicola → Pilosocereus catingicola
Pilosocereus arrabidae
Pilosocereus atroflavispinus → Pilosocereus pachycladus ssp. pachycladus
Pilosocereus aureispinus
Pilosocereus aurilanatus → Pilosocereus aurisetus ssp. aurilanatus
Pilosocereus aurisetus
Pilosocereus aurisetus ssp. **aurilanatus**
Pilosocereus aurisetus ssp. **aurisetus**
Pilosocereus aurisetus ssp. *densilanatus* → Pilosocereus aurisetus ssp. aurisetus
Pilosocereus aurisetus ssp. *supthutianus* → Pilosocereus aurisetus ssp. aurisetus
Pilosocereus aurisetus ssp. *werdermannianus* → Pilosocereus aurisetus ssp. aurisetus
Pilosocereus azulensis
Pilosocereus azureus → Pilosocereus pachycladus ssp. pachycladus
Pilosocereus backebergii → Pilosocereus lanuginosus
Pilosocereus bahamensis → Pilosocereus polygonus
Pilosocereus barbadensis → Pilosocereus royenii
Pilosocereus bradei → Cipocereus bradei
Pilosocereus brasiliensis
Pilosocereus brasiliensis ssp. **brasiliensis**
Pilosocereus brasiliensis ssp. **ruschianus**
Pilosocereus braunii → Pilosocereus gounellei ssp. zehntneri
Pilosocereus brooksianus → Pilosocereus polygonus
Pilosocereus carolinensis → Pilosocereus flavipulvinatus
Pilosocereus catingicola
Pilosocereus catingicola ssp. *arenicola* → Pilosocereus catingicola
Pilosocereus catingicola ssp. **catingicola**
Pilosocereus catingicola ssp. *hapalacanthus* → Pilosocereus catingicola ssp. salvadorensis
Pilosocereus catingicola ssp. *robustus* → Pilosocereus catingicola ssp. catingicola
Pilosocereus catingicola ssp. **salvadorensis**
Pilosocereus cenepequei → Pilosocereus pachycladus ssp. pachycladus
Pilosocereus chrysacanthus
Pilosocereus chrysostele
Pilosocereus claroviridis → Pilosocereus lanuginosus

ⓂⓊ*Pilosocereus coerulescens* → Pilosocereus aurisetus
Pilosocereus collinsii → ?Pilosocereus purpusii
Pilosocereus colombianus → Pilosocereus lanuginosus
Pilosocereus cometes → ?Pilosocereus leucocephalus
Pilosocereus cristalinensis → Pilosocereus machrisii
Ⓤ*Pilosocereus cuyabensis* → Pilosocereus machrisii
Pilosocereus cyaneus → Pilosocereus pachycladus ssp. pachycladus
Pilosocereus deeringii → Pilosocereus polygonus
Pilosocereus densiareolatus
Pilosocereus densivillosus → Pilosocereus machrisii
Pilosocereus diersianus
Pilosocereus flavipulvinatus
Pilosocereus flavipulvinatus ssp. *carolinensis* → Pilosocereus flavipulvinatus
Pilosocereus flexibilispinus
Pilosocereus floccosus
Pilosocereus floccosus ssp. **floccosus**
Pilosocereus floccosus ssp. **quadricostatus**
Pilosocereus fulvilanatus
Pilosocereus fulvilanatus ssp. **fulvilanatus**
Pilosocereus fulvilanatus ssp. **rosae**
Pilosocereus gaturianensis → Pilosocereus piauhyensis
Pilosocereus gaumeri → Pilosocereus royenii
Pilosocereus gironensis → Pilosocereus lanuginosus
Ⓜ*Pilosocereus glaucescens* → Pilosocereus pachycladus
Pilosocereus glaucochrous
Pilosocereus gounellei
Pilosocereus gounellei ssp. **gounellei**
Pilosocereus gounellei ssp. **zehntneri**
Pilosocereus gruberi → Cereus mortensenii
Pilosocereus guerreronis → Pilosocereus alensis
Pilosocereus hapalacanthus → Pilosocereus catingicola ssp. salvadorensis
Pilosocereus juaruensis → Pilosocereus machrisii
Pilosocereus kanukuensis → Pilosocereus oligolepis
Pilosocereus keyensis → Pilosocereus polygonus
Pilosocereus lanuginosus
Pilosocereus leucocephalus
Ⓤ*Pilosocereus lindaianus* → Pilosocereus machrisii
Pilosocereus lindanus → Pilosocereus machrisii
Pilosocereus luetzelburgii → Stephanocereus luetzelburgii
Pilosocereus machrisii
Pilosocereus magnificus
Pilosocereus maxonii → Pilosocereus leucocephalus
Pilosocereus millspaughii → Pilosocereus royenii
Pilosocereus minensis → Cipocereus minensis
Pilosocereus monoclonos → Pilosocereus royenii
Pilosocereus moritzianus → Pilosocereus lanuginosus
Pilosocereus mortensenii → Cereus mortensenii
Pilosocereus mucosiflorus → Pilosocereus piauhyensis
Pilosocereus multicostatus
Ⓜ*Pilosocereus nobilis* → Pilosocereus royenii
Pilosocereus oligolepis
Pilosocereus oligolepis ssp. *kanukuensis* → Pilosocereus oligolepis
Pilosocereus oreus → Pilosocereus pachycladus ssp. pachycladus
Pilosocereus pachycladus
Pilosocereus pachycladus ssp. **pachycladus**
Pilosocereus pachycladus ssp. **pernambucoensis**
Pilosocereus palmeri → Pilosocereus leucocephalus
Pilosocereus paraguayensis → Pilosocereus machrisii
Pilosocereus parvus → Pilosocereus machrisii
Pilosocereus pentaedrophorus

Pilosocereus pentaedrophorus ssp. **pentaedrophorus**
Pilosocereus pentaedrophorus ssp. **robustus**
Pilosocereus pernambucoensis → Pilosocereus pachycladus ssp. pernambucoensis
Pilosocereus piauhyensis
Pilosocereus piauhyensis ssp. *gaturianensis*
Pilosocereus piauhyensis ssp. *mucosiflorus*
Pilosocereus pleurocarpus → Cipocereus minensis ssp. pleurocarpus
Pilosocereus polygonus
Pilosocereus purpusii
Pilosocereus pusillibaccatus → Pilosocereus machrisii
Pilosocereus pusilliflorus → Cipocereus pusilliflorus
Pilosocereus quadricentralis
Pilosocereus quadricostatus → Pilosocereus floccosus ssp. quadricostatus
Pilosocereus rizzoanus → Pilosocereus vilaboensis
Pilosocereus robinii → Pilosocereus polygonus
Pilosocereus robustus → Pilosocereus catingicola ssp. catingicola
Pilosocereus rosae → Pilosocereus fulvilanatus ssp. rosae
Pilosocereus royenii
Pilosocereus rupicola → Pilosocereus catingicola ssp. salvadorensis
Pilosocereus ruschianus → Pilosocereus brasiliensis ssp. ruschianus
Pilosocereus salvadorensis → Pilosocereus catingicola ssp. salvadorensis
Pilosocereus sartorianus → Pilosocereus leucocephalus
Pilosocereus saudadensis → Pilosocereus machrisii
Pilosocereus schoebelii → Pilosocereus pachycladus ssp. pachycladus
Pilosocereus sergipensis → Pilosocereus catingicola ssp. salvadorensis
Pilosocereus splendidus → Pilosocereus pachycladus ssp. pachycladus
Pilosocereus sublanatus → ?Pilosocereus brasiliensis
Pilosocereus ×subsimilis (P. magnificus × P. floccosus ssp. quadricostatus)
Pilosocereus superbus → Pilosocereus pachycladus ssp. pachycladus
Pilosocereus superfloccosus → Pilosocereus gounellei ssp. zehntneri
Pilosocereus supthutianus → Pilosocereus aurisetus ssp. aurisetus
Pilosocereus swartzii → Pilosocereus royenii
Pilosocereus tehuacanus → Pilosocereus leucocephalus
Pilosocereus tillianus → Pilosocereus lanuginosus
Pilosocereus tuberculatus
Pilosocereus tuberculosus → Pilosocereus lanuginosus
Pilosocereus tweedyanus → Pilosocereus lanuginosus
Pilosocereus ulei
Pilosocereus urbanianus → Pilosocereus royenii
Pilosocereus vilaboensis
Pilosocereus werdermannianus → Pilosocereus aurisetus ssp. aurisetus
Pilosocereus zehntneri → Pilosocereus gounellei ssp. zehntneri

Piptanthocereus aethiops → Cereus aethiops
Piptanthocereus alacriportanus → Cereus sp.
Piptanthocereus bageanus → Cereus hildmannianus ssp. uruguayanus
Piptanthocereus cabralensis → Cereus jamacaru ssp. calcirupicola
Piptanthocereus calcirupicola → Cereus jamacaru ssp. calcirupicola
Piptanthocereus cipoensis → Cereus jamacaru ssp. calcirupicola
Piptanthocereus colosseus → ?Cereus lamprospermus
Piptanthocereus comarapanus → Cereus comarapanus
Piptanthocereus crassisepalus → Cipocereus crassisepalus
Piptanthocereus dayamii → Cereus stenogonus
Piptanthocereus goiasensis → Cereus jamacaru ssp. goiasensis
Piptanthocereus huilunchu → Cereus huilunchu
Piptanthocereus lanosus → Cereus lanosus
Piptanthocereus lindenzweigianus → Cereus spegazzinii
Piptanthocereus neonesioticus → Cereus hildmannianus
Piptanthocereus pachyrhizus → Cereus pachyrrhizus

Piptanthocereus sericifer → Cereus fernambucensis ssp. sericifer
Piptanthocereus stenogonus → Cereus stenogonus
®*Piptanthocereus uruguayanus* → Cereus hildmannianus ssp. uruguayanus
Piptanthocereus xanthocarpus → Cereus hildmannianus

Platyopuntia albisaetacens → Opuntia albisaetacens
Platyopuntia apurimacensis → Opuntia apurimacensis
®*Platyopuntia atroglobosa* → Opuntia nigrispina
Platyopuntia brachyacantha → Opuntia sulphurea ssp. brachyacantha
Platyopuntia brunneogemmia → Opuntia monacantha
Platyopuntia cardiosperma → Opuntia cardiosperma
Platyopuntia chilensis → Opuntia alcerrecensis
Platyopuntia cognata → Opuntia cognata
Platyopuntia conjungens → Opuntia conjungens
Platyopuntia cordobensis → Opuntia ficus-indica
Platyopuntia corrugata → Opuntia corrugata
Platyopuntia discolor → Opuntia discolor
Platyopuntia dumetorum → Opuntia dumetorum
Platyopuntia ianthinantha → Opuntia ianthinantha
Platyopuntia inaequilateralis → Opuntia inaequilateralis
Platyopuntia inamoena → Opuntia inamoena
Platyopuntia infesta → Opuntia infesta
Platyopuntia interjecta → Opuntia vitelliniflora ssp. interjecta
Platyopuntia kiska-loro → Opuntia anacantha
Platyopuntia limitata → Opuntia limitata
Platyopuntia microdisca → Opuntia microdisca
Platyopuntia nana → Opuntia pubescens
Platyopuntia nigrispina → Opuntia nigrispina
Platyopuntia orurensis → Opuntia orurensis
Platyopuntia pituitosa → Opuntia pituitosa
Platyopuntia pyrrhantha → Opuntia pyrrhantha
Platyopuntia quimilo → Opuntia quimilo
Platyopuntia quitensis → Opuntia quitensis
Platyopuntia retrorsa → Opuntia anacantha
Platyopuntia rubrogemmia → Opuntia viridirubra ssp. rubrogemmia
Platyopuntia salmiana → Opuntia salmiana
Platyopuntia saxatilis → Opuntia saxatilis
Platyopuntia soehrensii → Opuntia soehrensii
Platyopuntia spinibarbis → Opuntia sulphurea ssp. spinibarbis
Platyopuntia sulphurea → Opuntia sulphurea
Platyopuntia viridirubra → Opuntia viridirubra
Platyopuntia vitelliniflora → Opuntia vitelliniflora
®*Platyopuntia vulgaris* → Opuntia monacantha

Polaskia chende
Polaskia chichipe

Praecereus amazonicus → Praecereus euchlorus ssp. amazonicus
Praecereus apoloensis → Praecereus euchlorus ssp. amazonicus
Praecereus campinensis → Praecereus euchlorus ssp. euchlorus
Praecereus euchlorus
Praecereus euchlorus ssp. **amazonicus**
Praecereus euchlorus ssp. **diffusus**
Praecereus euchlorus ssp. **euchlorus**
Praecereus euchlorus ssp. **jaenensis**
Praecereus euchlorus ssp. **smithianus**
Praecereus jaenensis → Praecereus euchlorus ssp. jaenensis
Praecereus maritimus → Praecereus euchlorus ssp. diffusus
Praecereus saxicola

Praecereus smithianus → Praecereus euchlorus ssp. smithianus
Pseudoacanthocereus boreominarum → Pseudoacanthocereus brasiliensis
Pseudoacanthocereus brasiliensis
Pseudoacanthocereus sicariguensis

Pseudoespostoa melanostele → Espostoa melanostele
Pseudoespostoa nana → Espostoa nana

®*Pseudolobivia acanthoplegma* → Echinopsis cinnabarina
Pseudolobivia ancistrophora → Echinopsis ancistrophora
Pseudolobivia aurea → Echinopsis aurea
Pseudolobivia boyuibensis → Echinopsis boyuibensis
Pseudolobivia callichroma → Echinopsis callichroma
Pseudolobivia calorubra → Echinopsis obrepanda ssp. calorubra
®*Pseudolobivia carmineoflora* → Echinopsis obrepanda
Pseudolobivia ferox → Echinopsis ferox
Pseudolobivia hamatacantha → Echinopsis ancistrophora
Pseudolobivia kermesina → Echinopsis mamillosa
Pseudolobivia kratochviliana → Echinopsis ancistrophora ssp. arachnacantha
Pseudolobivia lecoriensis → Echinopsis ferox
Pseudolobivia leucorhodantha → Echinopsis ancistrophora
Pseudolobivia longispina → Echinopsis ferox
®*Pseudolobivia luteiflora* → Echinopsis aurea
Pseudolobivia obrepanda → Echinopsis obrepanda
Pseudolobivia orozasana → ?Echinopsis obrepanda
Pseudolobivia pelecyrhachis → Echinopsis ancistrophora
Pseudolobivia polyancistra
Pseudolobivia potosina → Echinopsis ferox
Pseudolobivia rojasii → ?Echinopsis obrepanda
Pseudolobivia toralapana → ?Echinopsis obrepanda
Pseudolobivia torrecillasensis → Echinopsis ancistrophora ssp. arachnacantha
®*Pseudolobivia wilkeae* → Echinopsis ferox

Pseudomammillaria kraehenbuehlii → Mammillaria kraehenbuehlii

Pseudomitrocereus fulviceps → Pachycereus fulviceps

Pseudonopalxochia conzattiana → Disocactus ackermannii

Pseudopilocereus arrabidae → Pilosocereus arrabidae
Pseudopilocereus atroflavispinus → Pilosocereus pachycladus ssp. pachycladus
Pseudopilocereus aurilanatus → Pilosocereus aurisetus ssp. aurilanatus
Pseudopilocereus aurisetus → Pilosocereus aurisetus ssp. aurisetus
Pseudopilocereus azureus → Pilosocereus pachycladus ssp. pachycladus
Pseudopilocereus bradei → Cipocereus bradei
Pseudopilocereus carolinensis → Pilosocereus flavipulvinatus
Pseudopilocereus catingicola → Pilosocereus catingicola ssp. catingicola
Pseudopilocereus chrysostele → Pilosocereus chrysostele
®*Pseudopilocereus cuyabensis* → Pilosocereus machrisii
Pseudopilocereus densiareolatus → Pilosocereus densiareolatus
Pseudopilocereus diersianus → Pilosocereus diersianus
Pseudopilocereus flavipulvinatus → Pilosocereus flavipulvinatus
Pseudopilocereus floccosus → Pilosocereus floccosus
Pseudopilocereus fulvilanatus → Pilosocereus fulvilanatus ssp. fulvilanatus
Pseudopilocereus gaturianensis → Pilosocereus piauhyensis
Pseudopilocereus glaucochrous → Pilosocereus glaucochrous
Pseudopilocereus gounellei → Pilosocereus gounellei ssp. gounellei
Pseudopilocereus hapalacanthus → Pilosocereus catingicola ssp. salvadorensis
Pseudopilocereus juaruensis → Pilosocereus machrisii

Pseudopilocereus luetzelburgii → Stephanocereus luetzelburgii
Pseudopilocereus machrisii → Pilosocereus machrisii
Pseudopilocereus magnificus → Pilosocereus magnificus
Pseudopilocereus mucosiflorus → Pilosocereus piauhyensis
Pseudopilocereus multicostatus → Pilosocereus multicostatus
⊗*Pseudopilocereus nobilis* → Pilosocereus royenii
Pseudopilocereus oligolepis → Pilosocereus oligolepis
Pseudopilocereus oreus → Pilosocereus pachycladus ssp. pachycladus
Pseudopilocereus pachycladus → Pilosocereus pachycladus ssp. pachycladus
Pseudopilocereus parvus → Pilosocereus machrisii
Pseudopilocereus pentaedrophorus → Pilosocereus pentaedrophorus ssp. pentaedrophorus
Pseudopilocereus pernambucoensis → Pilosocereus pachycladus ssp. pernambucoensis
Pseudopilocereus piauhyensis → Pilosocereus piauhyensis
Pseudopilocereus quadricostatus → Pilosocereus floccosus ssp. quadricostatus
Pseudopilocereus robustus → Pilosocereus catingicola ssp. catingicola
Pseudopilocereus rupicola → Pilosocereus catingicola ssp. salvadorensis
Pseudopilocereus ruschianus → Pilosocereus brasiliensis ssp. ruschianus
Pseudopilocereus salvadorensis → Pilosocereus catingicola ssp. salvadorensis
Pseudopilocereus saudadensis → Pilosocereus machrisii
Pseudopilocereus sergipensis → Pilosocereus catingicola ssp. salvadorensis
Pseudopilocereus splendidus → Pilosocereus pachycladus ssp. pachycladus
Pseudopilocereus superbus → Pilosocereus pachycladus ssp. pachycladus
Pseudopilocereus superfloccosus → Pilosocereus gounellei ssp. zehntneri
Pseudopilocereus tuberculatus → Pilosocereus tuberculatus
Pseudopilocereus ulei → Pilosocereus ulei
Pseudopilocereus vilaboensis → Pilosocereus vilaboensis
Pseudopilocereus werdermannianus → Pilosocereus aurisetus ssp. aurisetus
Pseudopilocereus zehntneri → Pilosocereus gounellei ssp. zehntneri

Pseudorhipsalis acuminata
Pseudorhipsalis alata
Pseudorhipsalis himantoclada
Pseudorhipsalis horichii
Pseudorhipsalis lankesteri
Pseudorhipsalis macrantha → Disocactus macranthus
Pseudorhipsalis ramulosa

Pseudozygocactus epiphylloides → Hatiora epiphylloides

Pterocactus araucanus
Pterocactus australis
Pterocactus decipiens → Pterocactus tuberosus
Pterocactus fischeri
Pterocactus gonjianii
Pterocactus hickenii
Pterocactus kuntzei → Pterocactus tuberosus
Pterocactus megliolii
Pterocactus pumilus → Pterocactus valentinii
Pterocactus reticulatus
Pterocactus skottsbergii → Pterocactus hickenii
Pterocactus tuberosus
Pterocactus valentinii

Pterocereus foetidus → Pachycereus gaumeri
Pterocereus gaumeri → Pachycereus gaumeri

Puna bonnieae → ?Opuntia subterranea
Puna clavarioides → Opuntia clavarioides
Puna subterranea → Opuntia subterranea

Pygmaeocereus akersii → Pygmaeocereus bylesianus
Pygmaeocereus bieblii
Pygmaeocereus bylesianus
Pygmaeocereus familiaris
Pygmaeocereus rowleyanus → Pygmaeocereus bylesianus

Pyrrhocactus andicola → Eriosyce curvispina
Pyrrhocactus aricensis → Eriosyce recondita ssp. iquiquensis
Pyrrhocactus armatus → Eriosyce curvispina
Pyrrhocactus aspillagae → Eriosyce aspillagae
Pyrrhocactus atrospinosus → Eriosyce strausiana
Pyrrhocactus atroviridis → Eriosyce crispa ssp. atroviridis
Pyrrhocactus bulbocalyx → Eriosyce bulbocalyx
Pyrrhocactus calderanus → Eriosyce taltalensis ssp. taltalensis
Pyrrhocactus carrizalensis → Eriosyce crispa ssp. atroviridis
Pyrrhocactus catamarcensis → ?Echinopsis sp.
Pyrrhocactus chaniarensis → Eriosyce heinrichiana
Pyrrhocactus chilensis → Eriosyce chilensis
Pyrrhocactus choapensis → Eriosyce curvispina
Pyrrhocactus chorosensis → Eriosyce heinrichiana
Pyrrhocactus coliguayensis → Eriosyce curvispina
Pyrrhocactus confinis → Eriosyce confinis
Pyrrhocactus crispus → Eriosyce crispa ssp. crispa
Pyrrhocactus curvispinus → Eriosyce curvispina
Pyrrhocactus deherdtianus → Eriosyce heinrichiana
Pyrrhocactus dimorphus → Eriosyce heinrichiana ssp. intermedia
Pyrrhocactus dubius → Eriosyce bulbocalyx
Pyrrhocactus echinus → Eriosyce taltalensis ssp. echinus
Pyrrhocactus engleri → Eriosyce engleri
Pyrrhocactus eriosyzoides → Eriosyce kunzei
Pyrrhocactus floccosus → Eriosyce taltalensis ssp. echinus
ⓤ*Pyrrhocactus floribundus* → Eriosyce recondita ssp. iquiquensis
Pyrrhocactus garaventae → Eriosyce garaventae
Pyrrhocactus glaucescens → Eriosyce taltalensis ssp. echinus
Pyrrhocactus gracilis → Eriosyce taltalensis ssp. taltalensis
Pyrrhocactus grandiflorus → Eriosyce curvispina
Pyrrhocactus horridus → Eriosyce curvispina
Pyrrhocactus huascensis → Eriosyce crispa ssp. crispa
Pyrrhocactus intermedius → Eriosyce taltalensis ssp. taltalensis
Pyrrhocactus iquiquensis → Eriosyce recondita ssp. iquiquensis
Pyrrhocactus jussieui → ?Eriosyce heinrichiana
Pyrrhocactus krausii → Eriosyce chilensis
Pyrrhocactus limariensis → Eriosyce limariensis
Pyrrhocactus lissocarpus → Eriosyce marksiana
ⓤ*Pyrrhocactus marayesensis* → Eriosyce bulbocalyx
Pyrrhocactus marksianus → Eriosyce marksiana
Pyrrhocactus megliolii → Eriosyce bulbocalyx
ⓤ*Pyrrhocactus melanacanthus* → Eriosyce villicumensis
Pyrrhocactus neohankeanus → Eriosyce taltalensis ssp. paucicostata
ⓤ*Pyrrhocactus occultus* → Eriosyce occulta
Pyrrhocactus odoriflorus → Eriosyce curvispina
Pyrrhocactus pachacoensis → Eriosyce strausiana
Pyrrhocactus pamaensis → Eriosyce curvispina
Pyrrhocactus paucicostatus → Eriosyce taltalensis ssp. paucicostata
Pyrrhocactus pilispinus → Eriosyce taltalensis ssp. pilispina
Pyrrhocactus platyacanthus → Eriosyce strausiana
Pyrrhocactus pulchellus → Eriosyce taltalensis ssp. taltalensis
Pyrrhocactus pygmaeus → Eriosyce taltalensis ssp. taltalensis
Pyrrhocactus reconditus → Eriosyce recondita

Pyrrhocactus residuus → Eriosyce recondita ssp. iquiquensis
Pyrrhocactus rupicola → Eriosyce taltalensis ssp. taltalensis
Pyrrhocactus sanjuanensis → Eriosyce strausiana
Pyrrhocactus saxifragus → Eriosyce recondita ssp. iquiquensis
Pyrrhocactus scoparius → Eriosyce taltalensis ssp. taltalensis
Pyrrhocactus setiflorus → Eriosyce strausiana
Pyrrhocactus setosiflorus → Eriosyce heinrichiana ssp. intermedia
Pyrrhocactus simulans → Eriosyce heinrichiana ssp. simulans
Pyrrhocactus strausianus → Eriosyce strausiana
Pyrrhocactus subaianus → Eriosyce garaventae
Pyrrhocactus taltalensis → Eriosyce taltalensis
Pyrrhocactus tenuis → Eriosyce taltalensis ssp. taltalensis
Pyrrhocactus totoralensis → Eriosyce crispa ssp. atroviridis
Pyrrhocactus transiens → Eriosyce taltalensis ssp. taltalensis
Pyrrhocactus transitensis → Eriosyce kunzei
Pyrrhocactus trapichensis → Eriosyce heinrichiana
Pyrrhocactus truncatipetalus → Eriosyce curvispina
Pyrrhocactus umadeave → Eriosyce umadeave
Pyrrhocactus vallenarensis → Eriosyce kunzei
Pyrrhocactus vertongenii → Eriosyce vertongenii
Pyrrhocactus vexatus → Eriosyce recondita ssp. recondita
Pyrrhocactus villicumensis → Eriosyce villicumensis
Pyrrhocactus vollianus → Eriosyce strausiana
Pyrrhocactus wagenknechtii → Eriosyce heinrichiana ssp. intermedia

Quiabentia chacoensis → Quiabentia verticillata
Quiabentia pereziensis → Quiabentia verticillata
Quiabentia pflanzii → Quiabentia verticillata
Quiabentia verticillata
Quiabentia zehntneri

Rathbunia alamosensis → Stenocereus alamosensis
Rathbunia beneckei → Stenocereus beneckei
Rathbunia chacalapensis → Stenocereus chacalapensis
Rathbunia chrysocarpa → Stenocereus chrysocarpus
Rathbunia deficiens → Stenocereus griseus
Rathbunia dumortieri → Stenocereus dumortieri
Rathbunia eichlamii → Stenocereus eichlamii
Rathbunia eruca → Stenocereus eruca
Rathbunia fimbriata → Stenocereus fimbriatus
Rathbunia fricii → Stenocereus fricii
Rathbunia grisea → Stenocereus griseus
Rathbunia gummosa → Stenocereus gummosus
Rathbunia kerberi → Stenocereus kerberi
Rathbunia laevigata → Stenocereus laevigatus
Rathbunia longispinus → Stenocereus eichlamii
Rathbunia martinezii → Stenocereus martinezii
Rathbunia montana → Stenocereus montanus
Rathbunia neosonorensis → Stenocereus alamosensis
Rathbunia pruinosa → Stenocereus pruinosus
Rathbunia queretaroensis → Stenocereus queretaroensis
Rathbunia quevedonis → Stenocereus quevedonis
Rathbunia sonorensis → Stenocereus alamosensis
Rathbunia standleyi → Stenocereus standleyi
Rathbunia stellata → Stenocereus stellatus
Rathbunia thurberi → Stenocereus thurberi
Rathbunia treleasei → Stenocereus treleasei
Rathbunia yunckeri → Stenocereus yunckeri

Rauhocereus riosaniensis
Rauhocereus riosaniensis ssp. **jaenensis**
Rauhocereus riosaniensis ssp. **riosaniensis**

Rebutia albiareolata → Rebutia pseudodeminuta
Rebutia albiflora
Rebutia albipilosa → Rebutia fiebrigii
Rebutia albispina → Rebutia nigricans
Rebutia albopectinata
Rebutia applanata → Rebutia steinmannii
Rebutia archibuiningiana → Rebutia spinosissima
Rebutia arenacea
Rebutia atrovirens → ?Rebutia pygmaea
Rebutia auranitida → Rebutia einsteinii
Rebutia aureiflora
Rebutia aureiflora ssp. *elegans* → Rebutia aureiflora
Rebutia brachyantha → Rebutia steinmannii
Rebutia brunescens
Rebutia brunneoradicata → Rebutia steinmannii
Rebutia buiningiana → Rebutia pseudodeminuta
Rebutia caineana
Rebutia cajasensis → Rebutia fiebrigii
Rebutia calliantha → Rebutia wessneriana
Rebutia camargoensis → Rebutia steinmannii
Rebutia canacruzensis → Rebutia pygmaea
Rebutia candiae → Rebutia arenacea
Rebutia canigueralii
Rebutia canigueralii ssp. **crispata**
Rebutia canigueralii ssp. **pulchra**
Rebutia caracarensis → Rebutia canigueralii
Rebutia cardenasiana
Rebutia carmeniana → Rebutia nigricans
Rebutia carminea → Rebutia minuscula
Rebutia christinae → Rebutia steinmannii
Rebutia chrysacantha → Rebutia minuscula
Rebutia cincinnata → Rebutia steinmannii
Rebutia cintiensis → Rebutia fiebrigii
Rebutia colorea → Rebutia pygmaea
Rebutia corroana → Rebutia neocumingii
Rebutia costata → Rebutia steinmannii
Rebutia crassa → Rebutia pygmaea
Rebutia cylindrica
Rebutia deminuta
Rebutia diersiana → Rebutia pygmaea
Rebutia donaldiana → Rebutia fiebrigii
Rebutia einsteinii
Rebutia elegantula → Rebutia pygmaea
Rebutia eos → Rebutia pygmaea
©*Rebutia euanthema* → Rebutia aureiflora
Rebutia eucaliptana → Rebutia steinmannii
Rebutia fabrisii
Rebutia famatinensis → Echinopsis famatinensis
Rebutia fidaiana
Rebutia fidaiana ssp. **cintiensis**
Rebutia fidaiana ssp. **fidaiana**
Rebutia fiebrigii
Rebutia flavistyla
Rebutia friedrichiana → Rebutia pygmaea
Rebutia froehlichiana → Rebutia spegazziniana
Rebutia fuauxiana → Rebutia sp.

Rebutia fulviseta
Rebutia fusca → Rebutia spegazziniana
Rebutia glomeriseta → Rebutia arenacea
Rebutia glomerispina → Rebutia steinbachii
Rebutia gonjianii
Rebutia graciliflora → Rebutia minuscula
Rebutia gracilispina → Rebutia pygmaea
Rebutia grandiflora → Rebutia minuscula
Rebutia grandilacea → Rebutia minuscula
Rebutia haagei → Rebutia pygmaea
Rebutia haefneriana → Rebutia pygmaea
Rebutia haseltonii → Rebutia caineana
Rebutia heliosa
Rebutia hoffmannii → Rebutia spinosissima
Rebutia huasiensis
®*Rebutia hyalacantha* → Rebutia marsoneri
Rebutia inflexiseta → Rebutia canigueralii
Rebutia iridescens → Rebutia pygmaea
Rebutia iscayachensis → Rebutia pygmaea
Rebutia ithyacantha → Rebutia fiebrigii
Rebutia jujuyana → Rebutia fiebrigii
Rebutia kariusiana → Rebutia minuscula
Rebutia kieslingii → Rebutia fiebrigii
Rebutia knizei → Rebutia pygmaea
Rebutia krainziana → Rebutia marsoneri
Rebutia kruegeri → Rebutia steinbachii ssp. kruegeri
Rebutia kupperiana → ?Rebutia pseudodeminuta
Rebutia lanosiflora → Rebutia pygmaea
Rebutia leucacantha → Rebutia steinmannii
Rebutia leucanthema
Rebutia major → Rebutia steinmannii
Rebutia mamillosa → Rebutia spegazziniana
Rebutia margarethae → Rebutia padcayensis
Rebutia marsoneri
Rebutia melanocentra → Rebutia steinmannii
Rebutia menesesii → Rebutia arenacea
Rebutia mentosa
Rebutia mentosa ssp. **mentosa**
Rebutia mentosa ssp. **purpurea**
Rebutia minuscula
Rebutia minuscula ssp. *grandiflora* → Rebutia minuscula
Rebutia minuscula ssp. *violaciflora* → Rebutia minuscula
Rebutia minutissima → Rebutia pseudodeminuta
Rebutia mixta → Rebutia pygmaea
Rebutia mixticolor → Rebutia pygmaea
Rebutia mudanensis → Rebutia pygmaea
Rebutia muscula → Rebutia fiebrigii
Rebutia narvaecensis
Rebutia nazarenoensis → Rebutia pygmaea
Rebutia neocumingii
Rebutia neocumingii ssp. **neocumingii**
Rebutia neocumingii ssp. **pilcomayensis**
Rebutia neocumingii ssp. **riograndensis**
Rebutia neocumingii ssp. **saipinensis**
Rebutia neocumingii ssp. **trollii**
Rebutia neumanniana
Rebutia nigricans
Rebutia nitida → Rebutia pseudodeminuta
Rebutia nogalesensis → Rebutia pseudodeminuta
Rebutia oculata → Rebutia aureiflora

Rebutia odontopetala → Rebutia pygmaea
Rebutia oligacantha
Rebutia orurensis → Rebutia pygmaea
Rebutia padcayensis
Rebutia pallida → Rebutia pygmaea
Rebutia parvula → Rebutia steinmannii
Rebutia patericalyx → Rebutia spegazziniana
Rebutia pauciareolata → Rebutia pygmaea
Rebutia paucicostata → Rebutia pygmaea
Rebutia pelzliana → Rebutia pygmaea
Rebutia perplexa
Rebutia poecilantha → Rebutia steinmannii
Rebutia polymorpha → Rebutia steinbachii
Rebutia polypetala → Rebutia pygmaea
Rebutia potosina → Rebutia steinmannii
Rebutia pseudodeminuta
Rebutia pseudominuscula → Rebutia deminuta
Rebutia pseudoritteri → Rebutia pygmaea
Rebutia pulchella → Rebutia fiebrigii
Rebutia pulchra → Rebutia canigueralii ssp. pulchra
Rebutia pulvinosa
Rebutia pygmaea
Rebutia raulii → Rebutia ritteri
Rebutia rauschii → Rebutia steinmannii
®*Rebutia rauschii* → Rebutia canigueralii
Rebutia ritteri
Rebutia robustispina → Rebutia pseudodeminuta
Rebutia rosalbiflora → Rebutia pygmaea
Rebutia rubiginosa → Rebutia spegazziniana
Rebutia rutiliflora → Rebutia pygmaea
Rebutia salpingantha → Rebutia pygmaea
Rebutia sanguinea → Rebutia pseudodeminuta
Rebutia sarothroides → Rebutia aureiflora
Rebutia schatzliana → Rebutia albopectinata
Rebutia senilis → Rebutia minuscula
Rebutia senilis ssp. *chrysacantha* → Rebutia minuscula
Rebutia simoniana
Rebutia singularis → Rebutia padcayensis
Rebutia spegazziniana
Rebutia spinosissima
Rebutia steinbachii
Rebutia steinbachii ssp. **kruegeri**
Rebutia steinbachii ssp. **steinbachii**
Rebutia steinbachii ssp. **tiraquensis**
Rebutia steinbachii ssp. **verticillacantha**
Rebutia steinmannii
Rebutia sumayana → Rebutia spegazziniana
Rebutia supthutiana → Rebutia albopectinata
Rebutia tamboensis → Rebutia fiebrigii
Rebutia taratensis → Rebutia steinbachii
Rebutia tarijensis → Rebutia spegazziniana
Rebutia tarvitaensis → Rebutia spegazziniana
Rebutia tilcarensis → Rebutia aureiflora
Rebutia tiraquensis → Rebutia steinbachii ssp. tiraquensis
Rebutia torquata → Rebutia pygmaea
Rebutia tropaeolipicta → Rebutia pygmaea
Rebutia tuberculata → Rebutia steinmannii
Rebutia tuberculato-chrysantha → Rebutia steinbachii
Rebutia tuberosa → Rebutia spegazziniana
Rebutia tunariensis → Rebutia steinbachii

Rebutia vallegrandensis → Rebutia fiebrigii
Rebutia villazonensis → Rebutia pygmaea
Rebutia violaciflora → Rebutia minuscula
Rebutia violacistaminea → Rebutia pygmaea
Rebutia violascens → Rebutia pygmaea
Rebutia vizcarrae → Rebutia steinbachii
Rebutia vulpina → Rebutia spegazziniana
Rebutia wahliana → Rebutia pseudodeminuta
Rebutia walteri → Rebutia spinosissima
Rebutia wessneriana
Rebutia wessneriana ssp. *beryllioides* → Rebutia wessneriana
Rebutia xanthocarpa
Rebutia yuncharasensis → Rebutia pygmaea
Rebutia yuquinensis → Rebutia pygmaea
Rebutia zecheri → Rebutia spegazziniana

Ⓤ*Reicheocactus floribundus* → Eriosyce recondita ssp. iquiquensis
Ⓤ*Reicheocactus neoreichei* → Eriosyce napina ssp. lembckei
Reicheocactus pseudoreicheanus → Echinopsis famatinensis

Rhipsalidopsis gaertneri → Hatiora gaertneri
Rhipsalidopsis rosea → Hatiora rosea

Rhipsalis aculeata → Lepismium aculeatum
Rhipsalis alboareolata → Rhipsalis teres
Rhipsalis angustissima → Pseudorhipsalis ramulosa
Rhipsalis asperula → Lepismium micranthum
Rhipsalis baccifera
Rhipsalis baccifera ssp. **baccifera**
Rhipsalis baccifera ssp. **erythrocarpa**
Rhipsalis baccifera ssp. *fasciculata* → Rhipsalis baccifera
Rhipsalis baccifera ssp. *fortdauphinensis* → Rhipsalis baccifera ssp. mauritiana
Rhipsalis baccifera ssp. **hileiabaiana**
Rhipsalis baccifera ssp. **horrida**
Rhipsalis baccifera ssp. **mauritiana**
Rhipsalis baccifera ssp. *rhodocarpa* → Rhipsalis baccifera
Rhipsalis baccifera ssp. **shaferi**
Rhipsalis bartlettii → Rhipsalis baccifera
Rhipsalis boliviana → Lepismium bolivianum
Ⓤ*Rhipsalis brevispina* → Lepismium brevispinum
Rhipsalis brevispina → Lepismium brevispinum
Rhipsalis burchellii
Rhipsalis campos-portoana
Rhipsalis capilliformis → Rhipsalis teres
Rhipsalis cassutha → Rhipsalis baccifera
Rhipsalis cassuthopsis → Rhipsalis baccifera
Rhipsalis cassytha → Rhipsalis baccifera
Rhipsalis cassythoides → Rhipsalis baccifera
Rhipsalis cereoides
Rhipsalis cereuscula
Rhipsalis chloroptera → Rhipsalis elliptica
Rhipsalis chrysantha → ?Rhipsalis dissimilis
Rhipsalis chrysocarpa → Rhipsalis puniceodiscus
Rhipsalis clavata
Rhipsalis clavellina → Rhipsalis teres
Rhipsalis coralloides → Rhipsalis baccifera ssp. horrida
Rhipsalis coriacea → Pseudorhipsalis ramulosa
Rhipsalis crenata → Lepismium crenatum
Ⓜ*Rhipsalis cribrata* → Rhipsalis teres
Rhipsalis crispata

Rhipsalis crispimarginata → Rhipsalis oblonga
Rhipsalis cuneata
Rhipsalis densiareolata → Rhipsalis lindbergiana
Rhipsalis dissimilis
Rhipsalis dusenii → Rhipsalis pachyptera
Rhipsalis elliptica
Rhipsalis epiphyllanthoides → Rhipsalis dissimilis
Rhipsalis erythrocarpa → Rhipsalis baccifera ssp. erythrocarpa
Rhipsalis ewaldiana
Rhipsalis fasciculata → Rhipsalis baccifera
Rhipsalis fastigiata → Rhipsalis grandiflora
Rhipsalis floccosa
Rhipsalis floccosa ssp. **floccosa**
Rhipsalis floccosa ssp. **hohenauensis**
Rhipsalis floccosa ssp. **oreophila**
Rhipsalis floccosa ssp. **pittieri**
Rhipsalis floccosa ssp. **pulvinigera**
Rhipsalis floccosa ssp. **tucumanensis**
Rhipsalis flosculosa → Rhipsalis floccosa ssp. pulvinigera
Rhipsalis gibberula → Rhipsalis floccosa ssp. pulvinigera
Rhipsalis goebeliana
Rhipsalis gonocarpa → Lepismium warmingianum
Rhipsalis grandiflora
Rhipsalis hadrosoma → Rhipsalis grandiflora
Rhipsalis heptagona → Rhipsalis baccifera
Rhipsalis herminiae → Hatiora herminiae
Rhipsalis heteroclada → Rhipsalis teres
Rhipsalis hoelleri
Rhipsalis hohenauensis → Rhipsalis floccosa ssp. hohenauensis
Rhipsalis horrida → Rhipsalis baccifera ssp. horrida
Rhipsalis houlletiana → Lepismium houlletianum
Rhipsalis hylaea → Rhipsalis baccifera
Rhipsalis incachacana → Lepismium incachacanum
Rhipsalis jamaicensis → Pseudorhipsalis ramulosa
Rhipsalis juengeri
Rhipsalis kirbergii → Rhipsalis micrantha
Rhipsalis leiophloea → Pseudorhipsalis ramulosa
Rhipsalis leucorhaphis → Lepismium lumbricoides
Rhipsalis lindbergiana
Rhipsalis linearis → Lepismium warmingianum
Rhipsalis loefgrenii → Lepismium lumbricoides
Rhipsalis lorentziana → Lepismium lorentzianum
Rhipsalis lumbricoides → Lepismium lumbricoides
Rhipsalis madagascariensis → Rhipsalis baccifera ssp. horrida
Rhipsalis maricaensis → Rhipsalis teres
Rhipsalis megalantha → Rhipsalis neves-armondii
Rhipsalis mesembryanthemoides
Rhipsalis mesembryanthoides → Rhipsalis mesembryanthemoides
Rhipsalis micrantha
Rhipsalis minutiflora → Rhipsalis baccifera
Rhipsalis miyagawae → Lepismium miyagawae
Rhipsalis monteazulensis → Rhipsalis floccosa ssp. oreophila
Rhipsalis myosurus → Lepismium cruciforme
Rhipsalis neves-armondii
Rhipsalis oblonga
Rhipsalis occidentalis
Rhipsalis olivifera
Rhipsalis ormindoi
Rhipsalis pacheco-leonis
Rhipsalis pacheco-leonis ssp. **catenulata**

Rhipsalis pacheco-leonis ssp. **pacheco-leonis**
Rhipsalis pachyptera
Rhipsalis paradoxa
Rhipsalis paradoxa ssp. **paradoxa**
Rhipsalis paradoxa ssp. **septentrionalis**
Rhipsalis paranganiensis → Lepismium paranganiense
Rhipsalis penduliflora → Rhipsalis cereuscula
Rhipsalis pentaptera
Rhipsalis pilocarpa
Rhipsalis pilosa → Rhipsalis baccifera ssp. horrida
Rhipsalis pittieri → Rhipsalis floccosa ssp. pittieri
Rhipsalis platycarpa → Rhipsalis sp.
Rhipsalis prismatica → Rhipsalis teres
Rhipsalis pulchra
Rhipsalis pulvinigera → Rhipsalis floccosa ssp. pulvinigera
Rhipsalis puniceodiscus
Rhipsalis purpusii → Pseudorhipsalis ramulosa
®*Rhipsalis quellebambensis* → Rhipsalis baccifera
Rhipsalis ramulosa → Pseudorhipsalis ramulosa
Rhipsalis rauhiorum → Rhipsalis micrantha
Rhipsalis rhombea → Rhipsalis sp.
Rhipsalis rigida → Rhipsalis dissimilis
Rhipsalis robusta → Rhipsalis pachyptera
Rhipsalis roseana → Rhipsalis micrantha
Rhipsalis russellii
Rhipsalis saxatilis → Rhipsalis dissimilis
Rhipsalis shaferi → Rhipsalis baccifera ssp. shaferi
Rhipsalis simmleri → Rhipsalis cereuscula
Rhipsalis spinescens → Rhipsalis dissimilis
Rhipsalis squamulosa → Lepismium cruciforme
Rhipsalis sulcata
Rhipsalis teres
Rhipsalis tonduzii → Rhipsalis micrantha
Rhipsalis triangularis → Rhipsalis sp.
Rhipsalis trigona
Rhipsalis tucumanensis → Rhipsalis floccosa ssp. tucumanensis
Rhipsalis undulata → Rhipsalis baccifera
Rhipsalis virgata → Rhipsalis teres
Rhipsalis warmingiana → Lepismium warmingianum
Rhipsalis wercklei → Rhipsalis micrantha

Rhodocactus antonianus → Pereskia weberiana
Rhodocactus autumnalis → Pereskia lychnidiflora
Rhodocactus bleo → Pereskia bleo
Rhodocactus colombianus → Pereskia guamacho
Rhodocactus corrugatus → Pereskia bleo
Rhodocactus cubensis → Pereskia zinniiflora
Rhodocactus grandifolius → Pereskia grandifolia ssp. grandifolia
Rhodocactus guamacho → Pereskia guamacho
Rhodocactus higueranus → Pereskiopsis spathulata
Rhodocactus lychnidiflorus → Pereskia lychnidiflora
Rhodocactus nicoyanus → Pereskia lychnidiflora
Rhodocactus portulacifolius → Pereskia portulacifolia
Rhodocactus sacharosa → Pereskia sacharosa
Rhodocactus saipinensis → Pereskia sacharosa
Rhodocactus tampicanus → Pereskia grandifolia ssp. grandifolia
Rhodocactus zinniiflorus → Pereskia zinniiflora

Ritterocereus deficiens → Stenocereus griseus
Ritterocereus eichlamii → Stenocereus eichlamii

Ritterocereus fimbriatus → Stenocereus fimbriatus
Ritterocereus griseus → Stenocereus griseus
Ritterocereus laevigatus → Stenocereus laevigatus
Ritterocereus pruinosus → Stenocereus pruinosus
Ritterocereus queretaroensis → Stenocereus queretaroensis
Ritterocereus standleyi → Stenocereus standleyi

®*Rodentiophila atacamensis* → Eriosyce rodentiophila
Rodentiophila megacarpa → Eriosyce rodentiophila

Rooksbya euphorbioides → Neobuxbaumia euphorbioides

Roseocactus fissuratus♦ → Ariocarpus fissuratus♦
Roseocactus intermedius♦ → Ariocarpus fissuratus♦
Roseocactus kotschoubeyanus♦ → Ariocarpus kotschoubeyanus♦
Roseocactus kotschoubeyanus ssp. *macdowellii* → Ariocarpus kotschoubeyanus
Roseocactus lloydii♦ → Ariocarpus fissuratus♦

Roseocereus tephracanthus → Harrisia tetracantha

Samaipaticereus corroanus
Samaipaticereus inquisivensis → Yungasocereus inquisivensis

®*Schlumbergera bridgesii* → Schlumbergera russelliana
Schlumbergera ×buckleyi (S. russelliana × S. truncata)
Schlumbergera candida → Schlumbergera microsphaerica ssp. candida
Schlumbergera ×exotica (S. opuntioides × S. truncata)
Schlumbergera gaertneri → Hatiora gaertneri
Schlumbergera kautskyi
Schlumbergera microsphaerica
Schlumbergera microsphaerica ssp. ***candida***
Schlumbergera obtusangula → Schlumbergera microsphaerica
Schlumbergera opuntioides
Schlumbergera orssichiana
Schlumbergera ×reginae (S. orssichiana × S. truncata)
Schlumbergera russelliana
Schlumbergera truncata
Schlumbergera truncata ssp. *kautskyi* → Schlumbergera kautskyi

Sclerocactus blainei → Sclerocactus spinosior ssp. blainei
Sclerocactus brevihamatus
Sclerocactus brevihamatus ssp. **brevihamatus**
Sclerocactus brevihamatus ssp. **tobuschii**
Sclerocactus brevispinus♦ → Sclerocactus glaucus♦
Sclerocactus cloverae → Sclerocactus parviflorus
Sclerocactus cloverae ssp. *brackii* → Sclerocactus parviflorus
Sclerocactus contortus → Sclerocactus parviflorus ssp. parviflorus
Sclerocactus erectocentrus♦
Sclerocactus glaucus♦
Sclerocactus havasupaiensis → Sclerocactus parviflorus ssp. havasupaiensis
Sclerocactus intermedius → Sclerocactus parviflorus ssp. intermedius
Sclerocactus intertextus
Sclerocactus johnsonii
Sclerocactus mariposensis♦
Sclerocactus mesae-verdae♦
Sclerocactus nyensis
Sclerocactus papyracanthus♦
Sclerocactus parviflorus
Sclerocactus parviflorus ssp. ***havasupaiensis***
Sclerocactus parviflorus ssp. ***intermedius***

Sclerocactus parviflorus ssp. **parviflorus**
Sclerocactus parviflorus ssp. ***terrae-canyonae***
Sclerocactus polyancistrus
Sclerocactus pubispinus♦
Sclerocactus scheeri
Sclerocactus schlesseri → Sclerocactus spinosior ssp. blainei
Sclerocactus sileri
Sclerocactus spinosior
Sclerocactus spinosior ssp. ***blainei***
Sclerocactus terrae-canyonae → Sclerocactus parviflorus ssp. terrae-canyonae
Sclerocactus uncinatus
Sclerocactus uncinatus ssp. **crassihamatus**
Sclerocactus uncinatus ssp. **uncinatus**
Sclerocactus uncinatus ssp. **wrightii**
Sclerocactus unguispinus
Sclerocactus warnockii
Sclerocactus wetlandicus♦ → Sclerocactus glaucus♦
Sclerocactus wetlandicus ssp. *ilseae♦* → Sclerocactus glaucus♦
Sclerocactus whipplei
Sclerocactus whipplei ssp. ***busekii***
Sclerocactus wrightiae♦

Selenicereus anthonyanus
Selenicereus atropilosus
Selenicereus boeckmannii
Selenicereus brevispinus
Selenicereus chontalensis
Selenicereus chrysocardium
Selenicereus coniflorus
Selenicereus donkelaarii
Selenicereus grandiflorus
Selenicereus hallensis → Selenicereus grandiflorus
Selenicereus hamatus
Selenicereus hondurensis
Selenicereus inermis
Selenicereus innesii
Selenicereus kunthianus → Selenicereus grandiflorus
Selenicereus macdonaldiae
Selenicereus maxonii → Selenicereus urbanianus
Selenicereus megalanthus
Selenicereus mirandae → Weberocereus glaber
Selenicereus murrillii
Selenicereus nelsonii
Selenicereus nycticalus → Selenicereus pteranthus
Selenicereus pringlei → Selenicereus coniflorus
Selenicereus pseudospinulosus → Selenicereus spinulosus
Selenicereus pteranthus
Selenicereus rizzinii → Selenicereus setaceus
Selenicereus rothii → Selenicereus macdonaldiae
Selenicereus setaceus
Selenicereus spinulosus
Selenicereus testudo
Selenicereus tricae
Selenicereus urbanianus
Selenicereus vagans
Selenicereus validus
Selenicereus vaupelii → Selenicereus boeckmannii
Selenicereus wercklei
Selenicereus wittii

Seticereus chlorocarpus → Browningia chlorocarpa
Seticereus humboldtii → Cleistocactus icosagonus
Seticereus icosagonus → Cleistocactus icosagonus
Seticereus roezlii → Cleistocactus roezlii

Seticleistocactus dependens → Cleistocactus dependens
Seticleistocactus piraymirensis → Cleistocactus piraymirensis

Setiechinopsis mirabilis → Echinopsis mirabilis

Siccobaccatus dolichospermaticus → Micranthocereus dolichospermaticus
Siccobaccatus estevesii → Micranthocereus estevesii
Siccobaccatus estevesii ssp. *grandiflorus* → Micranthocereus estevesii
Siccobaccatus estevesii ssp. *insigniflorus* → Micranthocereus estevesii

Soehrensia bruchii → Echinopsis bruchii
Soehrensia formosa → Echinopsis formosa
Soehrensia grandis → Echinopsis bruchii
Soehrensia ingens → Echinopsis bruchii
Soehrensia korethroides → Echinopsis korethroides
Soehrensia oreopepon → Echinopsis formosa
Soehrensia smrziana → Echinopsis smrziana
®*Soehrensia uebelmanniana* → ?Echinopsis formosa

Solisia pectinata♦ → Mammillaria pectinifera♦

Spegazzinia fidaiana → Rebutia fidaiana ssp. fidaiana

®*Stenocactus albatus* → Stenocactus vaupelianus
Stenocactus arrigens → Stenocactus crispatus
Stenocactus bustamantei → Stenocactus ochoterenanus
Stenocactus coptonogonus
Stenocactus crispatus
Stenocactus dichroacanthus → Stenocactus crispatus
Stenocactus hastatus
Stenocactus heteracanthus → ?Stenocactus ochoterenanus
Stenocactus lamellosus → Stenocactus crispatus
Stenocactus lancifer → Stenocactus crispatus
Stenocactus lloydii → Stenocactus multicostatus
Stenocactus multicostatus
Stenocactus obvallatus
Stenocactus ochoterenanus
Stenocactus pentacanthus → Stenocactus obvallatus
Stenocactus phyllacanthus
Stenocactus rectispinus
Stenocactus sulphureus
Stenocactus vaupelianus
Stenocactus zacatecasensis → Stenocactus multicostatus

Stenocereus alamosensis
Stenocereus aragonii
Stenocereus beneckei
Stenocereus chacalapensis
Stenocereus chrysocarpus
Stenocereus deficiens → Stenocereus griseus
Stenocereus dumortieri
Stenocereus eichlamii
Stenocereus eruca
Stenocereus fimbriatus
Stenocereus fricii

Stenocereus griseus
Stenocereus gummosus
®*Stenocereus hystrix* → Stenocereus fimbriatus
Stenocereus kerberi
Stenocereus laevigatus
Stenocereus littoralis → Stenocereus thurberi ssp. littoralis
Stenocereus longispinus → Stenocereus eichlamii
Stenocereus marginatus → Pachycereus marginatus
Stenocereus martinezii
Stenocereus montanus
®*Stenocereus peruvianus* → Stenocereus fimbriatus
Stenocereus pruinosus
Stenocereus queretaroensis
Stenocereus quevedonis
Stenocereus standleyi
Stenocereus stellatus
Stenocereus thurberi
Stenocereus thurberi ssp. **thurberi**
Stenocereus thurberi ssp. **littoralis**
Stenocereus treleasei
Stenocereus weberi → Pachycereus weberi
Stenocereus yunckeri

Stephanocereus leucostele
Stephanocereus luetzelburgii

Stetsonia coryne

Strombocactus denegrii♦ → Obregonia denegrii♦
Strombocactus disciformis♦
Strombocactus disciformis ssp. **disciformis**♦
Strombocactus disciformis ssp. **esperanzae**♦
Strombocactus jarmilae♦ → ?Strombocactus disciformis♦
Strombocactus laui♦ → Turbinicarpus laui♦
Strombocactus polaskii♦ → Turbinicarpus schmiedickeanus ssp. schwarzii♦
Strombocactus pulcherrimus♦ → Strombocactus disciformis ssp. esperanzae♦
Strombocactus roseiflorus♦ → Turbinicarpus roseiflorus♦

Strophocactus wittii → Selenicereus wittii

Submatucana aurantiaca → Matucana aurantiaca ssp. aurantiaca
Submatucana aureiflora → Matucana aureiflora
Submatucana calvescens → Matucana aurantiaca ssp. aurantiaca
Submatucana currundayensis → Matucana aurantiaca ssp. currundayensis
Submatucana formosa → Matucana formosa
Submatucana intertexta → Matucana intertexta
Submatucana madisoniorum → Matucana madisoniorum
Submatucana myriacantha → Matucana haynei ssp. myriacantha
Submatucana paucicostata → Matucana paucicostata
Submatucana ritteri → Matucana ritteri

Subpilocereus atroviridis → Cereus repandus
Subpilocereus grenadensis → Cereus repandus
Subpilocereus horrispinus → Cereus horrispinus
Subpilocereus mortensenii → Cereus mortensenii
Subpilocereus ottonis → Cereus horrispinus
Subpilocereus remolinensis → Cereus repandus
Subpilocereus repandus → Cereus repandus
Subpilocereus repandus ssp. *micracanthus* → Cereus repandus
Subpilocereus russelianus → Cereus fricii

Sulcorebutia alba → Rebutia canigueralii
Sulcorebutia albaoides → Rebutia canigueralii
Sulcorebutia albissima → Rebutia mentosa
Sulcorebutia arenacea → Rebutia arenacea
Sulcorebutia augustinii → Rebutia mentosa
Ⓤ*Sulcorebutia breviflora* → Rebutia caineana
Sulcorebutia brevispina → Rebutia canigueralii
Sulcorebutia caineana → Rebutia caineana
Sulcorebutia callecallensis → Rebutia canigueralii
Sulcorebutia candiae → Rebutia arenacea
Sulcorebutia canigueralii → Rebutia canigueralii
Sulcorebutia caracarensis → Rebutia canigueralii
Sulcorebutia cardenasiana → Rebutia cardenasiana
Sulcorebutia chilensis → Neowerdermannia chilensis ssp. chilensis
Sulcorebutia cintiensis → Rebutia fidaiana ssp. cintiensis
Sulcorebutia clavata → Rebutia steinbachii
Sulcorebutia cochabambina → Rebutia steinbachii
Sulcorebutia corroana → Rebutia neocumingii
Sulcorebutia crispata → Rebutia canigueralii ssp. crispata
Sulcorebutia croceareolata → Rebutia steinbachii
Sulcorebutia cylindrica → Rebutia cylindrica
Sulcorebutia erinacea → Rebutia neocumingii
Sulcorebutia fidaiana → Rebutia fidaiana ssp. fidaiana
Sulcorebutia fischeriana → Rebutia canigueralii
Sulcorebutia flavida → Rebutia mentosa
Sulcorebutia flavissima → Rebutia mentosa
Sulcorebutia formosa → Rebutia mentosa
Sulcorebutia frankiana → Rebutia canigueralii
Sulcorebutia glomeriseta → Rebutia arenacea
Sulcorebutia glomerispina → Rebutia steinbachii
Sulcorebutia haseltonii → Rebutia caineana
Sulcorebutia hediniana → Rebutia neocumingii
Ⓤ*Sulcorebutia hoffmanniana* → Rebutia steinbachii
Sulcorebutia inflexiseta → Rebutia canigueralii
Sulcorebutia krahnii → Rebutia steinbachii
Sulcorebutia kruegeri → Rebutia steinbachii ssp. kruegeri
Sulcorebutia lanata → Rebutia neocumingii ssp. riograndensis
Sulcorebutia lecoriensis → Rebutia fidaiana
Sulcorebutia lepida → Rebutia steinbachii
Sulcorebutia longigibba → Rebutia neocumingii ssp. pilcomayensis
Sulcorebutia losenickyana → Rebutia canigueralii
Sulcorebutia mariana → Rebutia steinbachii
Sulcorebutia markusii → Rebutia mentosa
Sulcorebutia menesesii → Rebutia arenacea
Sulcorebutia mentosa → Rebutia mentosa ssp. mentosa
Sulcorebutia mizquensis → Rebutia steinbachii
Sulcorebutia multispina → Rebutia neocumingii
Sulcorebutia muschii → Rebutia arenacea
Sulcorebutia neocorroana → Rebutia neocumingii
Sulcorebutia neocumingii → Rebutia neocumingii ssp. neocumingii
Sulcorebutia neumanniana → Rebutia neumanniana
Sulcorebutia nigrofuscata → Rebutia steinbachii
Sulcorebutia oenantha → Rebutia steinbachii
Sulcorebutia oligacantha → Rebutia oligacantha
Sulcorebutia pampagrandensis → Rebutia steinbachii
Sulcorebutia pasopayana → Rebutia canigueralii
Sulcorebutia perplexiflora → Rebutia canigueralii
Sulcorebutia pilcomayensis → Rebutia neocumingii ssp. pilcomayensis
Sulcorebutia platygona → Rebutia neocumingii

Sulcorebutia polymorpha → Rebutia steinbachii
Sulcorebutia pulchra → Rebutia canigueralii ssp. pulchra
Sulcorebutia pulquinensis → Rebutia neocumingii ssp. neocumingii
Sulcorebutia purpurea → Rebutia mentosa ssp. purpurea
Sulcorebutia rauschii → Rebutia canigueralii
Sulcorebutia riograndensis → Rebutia neocumingii ssp. riograndensis
Sulcorebutia ritteri → Rebutia canigueralii
Sulcorebutia rubroaurea → Rebutia canigueralii
Sulcorebutia sanguineotarijensis → Rebutia oligacantha
Sulcorebutia santiaginensis → Rebutia mentosa
Sulcorebutia steinbachii → Rebutia steinbachii ssp. steinbachii
Sulcorebutia steinbachii ssp. *australis* → Rebutia steinbachii
®*Sulcorebutia steinbachii* ssp. *kruegeri* → Rebutia steinbachii ssp. kruegeri
Sulcorebutia steinbachii ssp. *tiraquensis* → Rebutia steinbachii ssp. tiraquensis
Sulcorebutia steinbachii ssp. *verticillacantha* → Rebutia steinbachii ssp. verticillacantha
Sulcorebutia sucrensis → Rebutia neocumingii ssp. neocumingii
Sulcorebutia swobodae → Rebutia mentosa
Sulcorebutia tarabucoensis → Rebutia canigueralii
Sulcorebutia taratensis → Rebutia steinbachii
Sulcorebutia tarijensis → Rebutia oligacantha
Sulcorebutia tiraquensis → Rebutia steinbachii ssp. tiraquensis
Sulcorebutia torotorensis → Rebutia mentosa
Sulcorebutia torotorensis → Rebutia mentosa
Sulcorebutia totoralensis → Rebutia steinbachii
Sulcorebutia totorensis → Rebutia steinbachii
Sulcorebutia tuberculato-chrysantha → Rebutia steinbachii
Sulcorebutia tunariensis → Rebutia steinbachii
Sulcorebutia unguispina → Rebutia mentosa
Sulcorebutia vasqueziana → Rebutia canigueralii
Sulcorebutia verticillacantha → Rebutia steinbachii ssp. verticillacantha
Sulcorebutia vilcayensis → Rebutia fidaiana
Sulcorebutia vizcarrae → Rebutia steinbachii
Sulcorebutia vorwerkii → Neowerdermannia vorwerkii
Sulcorebutia westii → Rebutia fidaiana
®*Sulcorebutia xanthoantha* → Rebutia arenacea
Sulcorebutia zavaletae → Rebutia canigueralii

Tacinga atropurpurea → Tacinga funalis
Tacinga braunii
Tacinga funalis
Tacinga funalis ssp. *atropurpurea* → Tacinga funalis

®*Tephrocactus albiscoparius* → Opuntia boliviana
Tephrocactus alboareolatus → Opuntia zehnderi
Tephrocactus alexanderi → Opuntia alexanderi
Tephrocactus articulatus → Opuntia articulata
Tephrocactus asplundii → Opuntia boliviana
Tephrocactus atacamensis → Opuntia atacamensis
®*Tephrocactus atroglobosus* → Opuntia nigrispina
Tephrocactus atroviridis → Opuntia floccosa
Tephrocactus blancii → Opuntia blancii
Tephrocactus bolivianus → Opuntia boliviana
Tephrocactus camachoi → Opuntia camachoi
Tephrocactus catacanthus → Opuntia sp.
Tephrocactus chichensis → Opuntia chichensis
Tephrocactus coloreus → Opuntia colorea
®*Tephrocactus conoideus* → Opuntia conoidea
Tephrocactus corotilla → Opuntia corotilla
Tephrocactus crassicylindricus → Opuntia crassicylindrica

Tephrocactus crispicrinitus → Opuntia floccosa
Tephrocactus curvispinus → Opuntia sp.
Tephrocactus cylindrarticulatus → Opuntia dactylifera
Tephrocactus cylindrolanatus → Opuntia floccosa
Tephrocactus dactylifer → Opuntia dactylifera
Tephrocactus darwinii → Opuntia darwinii
Tephrocactus dimorphus → Opuntia sphaerica
Tephrocactus echinaceus → Opuntia boliviana
Tephrocactus ferocior → Opuntia chichensis
Tephrocactus flexispinus → Opuntia sp.
Tephrocactus flexuosus → Opuntia flexuosa
Tephrocactus floccosus → Opuntia floccosa
Tephrocactus fulvicomus → Opuntia fulvicoma
Tephrocactus geometricus → Opuntia alexanderi
Tephrocactus glomeratus → Opuntia glomerata
Tephrocactus heteromorphus → Opuntia heteromorpha
Tephrocactus hickenii → Opuntia darwinii
Tephrocactus hirschii → Opuntia hirschii
Tephrocactus ignescens → Opuntia ignescens
Tephrocactus kuehnrichianus → Opuntia sphaerica
Tephrocactus lagopus → Opuntia lagopus
Tephrocactus malyanus → Opuntia lagopus
Tephrocactus mandragora → Opuntia minuta
Tephrocactus melanacanthus → Opuntia boliviana
Ⓤ*Tephrocactus microclados* → Opuntia rossiana
Tephrocactus microsphaericus → Opuntia alexanderi
Tephrocactus minor → Opuntia backebergii
Tephrocactus minusculus → Opuntia minuscula
Tephrocactus minutus → Opuntia minuta
Tephrocactus mirus → Opuntia sphaerica
Tephrocactus mistiensis → Opuntia mistiensis
Tephrocactus molinensis → Opuntia molinensis
Tephrocactus muellerianus → Opuntia sphaerica
Tephrocactus multiareolatus → Opuntia sphaerica
Tephrocactus neuquensis → Opuntia darwinii
Tephrocactus nigrispinus → Opuntia nigrispina
Tephrocactus noodtiae → Opuntia dactylifera
Tephrocactus ovallei → Opuntia glomerata
Tephrocactus ovatus → Opuntia ovata
Tephrocactus ovatus → Opuntia aoracantha
Tephrocactus p(a)ediophilus → Opuntia aoracantha
Ⓤ*Tephrocactus parvisetus* → Opuntia sp.
Tephrocactus pentlandii → Opuntia pentlandii
Tephrocactus platyacanthus → Opuntia darwinii
Tephrocactus pseudo-udonis → Opuntia floccosa
Tephrocactus punta-caillan → Opuntia punta-caillan
Tephrocactus pyrrhacanthus → Opuntia pyrrhacantha
Tephrocactus rarissimus → Opuntia pentlandii
Tephrocactus rauhii → Opuntia floccosa
Tephrocactus reicheanus → Opuntia glomerata
Tephrocactus russellii → Opuntia ovata
Tephrocactus silvestris → Opuntia silvestris
Tephrocactus sphaericus → Opuntia sphaerica
Tephrocactus subinermis → Opuntia pentlandii
Tephrocactus subterraneus → Opuntia subterranea
Tephrocactus tarapacanus → Opuntia tarapacana
Tephrocactus udonis → Opuntia floccosa
Ⓤ*Tephrocactus variflorus* → Opuntia subterranea
Tephrocactus verticosus → Opuntia floccosa

Tephrocactus weberi → Opuntia weberi
Tephrocactus wilkeanus → Opuntia pentlandii
Tephrocactus yanganucensis → Opuntia yanganucensis
Tephrocactus zehnderi → Opuntia zehnderi

Thelocactus aguirreanus → Escobaria aguirreana
Thelocactus beguinii♦ → Turbinicarpus beguinii♦
Thelocactus bicolor
Thelocactus bicolor ssp. **bicolor**
Thelocactus bicolor ssp. **flavidispinus**
Thelocactus bicolor ssp. **schwarzii**
Thelocactus buekii → Thelocactus tulensis ssp. buekii
Thelocactus conothelos
Thelocactus conothelos ssp. **argenteus**
Thelocactus conothelos ssp. **aurantiacus**
Thelocactus conothelos ssp. **conothelos**
Thelocactus durangensis → Sclerocactus unguispinus
Thelocactus ehrenbergii → Thelocactus leucacanthus
Thelocactus flavidispinus → Thelocactus bicolor ssp. flavidispinus
Thelocactus fossulatus → Thelocactus hexaedrophorus
Thelocactus garciae
Thelocactus gielsdorfianus♦ → Turbinicarpus gielsdorfianus♦
Thelocactus hastifer
Thelocactus heterochromus
Thelocactus hexaedrophorus
Thelocactus hexaedrophorus ssp. **hexaedrophorus**
Thelocactus hexaedrophorus ssp. **lloydii**
Thelocactus horripilus♦ → Turbinicarpus horripilus♦
Thelocactus knuthianus♦ → Turbinicarpus knuthianus♦
Thelocactus krainzianus → ?Thelocactus tulensis
Thelocactus lausseri
Thelocactus leucacanthus
Thelocactus lloydii → Thelocactus hexaedrophorus ssp. lloydii
Thelocactus lophothele → Thelocactus rinconensis
Thelocactus macdowellii
Thelocactus mandragora♦ → Turbinicarpus mandragora♦
Thelocactus matudae → Thelocactus tulensis ssp. matudae
Thelocactus multicephalus → ?Thelocactus rinconensis
Thelocactus nidulans → Thelocactus rinconensis
Thelocactus phymatothelos → Thelocactus rinconensis
⊗*Thelocactus pottsii* → Thelocactus heterochromus
Thelocactus pseudopectinatus♦ → Turbinicarpus pseudopectinatus♦
Thelocactus rinconensis
Thelocactus rinconensis ssp. **hintonii**
Thelocactus rinconensis ssp. *nidulans* → Thelocactus rinconensis
Thelocactus rinconensis ssp. *phymatothelos* → Thelocactus rinconensis
Thelocactus rinconensis ssp. **rinconensis**
Thelocactus roseanus → Escobaria roseana
Thelocactus saueri♦ → Turbinicarpus saueri♦
Thelocactus saussieri → Thelocactus conothelos
Thelocactus schottii → Thelocactus bicolor
Thelocactus schwarzii → Thelocactus bicolor ssp. schwarzii
Thelocactus setispinus
Thelocactus subterraneus♦ → Turbinicarpus subterraneus♦
Thelocactus tulensis
Thelocactus tulensis ssp. **buekii**
Thelocactus tulensis ssp. **matudae**
Thelocactus tulensis ssp. **tulensis**
Thelocactus unguispinus → Sclerocactus unguispinus
Thelocactus viereckii♦ → Turbinicarpus viereckii♦

Thelocactus wagnerianus → Thelocactus bicolor
Thelocactus ysabelae♦ → Turbinicarpus ysabelae♦
Thelocephala aerocarpa → Eriosyce aerocarpa
Thelocephala duripulpa → Eriosyce napina ssp. lembckei
Thelocephala esmeraldana → Eriosyce esmeraldana
Thelocephala fankhauseri → Eriosyce tenebrica
Thelocephala fulva → Eriosyce odieri ssp. fulva
Thelocephala glabrescens → Eriosyce odieri ssp. glabrescens
Thelocephala krausii → Eriosyce krausii
®*Thelocephala lembckei* → Eriosyce napina ssp. lembckei
Thelocephala longirapa → Eriosyce krausii
Thelocephala malleolata → Eriosyce krausii
Thelocephala napina → Eriosyce napina
Thelocephala nuda → Eriosyce aerocarpa
Thelocephala odieri → Eriosyce odieri ssp. odieri
Thelocephala reichei → ?Eriosyce odieri
Thelocephala tenebrica → Eriosyce tenebrica

Thrixanthocereus blossfeldiorum → Espostoa blossfeldiorum
Thrixanthocereus cullmannianus → Espostoa senilis
Thrixanthocereus longispinus → Espostoa senilis
Thrixanthocereus senilis → Espostoa senilis

Torreyocactus conothelos → Thelocactus conothelos ssp. conothelos

Toumeya bradyi♦ → Pediocactus bradyi♦
Toumeya fickeisenii♦ → Pediocactus peeblesianus♦
Toumeya krainziana♦ → Turbinicarpus pseudomacrochele♦
Toumeya papyracantha♦ → Sclerocactus papyracanthus♦

Trichocereus angelesii → Echinopsis angelesiae
Trichocereus antezanae → Echinopsis antezanae
Trichocereus arboricola → Echinopsis arboricola
Trichocereus bertramianus → Echinopsis bertramiana
Trichocereus bridgesii → Echinopsis lageniformis
Trichocereus bruchii → Echinopsis bruchii
Trichocereus cabrerae → Echinopsis ×cabrerae
Trichocereus callianthus → Echinopsis sp.
Trichocereus camarguensis → Echinopsis camarguensis
Trichocereus candicans → Echinopsis candicans
Trichocereus catamarcensis → Echinopsis huascha
®*Trichocereus caulescens* → Echinopsis sp.
Trichocereus cephalomacrostibas → Echinopsis cephalomacrostibas
Trichocereus chalaensis → Echinopsis chalaensis
Trichocereus chiloensis → Echinopsis chiloensis
®*Trichocereus chuquisacanus* → Echinopsis sp.
Trichocereus clavatus → Echinopsis clavatus
Trichocereus conaconensis → Echinopsis conaconensis
Trichocereus coquimbanus → Echinopsis coquimbana
Trichocereus courantii → Echinopsis candicans
®*Trichocereus crassicostatus* → Echinopsis sp.
Trichocereus cuzcoensis → Echinopsis cuzcoensis
Trichocereus deserticola → Echinopsis deserticola
Trichocereus eremophilus → Echinopsis atacamensis ssp. pasacana
Trichocereus escayachensis → Echinopsis escayachensis
Trichocereus fabrisii → Echinopsis fabrisii
Trichocereus fascicularis → Weberbauerocereus weberbaueri
Trichocereus formosus → Echinopsis formosa
Trichocereus fulvilanus → Echinopsis deserticola

Trichocereus glaucus → Echinopsis glauca
Trichocereus grandiflorus → Echinopsis huascha
Trichocereus grandis → Echinopsis bruchii
Trichocereus herzogianus → Echinopsis tarijensis ssp. herzogianus
Trichocereus huascha → Echinopsis huascha
Trichocereus ingens → Echinopsis bruchii
Trichocereus knuthianus → Echinopsis knuthiana
Trichocereus korethroides → Echinopsis korethroides
Trichocereus lamprochlorus → Echinopsis lamprochlora
Trichocereus litoralis → Echinopsis litoralis
Trichocereus lobivioides → Echinopsis huascha
Trichocereus macrogonus → Echinopsis macrogona
Trichocereus manguinii → Echinopsis schickendantzii
Trichocereus narvaecensis → ?Echinopsis tarijensis
Trichocereus neolamprochlorus → Echinopsis candicans
Trichocereus nigripilis → Echinopsis sp.
Trichocereus orurensis → ?Echinopsis ferox
Trichocereus pachanoi → Echinopsis pachanoi
Trichocereus pasacana → Echinopsis atacamensis ssp. pasacana
Trichocereus peruvianus → Echinopsis peruviana ssp. peruviana
Trichocereus poco → Echinopsis tarijensis
Ⓤ*Trichocereus pseudocandicans* → Echinopsis candicans
Trichocereus puquiensis → Echinopsis peruviana ssp. puquiensis
Trichocereus quadratiumbonatus → Echinopsis quadratiumbonatus
Trichocereus randallii → Echinopsis formosa
Ⓤ*Trichocereus riomizquensis* → Echinopsis sp.
Trichocereus rivierei → Echinopsis atacamensis ssp. pasacana
Trichocereus rowleyi → Echinopsis huascha
Ⓤ*Trichocereus rubinghianus* → Echinopsis thelegonoides
Trichocereus santaensis → Echinopsis santaensis
Trichocereus santiaguensis → Echinopsis spachiana
Trichocereus schickendantzii → Echinopsis schickendantzii
Trichocereus schoenii → Echinopsis schoenii
Trichocereus scopulicola → Echinopsis scopulicola
Trichocereus serenanus → Echinopsis coquimbana
Trichocereus shaferi → Echinopsis schickendantzii
Trichocereus skottsbergii → Echinopsis skottsbergii
Trichocereus smrzianus → Echinopsis smrziana
Trichocereus spachianus → Echinopsis spachiana
Trichocereus spinibarbis → Echinopsis spinibarbis
Trichocereus strigosus → Echinopsis strigosa
Trichocereus tacaquirensis → Echinopsis tacaquirensis
Trichocereus tacnaensis → Echinopsis peruviana ssp. peruviana
Trichocereus taquimbalensis → Echinopsis tacaquirensis ssp. taquimbalensis
Trichocereus taratensis → Echinopsis taratensis
Trichocereus tarijensis → Echinopsis tarijensis
Trichocereus tarmaensis → Echinopsis tarmaensis
Trichocereus tenuispinus → Echinopsis bridgesii
Trichocereus terscheckii → Echinopsis terscheckii
Trichocereus terscheckioides → Echinopsis sp.
Trichocereus thelegonoides → Echinopsis thelegonoides
Trichocereus thelegonus → Echinopsis thelegona
Trichocereus torataensis → Echinopsis peruviana ssp. peruviana
Trichocereus totorensis → Echinopsis tarijensis ssp. totorensis
Trichocereus trichosus → Echinopsis trichosa
Trichocereus tulhuayacensis → Echinopsis tulhuayacensis
Trichocereus tunariensis → Echinopsis tunariensis
Trichocereus uyupampensis → Echinopsis uyupampensis
Trichocereus validus → ?Echinopsis uyupampensis

Trichocereus vasquezii → Echinopsis vasquezii
Trichocereus vatteri → Echinopsis vatteri
Trichocereus volcanensis → Echinopsis schickendantzii
Trichocereus vollianus → Echinopsis volliana
Trichocereus walteri → Echinopsis walteri
Trichocereus werdermannianus → Echinopsis terscheckii

Turbinicarpus alonsoi♦
Turbinicarpus beguinii♦
Turbinicarpus bonatzii♦
Turbinicarpus booleanus♦
Turbinicarpus dickisoniae♦ → Turbinicarpus schmiedickeanus ssp. dickisoniae♦
Turbinicarpus flaviflorus♦ → Turbinicarpus schmiedickeanus ssp. flaviflorus♦
¶*Turbinicarpus gautii* → ?Sclerocactus warnockii
Ⓜ*Turbinicarpus gautii* → Turbinicarpus beguinii♦
Turbinicarpus gielsdorfianus♦
Turbinicarpus gracilis♦ → Turbinicarpus schmiedickeanus ssp. gracilis♦
Turbinicarpus hoferi♦
Turbinicarpus horripilus♦
Turbinicarpus jauernigii♦
Turbinicarpus klinkerianus♦ → Turbinicarpus schmiedickeanus ssp. klinkerianus♦
Turbinicarpus knuthianus♦
Turbinicarpus krainzianus♦ → Turbinicarpus pseudomacrochele♦
Turbinicarpus laui♦
Turbinicarpus lophophoroides♦
Turbinicarpus lophophoroides ssp. *jauernigii*♦ → Turbinicarpus jauernigii♦
Turbinicarpus macrochele♦ → Turbinicarpus schmiedickeanus ssp. macrochele♦
Turbinicarpus mandragora♦
Turbinicarpus ×***mombergeri***♦ (T. laui × T. pseudopectinatus)
Ⓤ*Turbinicarpus polaskii*♦ → Turbinicarpus schmiedickeanus ssp. schwarzii♦
Turbinicarpus pseudomacrochele♦
Turbinicarpus pseudomacrochele ssp. *krainzianus*♦ → Turbinicarpus pseudomacrochele♦
Turbinicarpus pseudomacrochele ssp. **lausseri**♦
Turbinicarpus pseudomacrochele ssp. **pseudomacrochele**♦
Turbinicarpus pseudopectinatus♦
Turbinicarpus rioverdensis♦
Ⓤ***Turbinicarpus roseiflorus***♦ → ?Turbinicarpus hybr.
Turbinicarpus saueri♦
Turbinicarpus schmiedickeanus♦
Turbinicarpus schmiedickeanus ssp. **dickisoniae**♦
Turbinicarpus schmiedickeanus ssp. **flaviflorus**♦
Turbinicarpus schmiedickeanus ssp. **gracilis**♦
Turbinicarpus schmiedickeanus ssp. **klinkerianus**♦
Turbinicarpus schmiedickeanus ssp. **macrochele**♦
Turbinicarpus schmiedickeanus ssp. **schmiedickeanus**♦
Turbinicarpus schmiedickeanus ssp. **schwarzii**♦
Turbinicarpus schwarzii♦ → Turbinicarpus schmiedickeanus ssp. schwarzii♦
Turbinicarpus subterraneus♦
Turbinicarpus swobodae♦
Turbinicarpus valdezianus♦
Turbinicarpus viereckii♦
Turbinicarpus viereckii ssp. **major**♦
Turbinicarpus ysabelae♦
Turbinicarpus zaragozae♦

Uebelmannia buiningii♦
Uebelmannia flavispina♦ → Uebelmannia pectinifera ssp. flavispina♦
Uebelmannia gummifera♦
Uebelmannia gummifera ssp. *meninensis*♦
Uebelmannia meninensis♦

Uebelmannia pectinifera♦
Uebelmannia pectinifera ssp. **flavispina**♦
Uebelmannia pectinifera ssp. **horrida**♦
Uebelmannia pectinifera ssp. **pectinifera**♦

Utahia peeblesiana♦ → Pediocactus peeblesianus♦
Utahia sileri♦ → Pediocactus sileri♦

Vatricania guentheri → Espostoa guentheri

Weberbauerocereus albus → Weberbauerocereus longicomus
Weberbauerocereus cephalomacrostibas → Echinopsis cephalomacrostibas
Weberbauerocereus churinensis
Weberbauerocereus cuzcoensis
Weberbauerocereus horridispinus → Weberbauerocereus weberbaueri
Weberbauerocereus johnsonii
Weberbauerocereus longicomus
Weberbauerocereus rauhii
Weberbauerocereus seyboldianus → Weberbauerocereus weberbaueri
Weberbauerocereus torataensis
Weberbauerocereus weberbaueri
Weberbauerocereus winterianus

Weberocereus biolleyi
Weberocereus bradei
Weberocereus glaber
Weberocereus imitans
Weberocereus panamensis
Weberocereus rosei
Weberocereus tonduzii
Weberocereus trichophorus
Weberocereus tunilla

Weingartia aglaia → Rebutia steinbachii
Weingartia alba → Rebutia canigueralii
Weingartia albaoides → Rebutia canigueralii
Weingartia albaoides ssp. *subfusca* → Rebutia canigueralii
Weingartia albissima → Rebutia mentosa
Ⓜ*Weingartia ambigua* → Rebutia neocumingii
Weingartia ansaldoensis → Rebutia steinbachii
Weingartia arenacea → Rebutia arenacea
Weingartia attenuata → Rebutia neocumingii
Weingartia aureispina → Rebutia canigueralii
Weingartia backebergiana → Rebutia steinbachii
Weingartia brachygraphisa → Rebutia neocumingii
Ⓤ*Weingartia breviflora* → Rebutia caineana
Weingartia brevispina → Rebutia canigueralii
Weingartia buiningiana → Rebutia neocumingii
Weingartia caineana → Rebutia caineana
Weingartia callecallensis → Rebutia neocumingii
Weingartia candiae → Rebutia arenacea
Weingartia canigueralii → Rebutia canigueralii
Weingartia caracarensis → Rebutia canigueralii
Weingartia cardenasiana → Rebutia cardenasiana
Weingartia chilensis → Neowerdermannia chilensis ssp. chilensis
Weingartia cintiensis → Rebutia fidaiana ssp. cintiensis
Weingartia clavata → Rebutia steinbachii
Weingartia columnaris → Rebutia neocumingii
Weingartia corroana → Rebutia neocumingii
Weingartia crispata → Rebutia canigueralii ssp. crispata

Weingartia croceareolata → Rebutia steinbachii
Weingartia cylindrica → Rebutia cylindrica
Weingartia electracantha → Rebutia steinbachii
Weingartia erinacea → Rebutia neocumingii
Weingartia fidaiana → Rebutia fidaiana ssp. fidaiana
Weingartia fidaiana ssp. *cintiensis* → Rebutia fidaiana ssp. cintiensis
Weingartia flavida → Rebutia neocumingii
Weingartia flavissima → Rebutia mentosa
Weingartia formosa → Rebutia mentosa
Weingartia frankiana → Rebutia canigueralii
Weingartia glomeriseta → Rebutia arenacea
Weingartia glomerispina → Rebutia steinbachii
Weingartia gracilispina → Rebutia neocumingii
Weingartia haseltonii → Rebutia caineana
Weingartia hediniana → Rebutia neocumingii
Ⓤ*Weingartia hoffmanniana* → Rebutia steinbachii
Weingartia inflexiseta → Rebutia canigueralii
Weingartia kargliana → Rebutia neumanniana
Weingartia knizei → Rebutia neocumingii
Weingartia krahnii → Rebutia steinbachii
Weingartia kruegeri → Rebutia steinbachii ssp. kruegeri
Weingartia lanata → Rebutia neocumingii ssp. riograndensis
Weingartia lanata ssp. *longigibba* → Rebutia neocumingii ssp. pilcomayensis
Weingartia lanata ssp. *pilcomayensis* → Rebutia neocumingii ssp. pilcomayensis
Weingartia lanata ssp. *riograndensis* → Rebutia neocumingii ssp. riograndensis
Weingartia lecoriensis → Rebutia fidaiana
Weingartia lepida → Rebutia steinbachii
Weingartia longigibba → Rebutia neocumingii ssp. pilcomayensis
Weingartia losenickyana → Rebutia canigueralii
Weingartia mairanana → Rebutia neocumingii
Weingartia margarethae → Rebutia padcayensis
Weingartia markusii → Rebutia mentosa
Weingartia mataralensis → Rebutia neocumingii
Weingartia menesesii → Rebutia arenacea
Weingartia mentosa → Rebutia mentosa ssp. mentosa
Weingartia minima → Rebutia steinbachii
Weingartia miranda → Rebutia neocumingii
Weingartia mizquensis → Rebutia steinbachii
Weingartia multispina → Rebutia neocumingii
Weingartia muschii → Rebutia arenacea
Weingartia neglecta → Rebutia neocumingii
Weingartia neocumingii → Rebutia neocumingii ssp. neocumingii
Weingartia neocumingii ssp. *pulquinensis* → Rebutia neocumingii ssp. neocumingii
Weingartia neocumingii ssp. *sucrensis* → Rebutia neocumingii ssp. neocumingii
Weingartia neumanniana → Rebutia neumanniana
Weingartia nigro-fuscata → Rebutia steinbachii
Weingartia oenantha → Rebutia steinbachii
Weingartia oligacantha → Rebutia oligacantha
Weingartia pampagrandensis → Rebutia steinbachii
Weingartia pasopayana → Rebutia canigueralii
Weingartia perplexiflora → Rebutia canigueralii
Weingartia pilcomayensis → Rebutia neocumingii ssp. pilcomayensis
Weingartia platygona → Rebutia neocumingii
Weingartia polymorpha → Rebutia steinbachii
Weingartia pulchra → Rebutia canigueralii ssp. pulchra
Weingartia pulquinensis → Rebutia neocumingii ssp. neocumingii
Weingartia purpurea → Rebutia mentosa ssp. purpurea
Weingartia pygmaea → Rebutia neumanniana
Weingartia rauschii → Rebutia canigueralii

Weingartia riograndensis → Rebutia neocumingii ssp. riograndensis
Weingartia ritteri → Rebutia canigueralii
Weingartia rubro-aurea → Rebutia canigueralii
Weingartia saetosa → Rebutia neocumingii
Weingartia saipinensis → Rebutia neocumingii ssp. saipinensis
Weingartia sanguineo-tarijensis → Rebutia oligacantha
Weingartia saxatilis → Rebutia canigueralii
Weingartia steinbachii → Rebutia steinbachii ssp. steinbachii
Weingartia sucrensis → Rebutia neocumingii ssp. neocumingii
Weingartia tarabucina → Rebutia canigueralii
Weingartia tarabucoensis → Rebutia canigueralii
Weingartia taratensis → Rebutia steinbachii
Weingartia tarijensis → Rebutia oligacantha
Weingartia tiraquensis → Rebutia steinbachii ssp. tiraquensis
Weingartia torotorensis → Rebutia mentosa
Weingartia totoralensis → Rebutia steinbachii
Weingartia totorensis → Rebutia steinbachii
Weingartia trollii → Rebutia neocumingii ssp. trollii
Weingartia tuberculato-chrysantha → Rebutia steinbachii
Weingartia tunariensis → Rebutia steinbachii
Weingartia verticillacantha → Rebutia steinbachii ssp. verticillacantha
Weingartia vilcayensis → Rebutia fidaiana
Weingartia vizcarrae → Rebutia steinbachii
Weingartia vorwerkii → Neowerdermannia vorwerkii
Weingartia westii → Rebutia fidaiana
Weingartia zavaletae → Rebutia canigueralii

Werckleocereus glaber → Weberocereus glaber
Werckleocereus imitans → Weberocereus imitans
Werckleocereus tonduzii → Weberocereus tonduzii

Wigginsia acuata → Parodia erinacea
Wigginsia arechavaletae → Parodia neoarechavaletae
Wigginsia corynodes → Parodia sellowii
Wigginsia courantii → Parodia sellowii
Wigginsia erinacea → Parodia erinacea
Wigginsia fricii → Parodia sellowii
Wigginsia horstii → Parodia neohorstii
Wigginsia kovaricii → Parodia sp.
Wigginsia langsdorfii → Parodia langsdorfii
Wigginsia leprosorum → Parodia langsdorfii
Wigginsia leucocarpa → Parodia sellowii
Wigginsia longispina → Parodia langsdorfii
Wigginsia macracantha → Parodia sellowii
Wigginsia macrogona → Parodia sellowii
Wigginsia polyacantha → Parodia langsdorfii
Wigginsia prolifera → Parodia langsdorfii
Wigginsia schaeferiana → Parodia turbinata
Wigginsia sellowii → Parodia sellowii
Wigginsia sessiliflora → Parodia sellowii
Wigginsia stegmannii → Parodia sellowii
Wigginsia tephracantha → Parodia sellowii
Wigginsia turbinata → Parodia turbinata
Wigginsia vorwerkiana → Parodia sellowii

Wilcoxia albiflora → Echinocereus leucanthus
Wilcoxia kroenleinii → Echinocereus poselgeri
Wilcoxia lazaro-cardenasii → Peniocereus lazaro-cardenasii
Wilcoxia nerispina♦ → Echinocereus schmollii♦

Wilcoxia papillosa → Peniocereus viperinus
Wilcoxia poselgeri → Echinocereus poselgeri
Wilcoxia schmollii♦ → Echinocereus schmollii♦
Wilcoxia striata → Peniocereus striatus
Wilcoxia tamaulipensis → Echinocereus poselgeri
Wilcoxia tomentosa → Peniocereus viperinus
Wilcoxia tuberosa → Echinocereus poselgeri
Wilcoxia viperina → Peniocereus viperinus
Wilcoxia zopilotensis → Peniocereus zopilotensis

Wilmattea minutiflora → Hylocereus minutiflorus
Wilmattea venezuelensis → Hylocereus lemairei

Wittia amazonica → Disocactus amazonicus
Wittia himantoclada → Pseudorhipsalis himantoclada
Wittia panamensis → Disocactus amazonicus

Wittiocactus amazonicus → Disocactus amazonicus
Wittiocactus panamensis → Disocactus amazonicus

Yungasocereus inquisivensis
®*Yungasocereus microcarpus* → Yungasocereus inquisivensis

Zehntnerella chaetacantha → Facheiroa squamosa
Zehntnerella polygona → Facheiroa squamosa
Zehntnerella squamulosa → Facheiroa squamosa

Zygocactus truncatus → Schlumbergera truncata

Part II. Reference Data

1. List of genera of Cactaceae and their principal synonyms

Notes: Names of genera currently accepted for the purpose of the Checklist are printed in **bold** (= accepted) or ***bold italic*** (= provisionally accepted) type. Synonyms are printed in *italic*, and the genera to which they are referred indicated after an arrow (→). The probable affinity of provisionally accepted genera is indicated by a tilde (~) and the name of the related genus.

The number of taxa quoted for each genus ('Taxa: ') is the number of currently accepted species, plus (+) the total number of provisionally accepted species, heterotypic subspecies (both fully and provisionally accepted) and hybrids, if any. Total numbers of taxa are summarized in the Table on pp. 159–160.

Acanthocalycium Backeberg in Backeberg & Knuth, Kaktus ABC, 224, 412 (1936). Type: *Echinocactus spiniflorus* Schumann. Taxa: 1+2. Distr.: Argentina. ~ Echinopsis

Acanthocereus (Engelmann ex Berger) Britton & Rose, Contr. US Nat. Herb. 12: 432 (1909). Type: *Cereus baxaniensis* Karwinsky ex Pfeiffer. Taxa: 1+5. Distr.: Trop. America, Caribbean. Syn.: *Monvillea* Britton & Rose (1920), as to type.

Acantholobivia Backeberg, Cact. Jahrb. Deutsch. Kakt. Ges. 1941: 76 (1942). Type: *Lobivia tegeleriana* Backeberg. → Echinopsis

Acanthorhipsalis (Schumann) Britton & Rose, Cact. 4: 211 (1923). Type: *Rhipsalis monacantha* Grisebach. → Lepismium

Acanthorhipsalis Kimnach, Cact. Succ. J. (US) 55: 177 (1983), non (Schumann) Britton & Rose. Type: *Cereus micranthus* Vaupel. → Lepismium

Akersia Buining, Succulenta 1961: 25 (1961). Type: *A. roseiflora* Buining. → Cleistocactus

Ancistrocactus Britton & Rose, Cact. 4: 3 (1923). Type: *Echinocactus megarhizus* Rose. → Sclerocactus

Anisocereus Backeberg, Blätt. Kakt.-Forsch. 1938(6): [8, 21] (1938). Type: *Cereus lepidanthus* Eichlam. → Pachycereus

Aporocactus Lemaire, Illustr. Hort. 7: Misc. 67 (1860). Type: *A. flagelliformis* (Linnaeus) Lemaire. → Disocactus

Arequipa Britton & Rose, Cact. 3: 100 (1922). Type: *Echinocactus leucotrichus* Philippi. → Oreocereus

Arequipiopsis Kreuzinger & Buining, Fedde's Repert. Sp. Nov. 50: 20 (1941). Type: *Echinopsis hempelianus* Gürke. → Oreocereus

Ariocarpus Scheidweiler, Bull. Acad. Sci. Brux. 5: 491 (1838). Type: *A. retusus* Scheidweiler. Taxa: 6+2. Distr.: SW USA (S Texas), N Mexico. Syn.: *Roseocactus* Berger (1925); *Neogomesia* Castañeda (1941). Ref.: Anderson (1960–64).

Armatocereus Backeberg, Blätt. Kakt.-Forsch. 1938(6): [21] (1938). Type: *A. laetus* (Kunth) Backeberg. Taxa: 10+5. Distr.: W South America.

Arrojadoa Britton & Rose, Cact. 2: 170 (1920). Type: *Cereus rhodanthus* Gürke. Taxa: 4+2. Distr.: E Brazil.

Arthrocereus Berger, Kakteen: 337 (1929), *nom. cons.;* F. Knuth, Den Nye Kaktusbog: 111 (1930). Type: Monatsschr. Kakt.-Kunde 28: illus. opp. p. 62 (1918), as '*Cereus damazioi*', *typ. cons.* (→ *A. glaziovii* (Schumann) Taylor & Zappi). Taxa: 4+2. Distr.: Brazil.

Astrophytum Lemaire, Cact. Gen. Nov. Sp. Hort. Monv., 3 (1839). Type: *A. myriostigma* Lemaire. Taxa: 4. Distr.: SW USA (S Texas), Mexico.

Austrocactus Britton & Rose, Cact. 3: 44 (1922). Type: *A. bertinii* (Cels) Britton & Rose. Taxa: 3+2. Distr.: S Argentina, S Chile.

Austrocephalocereus Backeberg, Blätt. Kakt.-Forsch. 1938(6): [22] (1938). Type: *Cephalocereus purpureus* Gürke. Distr.: E Brazil. → Micranthocereus

Austrocylindropuntia Backeberg, Blätt. Kakt.-Forsch. 1938(6): [3, 21] (1938). Type: *Opuntia exaltata* Berger. → Opuntia

Aylostera Spegazzini, An. Soc. Cient. Argent. 96: 75 (1923). Type: *Echinopsis pseudominuscula* Spegazzini. → Rebutia

Aztekium Bödeker, Monatsschr. Deutsch. Kakt. Ges. 1: 52 (1929). Type: *A. ritteri* Bödeker. Taxa: 2. Distr.: NE Mexico.

Azureocereus Akers & Johnson, Cact. Succ. J. (US) 21: 133 (1949). Type: *A. nobilis* Akers & Johnson. → Browningia

Backebergia Bravo, An. Inst. Biol. Mex. 24: 230 (1953). Type: *Pilocereus chrysomallus* Lemaire. → Pachycereus

Bartschella Britton & Rose, Cact. 4: 57 (1923). Type: *Mammillaria schumannii* Hildmann. → Mammillaria

Bergerocactus Britton & Rose, Contr. US Nat. Herb. 12: 435 (1909). Type: *B. emoryi* (Engelmann) Britton & Rose. Taxa: 1. Distr.: SW USA (S California), NW Mexico (NW Baja California).

Binghamia Britton & Rose, Cact. 2: 167 (1920). Type: *Cephalocereus melanostele* Vaupel. → Espostoa

Bisnaga Orcutt, Cactography, 1 (1926). Type: *Echinocactus cornigerus* De Candolle. → Ferocactus

Blossfeldia Werdermann, Kakteenkunde 1937: 162 (1937). Type: *B. liliputana* Werdermann. Taxa: 1. Distr.: N Argentina, Bolivia. ~ Parodia

Bolivicereus Cárdenas, Cact. Succ. J. (US) 23: 91 (1951). Type: *B. samaipatanus* Cárdenas. → Cleistocactus

Bonifazia Standley & Steyermark, Publ. Field Mus. Nat. Hist. bot. ser. 23: 66 (1944). Type: *B. quezalteca* Standley & Steyermark. → Disocactus

Borzicactella Ritter, Kakteen in Suedamerika 4: 1385 (1981). Type: *Cleistocactus tenuiserpens* Rauh & Backeberg. → Cleistocactus

Borzicactus Riccobono, Boll. R. Ort. Palermo 8: 261 (1909). Type: *B. ventimigliae* Riccobono. → Cleistocactus

Brachycalycium Backeberg, Cact. Jahrb. Deutsch. Kakt. Ges. 1941: 76 (1942). Type: *B. tilcarense* (Backeberg) Backeberg. → Gymnocalycium

Brachycereus Britton & Rose, Cact. 2: 120 (1920). Type: *Cereus nesioticus* Schumann (LT: Backeberg 1959). Taxa: 1. Distr.: Ecuador (Galapagos Islands).

Brasilicactus Backeberg, Cact. Jahrb. Deutsch. Kakt. Ges. 1941: 76 (1942). Type: *Echinocactus graessneri* Schumann. → Parodia

Brasilicereus Backeberg, Blätt. Kakt.-Forsch. 1938(6): [8, 22] (1938). Type: *Cereus phaeacanthus* Guerke. Taxa: 2. Distr.: Brazil.

Brasiliopuntia (Schumann) Berger, Entwicklungslinien Kakt., 17 (1926). Type: *Cactus brasiliensis* Willdenow. → Opuntia

Brasiliparodia Ritter, Kakteen in Südamerika 1: 144 (1979). Type: *Parodia buenekeri* Buining. → Parodia

Bravocactus Doweld, Sukkulenty (Moscow) 1998(1): 22 (1998). Type: *Bravocactus horripilus* (Lemaire) Doweld. → Turbinicarpus

Browningia Britton & Rose, Cact. 2: 63 (1920). Type: *B. candelaris* (Meyen) Britton & Rose. Taxa: 5+6. Distr.: W Paraguay, Bolivia, Peru, N Chile. Syn.: *Gymnanthocereus* Backeberg (1937); *Azureocereus* Akers & Johnson (1949); *Castellanosia* Cárdenas (1951); *Gymnocereus* Backeberg (1959).

Buiningia Buxbaum in Krainz, Die Kakteen, Lfg. 46–47 (1971). Type: *Buiningia brevicylindrica* Buining. → Coleocephalocereus

Cactus Linnaeus, Species Plantarum, 466 (1753). Type: *C. mammillaris* Linnaeus (see note). → Mammillaria

Cactus sensu Britton & Rose, Cact. 3: 220 (1922). → Melocactus

Calymmanthium Ritter, Kakt. and. Sukk. 13: 25 (1962). Type: *C. substerile* Ritter. Taxa: 1. Distr.: N Peru.

Carnegiea Britton & Rose, J. New York Bot. Gard. 9: 187 (1908). Type: *C. gigantea* (Engelmann) Britton & Rose. Taxa: 1. Distr.: SW USA (Arizona), Mexico.

Castellanosia Cárdenas, Cact. Succ. J. (US) 23: 90 (1951). Type: *C. caineana* Cárdenas. → Browningia

Cephalocereus Pfeiffer, Allg. Gartenz. 6: 142 (1838). Type: *Cactus senilis* Haworth. Taxa: 3+2. Distr.: Mexico. Syn.: *Pilocereus* Lemaire (1839); *Haseltonia* Backeberg (1949); *Neodawsonia* Backeberg (1949).

Cephalocleistocactus Ritter, Succulenta 1959: 108 (1959). Type: *C. chrysocephalus* Ritter. Taxa: 0+1. Distr.: Bolivia. ~ Cleistocactus

Cereus Miller, Gard. Dict. Abr. ed. 4, [unpaged] (1754). Type: *C. hexagonus* (Linnaeus) Miller. Taxa: 23+16. Distr.: Caribbean, South America. Syn.: *Piptanthocereus* (Berger) Riccobono (1909); *Subpilocereus* Backeberg (1938); *Mirabella* Ritter (1979).

Chamaecereus Britton & Rose, Cact. 3: 48 (1922). Type: *Cereus silvestrii* Spegazzini. → Echinopsis

Chiapasia Britton & Rose, Cact. 4: 203 (1923). Type: *Epiphyllum nelsonii* Rose. → Disocactus

Chichipia Backeberg in Marnier-Lapostolle, Liste Cact., 12 (1950). Type: *Cereus chichipe* Gosselin. → Polaskia

Chilita Orcutt, Cactography, 2 (1926). Type: *Mammillaria grahamii* Engelmann. → Mammillaria

Cintia Knize & Riha, Kaktusy 31(2): 35–39 (1996). Type: *C. knizei* Riha. Taxa: 1. Distr.: Bolivia. ~ Rebutia?

Cipocereus Ritter, Kakteen in Südamerika 1: 54 (1979). Type: *Cipocereus pleurocarpus* Ritter. Taxa: 5+1. Distr.: Brazil (Minas Gerais). Syn.: *Floribunda* Ritter (1979).

Cleistocactus Lemaire, Illustr. Hort. 8: Misc. 35 (1861). Type: *Cereus baumannii* Lemaire. Taxa: 23+25. Distr.: W South America, S Brazil, N Argentina. Syn.: *Borzicactus* Riccobono (1909); *Loxanthocereus* Backeberg (1937); *Seticereus* Backeberg (1942); *Maritimocereus* Akers & Buining (1950); *Bolivicereus* Cárdenas (1951); *Akersia* Buining (1961); *Winteria* Ritter (1962), non Murray; *Seticleistocactus* Backeberg (1963); *Winterocereus* Backeberg (1966); *Hildewintera* Ritter (1966); *Borzicactella* Ritter (1981).

×***Cleistocana*** Rowley, Bradleya 12: 5 (1994). (*Cleistocactus* × *Matucana*). Taxa: 1. Distr.: Peru.

Cochemiea (K. Brandegee) Walton, Cactus Journal 2: 50 (1899). Type: *Mammillaria halei* K. Brandegee. → Mammillaria

Cochiseia Earle, Saguaroland Bull. no. 30: 61, 64–66 (1976). Type: *Cochiseia robbinsorum* Earle. → Escobaria

Coleocephalocereus Backeberg, Blätt. Kakt.-Forsch. 1938(6): [22] (1938). Type: *C. fluminensis* (Miquel) Backeberg. Taxa: 6+2. Distr.: E Brazil. Syn.: *Buiningia* Buxbaum (1971).

Coloradoa Boissevain & Davidson, Colorado Cacti, 54 (1940). Type: *C. mesae-verdae* Boissevain & Davidson. → Sclerocactus

Consolea Lemaire, Rev. Hort. 1862: 174 (1862). Type: *Opuntia rubescens* Salm-Dyck. → Opuntia

Copiapoa Britton & Rose, Cact. 3: 85 (1922). Type: *C. marginata* (Salm-Dyck) Britton & Rose. Taxa: 20+9. Distr.: N Chile. Syn.: *Pilocopiapoa* Ritter (1961).

Corryocactus Britton & Rose, Cact. 2: 66 (1920). Type: *C. brevistylus* (Schumann) Britton & Rose. Taxa: 12+23. Distr.: Bolivia, Peru, N Chile. Syn.: *Erdisia* Britton & Rose (1920).

Corynopuntia F. Knuth in Backeberg & Knuth, Kaktus ABC, 114, 410 (1936). Type: *Opuntia clavata* Engelmann. → Opuntia

Coryphantha (Engelmann) Lemaire, Les Cactées, 32 (1868). Type: *C. sulcata* (Engelmann) Britton & Rose (LT: Hunt & Benson 1976). Taxa: 41+16. Distr.: SW USA, Mexico. Syn.: *Lepidocoryphantha* Backeberg (1938); *Cumarinia* Buxbaum (1951).

Cryptocereus Alexander, Cact. Succ. J. (US) 22: 164 (1950). Type: *C. anthonyanus* Alexander. → Selenicereus

Cullmannia Distefano, Kakt. and. Sukk. 7: 8 (1956). Type: *Cereus viperinus* Weber. → Peniocereus

Cumarinia Buxbaum, Oesterr. Bot. Zeitschr. 98: 61 (1951). Type: *Coryphantha odorata* Bödeker. → Coryphantha

Cumulopuntia Ritter, Kakteen in Südamerika 2: 399 (1980). Type: *Opuntia ignescens* Vaupel. → Opuntia

Cylindropuntia (Engelmann) F. Knuth in Backeberg & Knuth, Kaktus ABC, 117 (1936). Type: *Opuntia imbricata* (Haworth) De Candolle. → Opuntia

Deamia Britton & Rose, Cact. 2: 212 (1920). Type: *Cereus testudo* Karwinsky ex Zuccarini. → Selenicereus

Delaetia Backeberg, Die Cact. 6: 3788 (1962). Type: *Delaetia woutersiana* Backeberg, *nom. inval.* → Eriosyce

Dendrocereus Britton & Rose, Cact. 2: 113 (1920). Type: *Cereus nudiflorus* Engelmann ex Sauvalle. Taxa: 1+1. Distr.: Cuba, Haiti. ~ Acanthocereus

Denmoza Britton & Rose, Cact. 3: 78 (1922). Type: *Echinocactus rhodacanthus* Salm-Dyck. Taxa: 1. Distr.: NW Argentina.

Digitorebutia Buining, Succulenta 22: 51 (1940). Type: *Rebutia haagei* Fric & Schelle. → Rebutia

Discocactus Pfeiffer, Allg. Gartenz. 5: 241 (1837). Type: *D. insignis* Pfeiffer. Taxa: 6+4. Distr.: E Bolivia, Brazil, N Paraguay.

Disocactus Lindley, Bot. Reg. 31: t. 9 (1845). Type: *D. biformis* (Lindley) Lindley. Taxa: 16. Distr.: S Mexico, Central America, Caribbean, N & W South America. Syn.: *Aporocactus* Lemaire (1860); *Wittia* Schumann (1903) non Pantocsek; *Heliocereus* (Berger) Britton & Rose (1909); *Mediocactus* Britton & Rose (1920), as to type; *Nopalxochia* Britton & Rose (1923); *Chiapasia* Britton & Rose (1923); *Lobeira* Alexander (1944); *Bonifazia* Standley & Steyermark (1944); *Pseudonopalxochia* Backeberg (1958); *Wittiocactus* Rauschert (1982).

Dolichothele (Schumann) Britton & Rose, Cact. 4: 61 (1923). Type: *Mammillaria longimamma* De Candolle. → Mammillaria

Eccremocactus Britton & Rose, Contr. US Nat. Herb. 16: 261 (1913). Type: *E. bradei* Britton & Rose. → Weberocereus

Echinocactus Link & Otto, Verh. Ver. Beförd. Gartenb. Preuss. Staaten, 420 (1827). Type: *E. platyacanthus* Link & Otto. Taxa: 6+1. Distr.: SW USA, Mexico. Syn.: *Homalocephala* Britton & Rose (1922); *Echinofossulocactus* Lawrence (1841) non Britton & Rose; *Emorycactus* Doweld (1996); *Meyerocactus* Doweld (1996).

Echinocereus Engelmann in Wislizenus, Mem. Tour N. Mex., 91 (1848). Type: *E. viridiflorus* Engelmann (LT: Britton & Rose). Taxa: 54+57. Distr.: SW USA, Mexico. Syn.: *Wilcoxia* Britton & Rose (1909); *Morangaya* Rowley (1974). Ref.: Taylor (1985, 1988, 1989, 1994).

Echinofossulocactus Lawrence in Loudon, Gard. Mag. 17: 317 (1841). Type: *E. helophorus* (Lemaire) Lawrence (LT: Hunt 1980). → Echinocactus

Echinofossulocactus sensu Britton & Rose, Cact. 3: 109 (1922). Type: *E. coptonogonus* Lemaire (LT: Britton & Rose). → Stenocactus

Echinomastus Britton & Rose, Cact. 3: 147 (1922). Type: *Echinocactus erectocentrus* J. Coulter. → Sclerocactus

Echinopsis Zuccarini, Abh. Bayer. Akad. Wiss. München 2: 675 (1837). Type: *E. eyriesii* (Turpin) Zuccarini (LT: Britton & Rose 1922). Taxa: 61+85. Distr.: Bolivia, Peru, S Brazil, S South America. Syn.: *Trichocereus* (Berger) Riccobono (1909); *Chamaecereus* Britton & Rose (1922); *Lobivia* Britton & Rose (1922); *Soehrensia* Backeberg (1938); *Hymenorebutia* Fric ex Buining (1939); *Setiechinopsis* (Backeberg) De Haas (1940); *Acantholobivia* Backeberg (1942); *Pseudolobivia* (Backeberg) Backeberg (1942); *Reicheocactus* Backeberg (1942); *Helianthocereus* Backeberg (1949); *Neolobivia* Y. Ito, (1950); *Leucostele* Backeberg (1953).

Emorycactus Doweld, Succulenta 75(6): 270 (1996). Type: *Echinocactus polycephalus* Engelmann & Bigelow. → Echinocactus

Encephalocarpus Berger, Kakteen, 331, 332 (1929). Type: *Ariocarpus strobiliformis* Werdermann. → Pelecyphora

Eomatucana Ritter, Kakt. and. Sukk. 16: 230 (1965). Type: *E. oreodoxa* Ritter. → Matucana

Epiphyllanthus Berger, Rep. Missouri Bot. Gard. 16: 84 (1905). Type: *Cereus obtusangulus* Schumann. → Schlumbergera

Epiphyllopsis (Berger) Backeberg & Knuth, Kaktus ABC, 158 (1936). Type: *Epiphyllum gaertneri* Schumann. → Hatiora

Epiphyllum Haworth, Syn. Pl. Succ., 197 (1812). Type: *E. phyllanthus* (Linnaeus) Haworth. Taxa: 8+11. Distr.: Trop. America, Caribbean. Syn.: *Phyllocactus* Link (1829).

Epiphyllum Pfeiffer, Enum. Cact., 123 (1837), non Haworth. Type: *Epiphyllum truncatum* Haworth. → Schlumbergera

Epithelantha Weber ex Britton & Rose, Cact. 3: 92 (1922). Type: *Mammillaria micromeris* Engelmann. Taxa: 1+5. Distr.: SW USA (Texas), NE Mexico. → Epithelantha. Ref.: Glass & Foster (1978).

Erdisia Britton & Rose, Cact. 2: 104 (1920). Type: *Cereus squarrosus* Vaupel. → Corryocactus

Eriocactus Backeberg, Cact. Jahrb. Deutsch. Kakt. Ges. 1941: 76 (1942). Type: *Echinocactus schumannianus* Nicolai. → Parodia

Eriocephala Backeberg, Blätt. Kakt.-Forsch. 1938(6): [7,21] (1938), non *Eriocephalus* Linnaeus. → Parodia

Eriocereus Riccobono, Boll. R. Ort. Palermo 8: 238 (1909). Type: *Cereus gracilis* Miller. → Harrisia

Eriosyce Philippi, Anal. Univ. Chile 41: 721 (1872). Type: *Echinocactus sandillon* Gay (→ *Eriosyce aurata* (Pfeiffer) Backeberg; *E. ceratistes* sensu Britton & Rose). Taxa: 34+14. Distr.: Chile. Syn.: *Neoporteria* Britton & Rose (1922); *Islaya* Backeberg (1934); *Pyrrhocactus* Backeberg & Knuth (1936); *Horridocactus* Backeberg (1938); *Neochilenia* Backeberg ex Doelz (1942); *Thelocephala* Y. Ito (1957); *Rodentiophila* Backeberg (1959); *Delaetia* Backeberg (1962).

Erythrorhipsalis Berger, Monatsschr. Kakt. 30: 4 (1920). Type: *Rhipsalis pilocarpa* Loefgren. → Rhipsalis

Escobaria Britton & Rose, Cact. 4: 53 (1923). Type: *E. tuberculosa* (Engelmann) Britton & Rose. Taxa: 18+12. Distr.: W USA, N Mexico. Syn.: *Neobesseya* Britton & Rose (1923); *Cochiseia* Earle (1976). Ref.: Taylor (1986).

Escontria Rose, Contr. US Nat. Herb. 10: 125 (1906). Type: *Cereus chiotilla* Weber. Taxa: 1. Distr.: Mexico

Espostoa Britton & Rose, Cact. 2: 60 (1920). Type: *E. lanata* (Kunth) Britton & Rose. Taxa: 9+7. Distr.: W South America. Syn.: *Binghamia* Britton & Rose (1920); *Pseudoespostoa* Backeberg (1934); *Thrixanthocereus* Backeberg (1937); *Vatricania* Backeberg (1950).

×***Espostocactus*** Mottram, Contrib. New Class. Cact. Fam., 38 (1990). (*Cleistocactus* × *Espostoa*). Taxa: 0+1. Distr.: Peru.

Espostoopsis Buxbaum, Krainz, Die Kakteen Lfg. 38–39 (1968). Type: *Cereus dybowskii* Roland-Gosselin. Taxa: 1. Distr.: E Brazil (Bahia). Syn.: *Gerocephalus* Ritter (1968).

Eulychnia Philippi, Florula Atacamensis, 23 (1860). Type: *E. breviflora* Philippi. Taxa: 5+2. Distr.: Chile, S Peru. Syn.: *Philippicereus* Backeberg (1942).

Facheiroa Britton & Rose, Cact. 2: 173 (1920). Type: *F. pubiflora* Britton & Rose. Taxa: 3+1. Distr.: Brazil. Syn.: *Zehntnerella* Britton & Rose (1920).

Ferocactus Britton & Rose, Cact. 3: 123 (1922). Type: *F. wislizeni* (Engelmann) Britton & Rose. Taxa: 23+16. Distr.: SW USA, Mexico. Syn.: *Bisnaga* Orcutt (1926). Ref.: Taylor (1984).

Floribunda Ritter, Kakteen in Südamerika 1: 58 (1979). Type: *Floribunda pusilliflora* Ritter. → Cipocereus

Frailea Britton & Rose, Cact. 3: 208 (1922). Type: *F. cataphracta* (Dams) Britton & Rose. Taxa: 11+14. Distr.: E Bolivia, S Brazil, Paraguay, Uruguay, N Argentina. ~ Parodia

Geohintonia Glass & Fitz Maurice, Cact. Suc. Mex. 37(1): 16–19 (1991). Type: *Geohintonia mexicana* Glass & Fitz Maurice. Taxa: 1. Distr.: NE Mexico.

Gerocephalus Ritter, Kakt. and. Sukk. 19: 156 (1968). Type: *Cereus dybowskii* Roland-Gosselin. → Espostoopsis

Glandulicactus Backeberg, Blätt. Kakt.-Forsch. 1938(6): [10, 22] (1938). Type: *Echinocactus uncinatus* Galeotti. → Sclerocactus

Grusonia F. Reichenbach ex Britton & Rose, Cact. 1: 215 (1919). Type: *Cereus bradtiana* J. Coulter. → Opuntia

Gymnanthocereus Backeberg, Blätt. Kakt.-Forsch. 1937(7): Nachtr. 15 (1937). Type: *Cactus chlorocarpus* Kunth. → Browningia

Gymnocactus Backeberg, Blätt. Kakt.-Forsch. 1938(6): [10, 22] (1938). Type: *Echinocactus saueri* Bödeker. → Turbinicarpus

Gymnocalycium Pfeiffer ex Mittler, Taschenbuch für Cactusliebhaber 2: 124 (1844) ('Gymnocalicium'); Pfeiffer, Abbild. Beschr. Cact. 2: (1845). Type: *G. denudatum* (Link & Otto) Pfeiffer (LT: Britton & Rose 1922). Taxa: 42+48. Distr.: Bolivia, S Brazil, S South America. Syn.: *Brachycalycium* Backeberg (1942).

Gymnocereus Backeberg, Die Cact. 2: 920 (1959). Type: *Cereus microspermus* Werdermann & Backeberg. → Browningia

Haageocereus Backeberg, Blätt. Kakt.-Forsch. 1934(6): [1] (1934). Type: *H. pseudomelanostele* (Werdermann & Backeberg) Backeberg. Taxa: 13+12. Distr.: Bolivia, Peru, N Chile. Syn.: *Peruvocereus* Akers (1947).

×***Haagespostoa*** Rowley, Nat. Cact. Succ. J. 37(3): 76 (1982). (*Espostoa* × *Haageocereus*). Taxa: 0+2. Distr.: Peru.

Hamatocactus Britton & Rose, Cact. 3: 104 (1922). Type: *Echinocactus setispinus* Engelmann. → Thelocactus

Harrisia Britton, Bull. Torrey Bot. Club 35: 561 (1908). Type: *H. gracilis* (Miller) Britton. Taxa: 14+7. Distr.: SE USA (Florida), Caribbean, Brazil, Bolivia, Paraguay, Uruguay, Argentina. Syn.: *Eriocereus* Riccobono (1909); *Roseocereus* Backeberg (1938).

Haseltonia Backeberg, Blätt. Sukk.-Kunde 1: 3 (1949). Type: *Pilocereus hoppenstedtii* Weber. → Cephalocereus

Hatiora Britton & Rose in Bailey, Standard Cycl. Hort. 3: 1432 (1915). Type: *Rhipsalis salicornioides* Haworth. Taxa: 5+2. Distr.: NE to S Brazil. Syn.: *Rhipsalidopsis* Britton & Rose (1923); *Epiphyllopsis* Backeberg & Knuth (1936); *Pseudozygocactus* Backeberg (1938). Ref.: Barthlott (1987), Barthlott & Taylor (1995).

Heliabravoa Backeberg, Cact. Succ. J. Gt. Brit. 18: 23 (1956). Type: *Cereus chende* Roland-Gosselin. → Polaskia

Helianthocereus Backeberg, Cact. Succ. J. Gt. Brit. 11: 53 (1949). Type: *Trichocereus poco* Backeberg. → Echinopsis

Heliocereus (Berger) Britton & Rose, Contr. US Nat. Herb. 12: 433 (1909). Type: *Cactus speciosus* Cavanilles. Distr.: Mexico, Guatemala. → Disocactus

Hertrichocereus Backeberg, Cact. Succ. J. (US) 22: 153 (1950). Type: *Cereus beneckei* Ehrenberg. → Stenocereus

Hildewintera Ritter, Kakt. and. Sukk. 17: 11 (1966). Type: *Winteria auresipina* Ritter. → Cleistocactus

Homalocephala Britton & Rose, Cact. 3: 181 (1922). Type: *Echinocactus texensis* Hopffer. → Echinocactus

Horridocactus Backeberg, Blätt. Kakt.-Forsch. 1938(6): [7] (1938). Type: *Cactus horridus* Colla. → Eriosyce

Hylocereus (Berger) Britton & Rose, Contr. US Nat. Herb. 12: 428 (1909). Type: *H. triangularis* (Linnaeus) Britton & Rose. Taxa: 11+7. Distr.: Mexico, Central and N South America, Caribbean. Syn.: *Wilmattea* Britton & Rose (1920).

Hymenorebutia Fric ex Buining, Succulenta 21: 101 (1939). Type: *H. kreuzingeri* Fric ex Buining. → Echinopsis

Islaya Backeberg, Blätt. Kakt.-Forsch. 1934(10): [3] (1934). Type: *Islaya minor* Backeberg. → Eriosyce

Isolatocereus (Backeberg) Backeberg, Cact. Jahrb. Deutsch. Kakt. Ges. 1941: 76 (1942). Type: *Cereus dumortieri* Scheidweiler. → Stenocereus

Jasminocereus Britton & Rose, Cact. 2: 146 (1920). Type: *J. galapagensis* (F.A.C. Weber) Britton & Rose (→ *J. thouarsii* (Weber) Backeberg). Taxa: 1. Distr.: Ecuador (Galapagos Isls.). Ref.: Wiggins & Porter, Fl. Galapagos Isls. 534–537 (1971).

Kadenicarpus Doweld, Sukkulenty (Moscow) 1998(1): 22 (1998). Type: *Kadenicarpus pseudomacrochele* (Backeberg) Doweld. → Turbinicarpus

Krainzia Backeberg, Blätt. Kakt.-Forsch. 1938(6): [11, 21] (1938). Type: *Neomammillaria longiflora* Britton & Rose. → Mammillaria

Lasiocereus Ritter, Succulenta 48(8): (1966). Type: *Lasiocereus rupicola* Ritter. Taxa: 2. Distr.: Peru. ~ Haageocereus?

Lemaireocereus Britton & Rose, Contr. US Nat. Herb. 12: 424 (1909). Type: *Cereus hollianus* Weber. → Pachycereus

Leocereus Britton & Rose, Cact. 2: 108 (1920). Type: *L. bahiensis* Britton & Rose. Taxa: 1. Distr.: E Brazil.

Lepidocoryphantha Backeberg, Blätt. Kakt.-Forsch. 1938(6): [10, 22] (1938). Type: *Mammillaria macromeris* Engelmann. → Coryphantha

Lepismium Pfeiffer, Allg. Gartenz. 3: 315 (1835). Type: *L. commune* Pfeiffer (LT: Britton & Rose). Taxa: 14+1. Distr.: Brazil, Paraguay, Peru, Bolivia, Uruguay, Argentina. Syn.: *Pfeiffera* Salm-Dyck (1845); *Acanthorhipsalis* (Schumann) Britton & Rose (1923); *Acanthorhipsalis* Kimnach (1983); *Lymanbensonia* Kimnach (1984). Ref.: Barthlott (1987), Barthlott & Taylor (1995).

Leptocereus (Berger) Britton & Rose, Contr. US Nat. Herb. 12: 433 (1909). Type: *L. assurgens* (Grisebach) Britton & Rose. Taxa: 4+11. Distr.: Caribbean. Syn.: *Neoabbottia* Britton & Rose (1921).

Leptocladodia Buxbaum, Oesterr. Bot. Zeitschr. 101: 601 (1954). Type: *Mammillaria elongata* De Candolle. → Mammillaria

Leuchtenbergia Hooker, Curtis's Bot. Mag. 74: t.4393 (1848). Type: *L. principis* Hooker. Taxa: 1. Distr.: N Mexico.

Leucostele Backeberg, Kakt. and. Sukk. 4: 40 (1953). Type: *L. rivierei* Backeberg. → Echinopsis

Lobeira Alexander, Cact. Succ. J. (US) 16: 177 (1944). Type: *L. macdougallii* Alexander. → Disocactus

Lobivia Britton & Rose, Cact. 3: 49 (1922). Type: *Echinocactus pentlandii* Hooker. → Echinopsis

Lophocereus (Berger) Britton & Rose, Contr. US Nat. Herb. 12: 426 (1909). Type: *Cereus schottii* Engelmann. → Pachycereus

Lophophora J. Coulter, Contr. US Nat. Herb. 3: 131 (1894). Type: *L. williamsii* (Lemaire) J. Coulter. Taxa: 2. Distr.: SW USA (S Texas), N Mexico.

Loxanthocereus Backeberg, Cact. Jahrb. Deutsch. Kakt. Ges. 1937: 24 (1937). Type: *Cereus acanthurus* Vaupel. → Cleistocactus

Lymanbensonia Kimnach, Cact. Succ. J. (US) 56: 101 (1984). Type: *L. micrantha* (Vaupel) Kimnach. → Lepismium

Machaerocereus Britton & Rose, Cact. 2: 114 (1920). Type: *Cereus eruca* Brandegee. → Stenocereus

Maihuenia (Philippi ex F.A.C. Weber) Schumann, Gesamtb. Kakt., 754 (1898). Type: *Opuntia poeppigii* Otto ex Pfeiffer (LT). Taxa: 2. Distr.: Argentina and Chile. Ref.: Leuenberger (1997).

Maihueniopsis Spegazzini, An. Soc. Sci. Argent. 99: 86 (1925). Type: *Maihueniopsis molfinoi* Spegazzini. → Opuntia

Malacocarpus Salm-Dyck, Cact. Hort. Dyck. 1849: 24, 141 (1850), non Fischer & Meyer. Type: *Echinocactus corynodes* Pfeiffer. → Parodia

Mamillopsis Morren ex Britton & Rose, Cact. 4: 19 (1923). Type: *Mammillaria senilis* Loddiges. → Mammillaria

Mammillaria Haworth, Syn. Pl. Succ., 177 (1812), *nom. cons.* Type: *M. simplex* Haworth (→ *Cactus mammillaris* Linnaeus; *M. mammillaria* (Linnaeus) Karsten). Taxa: 145+119. Distr.: SW USA, Mexico, Central America, Caribbean, Colombia, Venezuela. Syn.: *Cactus* Linnaeus (1753); *Cochemiea* (K. Brandegee) Walton (1899); *Bartschella* Britton & Rose (1923); *Dolichothele* (Schumann) Britton & Rose (1923); *Mamillopsis* Morren ex Britton & Rose (1923); *Phellosperma* Britton & Rose (1923); *Solisia* Britton & Rose (1923); *Neomammillaria* Britton & Rose (1923); *Porfiria* Bödeker (1926); *Chilita* Orcutt (1926); *Krainzia* Backeberg (1938); *Oehmea* Buxbaum (1951); *Pseudomammillaria* Buxbaum (1951); *Leptocladodia* Buxbaum (1954). Ref.: Hunt (1983–87, 1989–92, 1996–97).

Mammilloydia Buxbaum, Bot. Stud. 12: 64,65 (1951). Type: *Mammillaria candida* Scheidweiler. Taxa: 1. Distr.: N Mexico. → Mammillaria

Marenopuntia Backeberg, Desert Plant Life 22(3): 27 (1950). Type: *Opuntia marenae* Parsons. → Opuntia

Marginatocereus (Backeberg) Backeberg, Cact. Jahrb. Deutsch. Kakt. Ges. 1941: 77 (1942). Type: *Cereus marginatus* De Candolle. → Pachycereus

Maritimocereus Akers & Buining, Succulenta 1950: 49 (1950). Type: *M. gracilis* Akers & Buining. → Cleistocactus

Marniera Backeberg, Cact. Succ. J. (US) 22: 153 (1950). Type: *Phyllocactus macropterus* Lemaire. → Selenicereus

Marshallocereus Backeberg, Cact. Succ. J. (US) 22: 154 (1950). Type: *Cereus aragonii* Weber. → Stenocereus

Matucana Britton & Rose, Cact. 3: 102 (1922). Type: *Echinocactus haynei* Otto. Taxa: 15+6. Distr.: Peru. Syn.: *Submatucana* Backeberg (1959); *Eomatucana* Ritter (1965). Ref.: Bregman et al. (1986–90); Bregman (1996).

Mediocactus Britton & Rose, Cact. 2: 210 (1920). Type: *Cereus coccineus* Salm-Dyck (→ Disocactus; cf. Hunt 1989); *Mediocactus* sensu Britton & Rose. → Selenicereus

Mediolobivia Backeberg, Blätt. Kakt.-Forsch. 1934(2): [3] (1934). Type: *Rebutia aureiflora* Backeberg. → Rebutia

Melocactus Link & Otto, Verh. Ver. Beförd. Gartenb. Preuss. Staaten, 417 (1827), *nom. cons.* Type: *Cactus melocactus* Linnaeus (= *M. caroli-linnaei* Taylor). Taxa: 29+16. Distr.: Mexico, Central America, Caribbean, N & W South America, E Brazil. Syn.: *Cactus* sensu Britton & Rose (1922). Ref.: Taylor (1991).

Meyerocactus Doweld, Succulenta 75(6): 271 (1996). Type: *Echinocactus horizonthalonius* Lemaire. → Echinocactus

Micranthocereus Backeberg, Blätt. Kakt.-Forsch. 1938(6): [22] (1938). Type: *Cereus polyanthus* Werdermann. Taxa: 9. Distr.: E Brazil. Syn.: *Austrocephalocereus* Backeberg (1938); *Siccobaccatus* Braun & Esteves (1990).

Mila Britton & Rose, Cact. 3: 211 (1922). Type: *M. caespitosa* Britton & Rose. Taxa: 1+3. Distr.: Peru. Ref.: Donald (1978).

Mirabella Ritter, Kakteen in Südamerika 1: 108 (1979). Type: *Acanthocereus albicaulis* Britton & Rose. → Cereus

Mitrocereus (Backeberg) Backeberg, Cact. Jahrb. Deutsch. Kakt. Ges. 1941: 77 (1942). Type: *Pilocereus chrysomallus* Lemaire. → Pachycereus

Monvillea Britton & Rose, Cact. 2: 21 (1920). Type: *M. cavendishii* (Monville) Britton & Rose (→ Acanthocereus); *Monvillea* sensu Britton & Rose → Praecereus

Morangaya Rowley, Ashingtonia 1: 44 (1974). Type: *Cereus pensilis* K. Brandegee. → Echinocereus

Morawetzia Backeberg, Cact. Jahrb. Deutsch. Kakt. Ges. 1936: 73 (1936). Type: *M. doelziana* Backeberg. → Oreocereus

×***Myrtgerocactus*** Moran, Cact. Succ. J. (US) 34: 186 (1962). (*Bergerocactus* × *Myrtillocactus*). Taxa: 1. Distr.: Mexico (Baja California).

Myrtillocactus Console, Boll. R. Ort. Bot. Palermo 1: 8 (1897). Type: *M. geometrizans* (Martius) Console. Taxa: 4. Distr.: Mexico.

Navajoa Croizat, Cact. Succ. J. (US) 15: 89 (1943). Type: *Navajoa peeblesiana* Croizat. → Pediocactus

Neoabbottia Britton & Rose, Smiths. Misc. Coll. 72: 2 (1921). Type: *Cactus paniculatus* Lamarck. → Leptocereus

Neobesseya Britton & Rose, Cact. 4: 51 (1923). Type: *Mammillaria missouriensis* Sweet. → Escobaria

Neobinghamia Backeberg, Cact. Succ. J. (US) 22(5): 154 (1950), gen. hybr. Type: *Binghamia climaxantha* Werdermann. → ×Haagespostoa

Neobuxbaumia Backeberg, Blätt. Kakt.-Forsch. 1938(6): [8,21] (1938). Type: *Pilocereus tetetzo* Weber. Taxa: 9. Distr.: Mexico. Syn.: *Rooksbya* (Backeberg) Backeberg (1959).

Neocardenasia Backeberg, Blätt. Sukk.-Kunde 1: 2 (1949). Type: *Neocardenasia herzogiana* Backeberg. → Neoraimondia

Neochilenia Backeberg ex Doelz, Fedde's Repert. Sp. Nov. 51: 60 (1942). Type: *Echinocactus jussieui* Monville. → Eriosyce

Neodawsonia Backeberg, Blätt. Sukk.-Kunde 1: 4 (1949). Type: *Cephalocereus apicicephalium* Dawson. → Cephalocereus

Neoevansia Marshall in Marshall & Bock, Cactaceae, 84 (1941). Type: *Cereus diguetii* Weber. → Peniocereus

Neogomesia Castañeda, Cact. Succ. J. (US) 13: 98 (1941). Type: *N. agavoides* Castañeda. → Ariocarpus

Neolloydia Britton & Rose, Bull. Torrey Bot. Club 49: 251 (1922). Type: *N. conoidea* (De Candolle) Britton & Rose. Taxa: 1+1. Distr.: SW USA (S Texas), N Mexico. Ref.: Anderson (1986).

Neolobivia (Backeberg) Y. Ito, Bull. Takarazuka Insectarium 71: 18 (1950). → Echinopsis

Neomammillaria Britton & Rose, Cact. 4: 65 (1923). Type: *Mammillaria simplex* Haworth. → Mammillaria

Neoporteria Britton & Rose, Cact. 3: 94 (1922). Type: *N. subgibbosa* (Haworth) Britton & Rose. Distr.: S Peru, NW Argentina, Chile. → Eriosyce

Neoraimondia Britton & Rose, Cact. 2: 181 (1920). Type: *N. macrostibas* (Schumann) Britton & Rose. Taxa: 2+1. Distr.: Bolivia, Peru. Syn.: *Neocardenasia* Backeberg (1949).

Neowerdermannia Fric, Kaktusar 1(11): 85 (1930). Type: *N. vorwerkii* Fric. Taxa: 2+1. Distr.: Bolivia, N Argentina, N Chile.

Nopalea Salm-Dyck, Cact. Hort. Dyck. 1849: 63,233 (1850). Type: *Cactus cochenillifer* Linnaeus. → Opuntia

Nopalxochia Britton & Rose, Cact. 4: 204 (1923). Type: *Cactus phyllanthoides* De Candolle. → Disocactus

Normanbokea Kladiwa & Buxbaum, in Krainz, Die Kakteen Lfg. 40–41 (1969). Type: *Pelecyphora valdeziana* Moeller. → Turbinicarpus

Notocactus (Schumann) Fric, Price list for 1928 [unpaged] (1928). → Parodia

Nyctocereus (Berger) Britton & Rose, Contr. US Nat. Herb. 12: 423 (1909). Type: *Cereus serpentinus* (Lagasca & Rodriguez) De Candolle. → Peniocereus

Obregonia Fric, Zivot v Prirode 29(2): 1–4 (1925). Type: *O. denegrii* Fric. Taxa: 1. Distr.: NE Mexico.

Oehmea Buxbaum, Sukkulentenkunde 4: 17 (1951). Type: *Neomammillaria nelsonii* Britton & Rose. → Mammillaria

Opuntia Miller, Gard. Dict. Abr. ed. 4, [] (1754). Type: *Cactus opuntia* Linnaeus (LT: Britton & Brown 1913). Taxa: 161+195. Distr.: S Canada, USA, Mexico, Caribbean, Central & South America. Syn.: *Nopalea* Salm-Dyck (1850); *Consolea* Lemaire (1862); *Tephrocactus* Lemaire (1868); *Grusonia* F. Reichenbach ex Britton & Rose (1919); *Maihueniopsis* Spegazzini (1925); *Brasiliopuntia* (Schumann) Berger (1926); *Cylindropuntia* (Engelmann) F. Knuth (1936); *Corynopuntia* F. Knuth (1936); *Austrocylindropuntia* Backeberg (1938); *Marenopuntia* Backeberg (1950); *Platyopuntia* (Engelmann) Ritter (1979); *Cumulopuntia* Ritter (1980); *Puna* Kiesling (1982).

Oreocereus (Berger) Riccobono, Boll. R. Ort. Palermo 8: 258 (1909). Type: *O. celsianus* (Lemaire) Riccobono. Taxa: 5+4. Distr.: Bolivia, Peru, NW Argentina, N Chile. Syn.: *Arequipa* Britton & Rose (1922); *Morawetzia* Backeberg (1936); *Arequipiopsis* Kreuzinger & Buining (1941).

Oroya Britton & Rose, Cact. 3: 102 (1922). Type: *Echinocactus peruvianus* Schumann. Taxa: 1+1. Distr.: Peru.

Ortegocactus Alexander, Cact. Succ. J. (US) 33: 39 (1961). Type: *Ortegocactus macdougallii* Alexander. Taxa: 1. Distr.: S Mexico.

×***Pacherocactus*** Rowley, Nat. Cact. Succ. J. 37(3): 78 (1982). (*Bergerocactus* × *Pachycereus*). Taxa: 1. Distr.: Mexico (Baja California).

Pachycereus (Berger) Britton & Rose, Contr. US Nat. Herb. 12: 420 (1909). Type: *P. pringlei* (S. Watson) Britton & Rose. Taxa: 12. Distr.: SW USA (S Arizona), Mexico, Guatemala, Honduras. Syn.: *Lemaireocereus* Britton & Rose (1909); *Lophocereus* (Berger) Britton & Rose (1909); *Anisocereus* Backeberg (1938); *Marginatocereus* (Backeberg) Backeberg (1942); *Mitrocereus* (Backeberg) Backeberg (1942); *Backebergia* Bravo (1953); *Pterocereus* MacDougall & Miranda (1954); *Pseudomitrocereus* Bravo & Buxbaum (1961).

Parodia Spegazzini, An. Soc. Cient. Argent. 96: 70 (1923). Type: *P. microsperma* (F.A.C. Weber) Spegazzini. Taxa: 60+29. Distr.: Brazil, Paraguay, Bolivia, Uruguay, Argentina. Syn.: *Malacocarpus* Salm-Dyck (1850) non Fischer & Meyer; *Notocactus* (Schumann) Fric (1928); *Eriocephala* Backeberg (1938); *Brasilicactus* Backeberg (1942); *Eriocactus* Backeberg (1942); *Wigginsia* D.M. Porter (1964); *Brasiliparodia* Ritter (1979).

Pediocactus Britton & Rose, Britton & Brown, Ill. Fl. ed. 2, 2: 569 (1913). Type: *P. simpsonii* (Engelmann) Britton & Rose. Taxa: 6+10. Distr.: SW USA. Syn.: *Utahia* Britton & Rose (1922); *Navajoa* Croizat (1943); *Pilocanthus* B.W. Benson & Backeberg (1957).

Pelecyphora Ehrenberg, Bot. Zeit. 1: 737 (1843). Type: *P. aselliformis* Ehrenberg. Taxa: 2. Distr.: N Mexico. Syn.: *Encephalocarpus* Berger (1929).

Peniocereus (Berger) Britton & Rose, Contr. US Nat. Herb. 12: 428 (1909). Type: *P. greggii* (Engelmann) Britton & Rose. Taxa: 13+5. Distr.: SW USA, Mexico and Central America. Syn.: *Nyctocereus* (Berger) Britton & Rose (1909); *Neoevansia* Marshall (1941); *Cullmannia* Distefano (1956). Ref.: Sánchez-Mejorada (1974).

Pereskia Miller, Gard. Dict. Abr. ed. 4, [unpaged] (1754). Type: *P. aculeata* Miller. Taxa: 16+3. Distr.: S Mexico, Cent. America, Caribbean, trop. South America. Syn.: *Rhodocactus* (Berger) F. Knuth (1936). Ref.: Leuenberger (1986).

Pereskiopsis Britton & Rose, Smiths. Misc. Coll. 50: 331 (1907). Type: *P. porteri* (K. Brandegee) Britton & Rose. Taxa: 6+1. Distr.: Mexico, Guatemala.

Peruvocereus Akers, Cact. Succ. J. (US) 19: 67 (1947). Type: *Peruvocereus salmonoideus* Akers. → Haageocereus

Pfeiffera Salm-Dyck, Cact. Hort. Dyck. 1844: 40 (1845). Type: *P. cereiformis* Salm-Dyck (→ *P. ianthothele* (Monville) Weber). → Lepismium

Phellosperma Britton & Rose, Cact. 4: 60 (1923). Type: *Mammillaria tetrancistra* Engelmann. → Mammillaria

Philippicereus Backeberg, Cact. Jahrb. Deutsch. Kakt. Ges. 1941: 75 (1942). Type: *Cereus castaneus* Schumann. → Eulychnia

Phyllocactus Link, Handb. Erkenn. Gewächse 2: 10 (1829). Type: *Cactus phyllanthus* Linnaeus. → Epiphyllum

Pierrebraunia Esteves Pereira, Cact. Succ. J. (US) 69: 296–302 (1997). Type: *Pierrebraunia bahiensis* (Braun & Esteves Pereira) Esteves Pereira. → Arrojadoa

Pilocanthus B.W. Benson & Backeberg, Kakt. and. Sukk. 8: 187 (1957). Type: *Pediocactus paradinei* B.W. Benson. → Pediocactus

Pilocereus Lemaire, Cact. Gen. Nov. Sp. Hort. Monv., 6 (1839). Type: *Cereus senilis* Haworth. → Cephalocereus

Pilocereus Schumann, Engler & Prantl, Nat. Pflanzenfam. 3, 6A: 179 (1894), non Lemaire. Type: *Pilocereus leucocephalus* Lemaire. → Pilosocereus

Pilocopiapoa Ritter, Kakt. and. Sukk. 13: 20 (1961). Type: *Pilocopiapoa solaris* Ritter. → Copiapoa

Pilosocereus Byles & Rowley, Cact. Succ. J. Gr. Brit. 19: 66 (1957). Type: *Pilocereus leucocephalus* Poselger. Taxa: 34+9. Distr.: Mexico, Caribbean, N & W South America, E Brazil, Paraguay. Syn.: *Pilocereus* Schumann (1894), non Lemaire; *Pseudopilocereus* Buxbaum (1968).

Piptanthocereus (Berger) Riccobono, Boll. R. Ort. Palermo 8: 255 (1909). Type: [?]. → Cereus

Platyopuntia (Engelmann) Ritter, Kakteen in Südamerika 1: 31 (1979). [Based on subgen. *Platopuntia* Engelmann.] → Opuntia

Polaskia Backeberg, Blätter f. Sukkulentenkunde 1: 4 (1949). Type: *P. chichipe* Backeberg. Taxa: 2. Distr.: Mexico. Syn.: *Chichipia* Backeberg (1950); *Heliabravoa* Backeberg (1956).

Porfiria Bödeker, Zeitschr. Sukkulentenk. 2: 210 (1926). Type: *Porfiria coahuilensis* Bödeker. → Mammillaria

Praecereus Buxbaum, Beitr. Biol. Pflanzen 44: 273 (1968). Type: *Cephalocereus smithianus*. Taxa: 2+4. Distr.: Trop. S America. Syn.: *Monvillea* Britton & Rose (1920), excl. type.

Pseudoacanthocereus Ritter, Kakteen in Südamerika 1: 47 (1979). Type: *Acanthocereus brasiliensis* Britton & Rose. Taxa: 2. Distr.: Colombia, Venezuela, Brazil.

Pseudoespostoa Backeberg, Blätt. Kakt.-Forsch. 1934(10): [1] (1934). Type: *Cephalocereus melanostele* Vaupel. → Espostoa

Pseudolobivia (Backeberg) Backeberg, Cact. Jahrb. Deutsch. Kakt. Ges. 1941: 76 (1942). Type: *Echinopsis ancistrophora* Spegazzini. → Echinopsis

Pseudomammillaria Buxbaum, Bot. Stud. 12: 84 (1951). Type: *Mammillaria camptotricha* Dams. → Mammillaria

Pseudomitrocereus Bravo & Buxbaum, Bot. Stud. 12: 49, 53 (1961). Type: *Pilocereus fulviceps* Weber. → Pachycereus

Pseudonopalxochia Backeberg, Die Cact. 1: 69 (1958). Type: *Nopalxochia conzattiana* MacDougall. → Disocactus

Pseudopilocereus Buxbaum, Beitr. Biol. Pflanzen 44: 249 (1968). Type: *Pilocereus arrabidae* Lemaire. → Pilosocereus

Pseudorhipsalis Britton & Rose, Cact. 4: 213–214 (1923). Type: *Cactus alatus* Swartz. Taxa: 4+2. Distr.: Trop. America, Caribbean.

Pseudozygocactus Backeberg, Blätt. Kakt.-Forsch. 1938(6): [5, 21] (1938). Type: *Rhipsalis epiphylloides* Campos-Porto & Werdermann. → Hatiora

Pterocactus Schumann, Monatsschr. Kakt. 7: 6 (1897). Type: *P. kuntzei* Schumann. Taxa: 9. Distr.: Argentina. Ref.: Kiesling, (1982).

Pterocereus MacDougall & Miranda, Ceiba 4: 135 (1954). Type: *Pterocereus foetidus* MacDougall & Miranda. → Pachycereus

Puna Kiesling, Hickenia 1: 289 (1982). Type: *Opuntia clavarioides* Pfeiffer. → Opuntia

Pygmaeocereus Johnson & Backeberg, Nat. Cact. Succ. J. 12: 86 (1957). Type: *P. bylesianus* Andreae & Backeberg. Taxa: 1+2. Distr.: Peru. ~ Echinopsis

Pyrrhocactus Berger, Kakteen, 345 (1929). Type: [not designated by Berger] *Echinocactus strausianus* Schumann. → Eriosyce

Quiabentia Britton & Rose, Cact. 4: 252 (1923). Type: *Pereskia zehntneri* Britton & Rose. Taxa: 2. Distr.: E Brazil, W Paraguay, S Bolivia, NW Argentina.

Rapicactus Buxbaum & Oehme, Cact. Jahrb. Deutsch. Kakt. Ges. 1942: 24 (1942). Type: *Rapicactus subterraneus* Buxbaum & Oehme. → Turbinicarpus

Rathbunia Britton & Rose, Contr. US Nat. Herb. 12: 414 (1909). Type: *Cereus sonorensis* Runge. → Stenocereus

Rauhocereus Backeberg, Descr. Cact. Nov., 5 (1957). Type: *Rauhocereus riosaniensis* Backeberg. Taxa: 1+1. Distr.: Peru. ~ Browningia

Rebutia Schumann, Monatsschr. Kakteenk. 5: 102 (1895). Type: *R. minuscula* Schumann. Taxa: 24+28. Distr.: Bolivia, Peru, N Argentina. Syn.: *Aylostera* Spegazzini (1923); *Mediolobivia* Backeberg (1934); *Weingartia* Werdermann (1937); *Digitorebutia* Buining (1940); *Sulcorebutia* Backeberg (1951).

Reicheocactus Backeberg, Cact. Jahrb. Deutsch. Kakt. Ges. 1941: 76 (1942). Type: *R. pseudoreicheanus* Backeberg. → Echinopsis

Rhipsalidopsis Britton & Rose, Cact. 4: 209 (1923). Type: *Rhipsalis rosea* Lagerheim. → Hatiora

Rhipsalis Gaertner, Fruct. Sem. 1: 137 (1788). Type: *R. cassutha* Gaertner. Taxa: 35+16. Distr.: Trop. and subtrop. America, trop. Africa, Madagascar, Mascarenes, Sri Lanka. Syn.: *Erythrorhipsalis* Berger (1920). Ref.: Barthlott & Taylor (1995).

Rhodocactus (Berger) F. Knuth in Backeberg & Knuth, Kaktus ABC, 48, 96 (1936). Type: *Pereskia grandifolia* Haworth. → Pereskia

Ritterocereus Backeberg, Cact. Jahrb. Deutsch. Kakt. Ges. 1941: 76 (1942). Type: *Lemaireocereus standleyi* Gonzalez Ortega. → Stenocereus

Rodentiophila Backeberg, Die Cact. 3: 1799 (1959), *nom. inval.* → Eriosyce

Rooksbya (Backeberg) Backeberg, Die Cact. 3: 2165 (1959). → Neobuxbaumia

Roseocactus Berger, J. Wash. Acad. Sci. 15: 45 (1925). Type: *Ariocarpus fissuratus* Schumann. → Ariocarpus

Roseocereus Backeberg, Blätt. Kakt.-Forsch. 1938(6): [21] (1938). Type: *Cereus tetracanthus* Labouret (*R. tephracanthus* Backeberg). Ref.: Hunt (1987: 92). → Harrisia

Samaipaticereus Cárdenas, Cact. Succ. J. (US) 24: 141 (1952). Type: *S. corroanus* Cárdenas. Taxa: 1. Distr.: Bolivia.

Schlumbergera Lemaire, Illustr. Hort. 5: misc. 24 (1858). Type: *S. epiphylloides* Lemaire (→ *S. russelliana* (Gardner ex Hooker) Britton & Rose). Taxa: 6+4. Distr.: SE Brazil. Syn.: *Epiphyllum* Pfeiffer (1837); *Zygocactus* Schumann (1890); *Epiphyllanthus* Berger (1905). Ref.: McMillan & Horobin, Succulent Pl. Res. 4 (1995).

Sclerocactus Britton & Rose, Cact. 3: 212 (1922). Type: *Echinocactus polyancistrus* Engelmann & Bigelow. Taxa: 18+10. Distr.: SW USA, N Mexico. Syn.: *Toumeya* Britton & Rose (1922); *Echinomastus* Britton & Rose (1922); *Ancistrocactus* Britton & Rose (1923); *Glandulicactus* Backeberg (1938); *Coloradoa* Boissevain & Davidson (1940).

Selenicereus (Berger) Britton & Rose, Contr. US Nat. Herb. 12: 429 (1909). Type: *S. grandiflorus* (Linnaeus) Britton & Rose. Taxa: 18+9. Distr.: Trop. America, Caribbean. Syn.: *Strophocactus* Britton & Rose (1913); *Deamia* Britton & Rose (1920); *Cryptocereus* Alexander (1950); *Marniera* Backeberg (1950). Ref.: Hunt (1989).

Seticereus Backeberg, Cact. Jahrb. Deutsch. Kakt. Ges. 1941: 75 (1942). Type: *Cactus icosagonus* Kunth. → Cleistocactus

Seticleistocactus Backeberg, Descr. Cact. Nov. III, 13 (1963). Type: *Cleistocactus piraymirensis* Cárdenas. → Cleistocactus

Setiechinopsis (Backeberg) De Haas, Succulenta 22: 9 (1940). Type: *Echinopsis mirabilis* Spegazzini. → Echinopsis

Siccobaccatus Braun & Esteves Pereira, Succulenta 69: 6 (1990). Type: *Austrocephalocereus dolichospermaticus* Buining & Brederoo. → Micranthocereus

Soehrensia Backeberg, Blätt. Kakt.-Forsch. 1938(6): [21] (1938). Type: *Lobivia bruchii* Britton & Rose. → Echinopsis

Solisia Britton & Rose, Cact. 4: 64 (1923). Type: *Pelecyphora pectinata* Stein. → Mammillaria

Stenocactus (Schumann) A.W. Hill, Index Kewensis, suppl. 8: 228 (1933). Type: *Echinocactus coptonogonus* Lemaire (LT: Byles 1956). Taxa: 7+3. Distr.: Mexico. Syn.: *Echinofossulocactus* sensu Britton & Rose (1922) non Lawrence.

Stenocereus (Berger) Riccobono, Boll. R. Ort. Palermo 8: 253 (1909), *nom. cons.* Type: *S. stellatus* (Pfeiffer) Riccobono. Taxa: 19+6. Distr.: SW USA (Arizona), Mexico, Caribbean, N South America. Syn.: *Rathbunia* Britton & Rose (1909); *Machaerocereus* Britton & Rose (1920); *Isolatocereus* (Backeberg) Backeberg (1942); *Ritterocereus* Backeberg (1942); *Hertrichocereus* Backeberg (1950); *Marshallocereus* Backeberg (1950). Ref.: Gibson & Horak, (1979).

Stephanocereus Berger, Entwicklungslinien Kakt., 97 (1926). Type: *S. leucostele* (Guerke) A. Berger. Taxa: 2. Distr.: Brazil (Bahia).

Stetsonia Britton & Rose, Cact. 2: 64 (1920). Type: *Cereus coryne* Salm-Dyck. Taxa: 1. Distr.: Paraguay, Argentina, Bolivia.

Strombocactus Britton & Rose, Cact. 3: 106 (1922). Type: *S. disciformis* (De Candolle) Britton & Rose. Taxa: 1+1. Distr.: E Mexico (Hidalgo and Querétaro).

Strophocactus Britton & Rose, Contr. US Nat. Herb. 16: 262 (1913). Type: *Cereus wittii* Schumann. → Selenicereus

Submatucana Backeberg, Die Cact. 2: 1059 (1959). Type: *Echinocactus aurantiacus* Vaupel. → Matucana

Subpilocereus Backeberg, Blätt. Kakt.-Forsch. 1938(6): [22] (1938). Type: *Cereus russel(l)ianus* Otto non Gardner ex Hooker (→ *C. fricii*). → Cereus

Sulcorebutia Backeberg, Cact. Succ. J. Gt. Brit. 13: 96 (1951). Type: *Rebutia steinbachii* Werdermann. → Rebutia

Tacinga Britton & Rose, Cact. 1: 39 (1919). Type: *T. funalis* Britton & Rose. Taxa: 2. Distr.: E Brazil.

Tephrocactus Lemaire, Les Cactées, 88 (1868). Type: *Opuntia diademata* Lemaire [LT: see Byles 1957: 30]. → Opuntia

Thelocactus (Schumann) Britton & Rose, Bull. Torrey Bot. Club 49: 251 (1922). Type: *T. hexaedrophorus* (Lemaire) Britton & Rose. Taxa: 10+11. Distr.: SW USA (Texas), Mexico. Syn.: *Hamatocactus* Britton & Rose (1922); *Torreyocactus* Doweld (1998). Ref.: Anderson (1987).

Thelocephala Y. Ito, Expl. Diagr., 292 (1957). Type: *Neoporteria napina* Backeberg. → Eriosyce

Thrixanthocereus Backeberg, Blätter Kakteenforsch. 1937: Nachtr. 15 (1937). Type: *Cephalocereus blossfeldiorum* Werdermann. → Espostoa

Torreyocactus Doweld, Sukkulenty (Moscow) 1998(1): 19 (1998). Type: *Torreyocactus conothelos* (Regel & Klein) Doweld. → Thelocactus

Toumeya Britton & Rose, Cact. 3: 91 (1922). Type: *Mammillaria papyracantha* Engelmann. → Sclerocactus

Trichocereus (Berger) Riccobono, Boll. R. Ort. Palermo 8: 236 (1909). Type: *Cereus macrogonus* Otto. → Echinopsis

Turbinicarpus (Backeberg) Buxbaum & Backeberg, Cact. Jahrb. Deutsch. Kakt. Ges. 1937: 27 (1937). Type: *Echinocactus schmiedeckeanus* Bödeker. Taxa: 16+17. Distr.: Mexico. Syn.: *Gymnocactus* Backeberg (1938); *Rapicactus* Buxbaum & Oehme (1942); *Normanbokea* Kladiwa & Buxbaum (1969); *Bravocactus* Doweld (1998); *Kadenicarpus* Doweld (1998).

Uebelmannia Buining, Succulenta (NL) 46: 159 (1967). Type: *Parodia gummifera* Backeberg & Voll. Taxa: 3+2. Distr.: E Brazil (Minas Gerais).

Utahia Britton & Rose, Cact. 3: 215 (1922). Type: *Echinocactus sileri* Engelmann. → Pediocactus

Vatricania Backeberg, Cact. Succ. J. (US) 22: 154 (1950). Type: *Cephalocereus guentheri* Kupper. → Espostoa

Weberbauerocereus Backeberg, Cact. Jahrb. Deutsch. Kakt. Ges. 1941: 75 (1942). Type: *Cereus fascicularis* Meyen. Taxa: 7+1. Distr.: Peru. ~ Echinopsis

Weberocereus Britton & Rose, Contr. US Nat. Herb. 12: 431 (1909). Type: *W. tunilla* (F.A.C. Weber) Britton & Rose. Taxa: 9. Distr.: Mexico, Central America, W South America. Syn.: *Werckleocereus* Britton & Rose (1909); *Eccremocactus* Britton & Rose (1913).

Weingartia Werdermann, Kakteenkunde 1937: 20,21 (1937). Type: *Echinocactus fidaianus* Backeberg. → Rebutia

Werckleocereus Britton & Rose, Contr. US Nat. Herb. 12: 432 (1909). Type: *Cereus tonduzii* Weber. → Weberocereus
Wigginsia D.M. Porter, Taxon 13: 210 (1964). Type: *Echinocactus corynodes* Pfeiffer. → Parodia
Wilcoxia Britton & Rose, Contr. US Nat. Herb. 12: 434 (1909). Type: *Echinocereus poselgeri* Lemaire. → Echinocereus
Wilmattea Britton & Rose, Cact. 2: 195 (1920). Type: *Cereus minutiflorus* Vaupel. → Hylocereus
Winteria Ritter, Kakt. and. Sukk. 13: 4 (1962), non Murray. Type: *Winteria aureispina*. → Cleistocactus
Winterocereus Backeberg, Kakteenlexikon, 455 (1966). Type: *Winteria aureispina* Ritter. → Cleistocactus
Wittia Schumann, Monatsschr. Kakt. 13: 117 (1903), non Pantocsek. Type: *W. amazonica* Schumann. → Disocactus
Wittiocactus Rauschert, Taxon 31: 558 (1982) [*nom. nov.* for *Wittia* Schumann]. → Disocactus
Yungasocereus Ritter, Kakteen in Südamerika 2: 668 (1980). Type: *Samaipaticereus inquisivensis* Cárdenas. Taxa: 1. Distr.: Bolivia. ~ Cleistocactus
Zehntnerella Britton & Rose, Cact. 2: 176 (1920). Type: *Zehntnerella squamulosa* Britton & Rose. → Facheiroa
Zygocactus Schumann, Martius, Fl. Bras. 4(4): 224 (1890). Type: *Epiphyllum truncatum* Haworth. → Schlumbergera

2. Summary table of accepted and provisionally accepted taxa

Columns: (1) Accepted species; (2) Accepted heterotypic subspecies; (3) Provisionally accepted species and hybrids; (4) Provisionally accepted heterotypic subspecies; (5) Total accepted taxa.

Genus	(1)	(2)	(3)	(4)	(5)
Acanthocalycium	1	0	2	0	3
Acanthocereus	1	0	5	0	6
Ariocarpus	6	2	0	0	8
Armatocereus	10	2	3	0	15
Arrojadoa	4	1	1	0	6
Arthrocereus	4	2	0	0	6
Astrophytum	4	0	0	0	4
Austrocactus	3	0	2	0	5
Aztekium	2	0	0	0	2
Bergerocactus	1	0	0	0	1
Blossfeldia	1	0	0	0	1
Brachycereus	1	0	0	0	1
Brasilicereus	2	0	0	0	2
Browningia	5	0	6	0	11
Calymmanthium	1	0	0	0	1
Carnegiea	1	0	0	0	1
Cephalocereus	3	0	2	0	5
Cephalocleistocactus	0	0	1	0	1
Cereus	23	3	12	1	39
Cintia	1	0	0	0	1
Cipocereus	5	1	0	0	6
Cleistocactus	33	5	16	4	58
×*Cleistocana*	0	0	1	0	1
Coleocephalocereus	6	2	0	0	8
Copiapoa	20	4	5	0	29
Corryocactus	12	1	22	0	35
Coryphantha	41	3	13	0	57
Dendrocereus	1	0	1	0	2
Denmoza	1	0	0	0	1
Discocactus	6	3	1	0	10
Disocactus	16	0	0	0	16
Echinocactus	6	1	0	0	7
Echinocereus	54	32	7	8	101
Echinopsis	61	2	69	14	146
Epiphyllum	8	0	11	0	19
Epithelantha	1	4	1	0	6
Eriosyce	34	13	1	0	48
Escobaria	18	4	7	1	30

Genus	(1)	(2)	(3)	(4)	(5)
Escontria	1	0	0	0	1
Espostoa	9	0	7	0	16
×*Espostocactus*	0	0	1	0	1
Espostoopsis	1	0	0	0	1
Eulychnia	5	0	2	0	7
Facheiroa	3	1	0	0	4
Ferocactus	23	10	6	0	39
Frailea	11	3	6	5	25
Geohintonia	1	0	0	0	1
Gymnocalycium	42	2	28	18	90
Haageocereus	13	4	8	0	25
×*Haagespostoa*	0	0	2	0	2
Harrisia	14	0	6	1	21
Hatiora	5	1	1	0	7
Hylocereus	11	0	7	0	18
Jasminocereus	1	0	0	0	1
Lasiocereus	2	0	0	0	2
Leocereus	1	0	0	0	1
Lepismium	14	0	1	0	15
Leptocereus	4	0	11	0	15
Leuchtenbergia	1	0	0	0	1
Lophophora	2	0	0	0	2
Maihuenia	2	0	0	0	2
Mammillaria	145	91	28	0	264
Mammilloydia	1	0	0	0	1
Matucana	15	4	2	0	21
Melocactus	29	11	5	0	45
Micranthocereus	9	0	0	0	9
Mila	3	0	1	0	4
×*Myrtgerocactus*	0	0	1	0	1
Myrtillocactus	4	0	0	0	4
Neobuxbaumia	9	0	0	0	9
Neolloydia	1	0	1	0	2
Neoraimondia	2	1	0	0	3
Neowerdermannia	2	1	0	0	3
Obregonia	1	0	0	0	1
Opuntia	161	0	195	5	361
Oreocereus	5	0	4	0	9
Oroya	1	0	1	0	2
Ortegocactus	1	0	0	0	1
×*Pacherocactus*	0	0	1	0	1
Pachycereus	12	0	0	0	12
Parodia	60	23	6	0	89
Pediocactus	6	0	2	8	16
Pelecyphora	2	0	0	0	2
Peniocereus	13	0	5	0	18
Pereskia	16	2	1	0	19
Pereskiopsis	6	0	1	0	7
Pilosocereus	34	8	1	0	43
Polaskia	2	0	0	0	2
Praecereus	2	4	0	0	6
Pseudoacanthocereus	2	0	0	0	2
Pseudorhipsalis	4	0	2	0	6
Pterocactus	9	0	0	0	9
Pygmaeocereus	2	0	1	0	3
Quiabentia	2	0	0	0	2
Rauhocereus	1	1	0	0	2
Rebutia	24	11	17	0	52
Rhipsalis	36	12	0	0	48
Samaipaticereus	1	0	0	0	1
Schlumbergera	6	2	1	1	10
Sclerocactus	18	3	2	5	28
Selenicereus	18	0	9	0	27
Stenocactus	7	0	3	0	10
Stenocereus	19	1	5	0	25
Stephanocereus	2	0	0	0	2
Stetsonia	1	0	0	0	1
Strombocactus	1	1	0	0	2
Tacinga	2	0	0	0	2
Thelocactus	10	9	2	0	21
Turbinicarpus	16	8	9	0	33
Uebelmannia	3	2	0	0	5
Weberbauerocereus	7	0	1	0	8
Weberocereus	9	0	0	0	9
Yungasocereus	1	0	0	0	1
Totals	1306	301	582	71	2260

3. Bibliography

Adams, C.D. (1972). Flowering Plants of Jamaica [Cactaceae: 271–276].
Alain & Leon [Sauget, J.S. & Liogier, A.H.] (1953). Flora de Cuba 3. [Cactaceae: 357–383].
Anderson, E.F. (1960–64). A revision of *Ariocarpus* (Cactaceae). Amer. J. Bot. 47: 582–589; ibid. 49: 615–622; ibid. 50: 724–732; ibid. 51: 144–151.
——— (1967). A study of the proposed genus *Obregonia* (Cactaceae). Amer. J. Bot. 54: 897–903.
——— (1980). Peyote, the divine cactus. Tucson: University of Arizona Press.
——— (1986). A revision of the genus *Neolloydia* B. & R. (Cactaceae). Bradleya 4: 1–28, with figs.
——— (1987). A revision of the genus *Thelocactus* B. & R. (Cactaceae). Bradleya 5: 49–76.
——— & Boke, N.H. (1969). The genus *Pelecyphora* (Cactaceae): Resolution of a controversy. Amer. J. Bot. 56: 314–326.
——— & Skillman (1984). A comparison of *Aztekium* and *Strombocactus* (Cactaceae). Syst. Bot. 9: 42–49.
——— & Walkington, D.L. (1971). Cactaceae. In Wiggins & Porter, Flora of the Galapagos Islands, 533–546.
——— & Fitz Maurice, W.A. (1997). *Ariocarpus* revisited. Haseltonia 5: 1–20.
Arias Montes, S. (1996, ined.). Revisión taxonómica del género *Pereskiopsis* Britton & Rose (Cactaceae). Tésis que para obtenir el grado de Maestro en Ciencias (Biología), Facultad de Ciencias, Universidad Nacional Autónoma de México.
———, Gama López, S. & Guzmán Cruz, U. (1997). Fasciculo 14. Cactaceae A.L. Juss. In Davila Aranda, P.D. et al., Flora del Valle de Tehuacán-Cuicatlán. Mexico D.F.: Instituto de Biología, UNAM.
Backeberg, C. (1958–62). Die Cactaceae. 6 vols. Jena: Gustav Fischer Verlag.
Backeberg, C. (1976). Kakteenlexikon, ed. 3. Jena: Gustav Fischer Verlag.
———, trs. Glass, L. (1977). Cactus Lexicon. Poole: Blandford Press.
Barthlott, W. (1987). New names in Rhipsalidinae (Cactaceae). Bradleya 5: 97 100.
——— (1991). In Hunt & Taylor eds, Notes on miscellaneous genera of Cactaceae. Bradleya 9: 81–92.
——— & Hunt D. (1993). Cactaceae. In Kubitzki, K. (ed), The Families and Genera of Vascular Plants 2: 161–197. Berlin etc: Springer Verlag.
——— & Taylor N.P. (1995). Notes towards a Monograph of Rhipsalideae (Cactaceae). Bradleya 13: 43–79.
Benson, L. (1982). The Cacti of the United States and Canada. Stanford: Stanford University Press.
Blum, W., Lange, M., Rischer, W. & Rutow, J. (1998). *Echinocereus*. Monographie. Publisher not stated [ISBN 3-00-001910-3].
Brako, L. & Zarucchi, J.L. (1993). Catalogue of the Flowering Plants and Gymnosperms of Peru. Monogr. Syst. Bot. 45. [Cactaceae: 265–309]. St Louis: Missouri Botanical Garden.
Braun, P.J. (1988). On the taxonomy of Brazilian Cereeae (Cactaceae). Bradleya 6: 85–99.
——— & Esteves Pereira, E. (1987–89). Revision der Gattung *Facheiroa* Britton & Rose (Cactaceae). Kakt. and. Sukk. 38–40. [Series in 7 parts].
Bravo-Hollis, H. (1978). Las Cactáceas de México, ed. 2, vol. 1. Mexico: Universidad Nacional Autónoma de México.
——— & Sánchez-Mejorada, H. (1983). Datos Preliminares Acerca de las Cactáceas en Mesoamérica. Cact. Suc. Mex. 28: 37–41, 60–71, 85–96.
——— & ——— (1991). Las Cáctaceas de México, ed. 2, vols. 2 and 3. Mexico City: Universidad Autónoma Nacional de México.
Breedlove, D.E. (1986). Listados Florísticos de México. IV. Flora de Chiapas. [Cactaceae: 65–66]. Mexico: Universidad Nacional Autónoma de México, Inst. de Biología.
Bregman, R. (1996). The genus *Matucana*. Biology and systematics of fascinating Peruvian cacti. Rotterdam: A.A. Balkema.
—— et al. (1986–90). Het geslacht *Matucana* Br. & R. Succulenta 65–69. [Revision in 32 parts].

Britton, N.L. & Rose, J.N. (1919–23). The Cactaceae. 4 vols. Washington: Carnegie Institution.
Cheesman, E.E. (1940). Flora of Trinidad & Tobago 1 [Cactaceae: 453–462].
Correll, D.S. & Correll, H.B. (1982). Flora of the Bahama Archipelago [Cactaceae: 1000–1016].
D'Arcy, W.G. (1987). Flora of Panama. Checklist and index. Monogr. Syst. Bot. Missouri Bot. Gard. 17 [Cactaceae: 207–209].
Donald, J.D. (1975). Occasional generic reviews no. 2. *Oroya* Br. & R. Ashingtonia 2: 136–141.
——— (1975–79). Systematics of *Rebutia*. Ashingtonia 1–3. [Series in several parts]
——— (1978). Occasional generic review no. 3. *Mila* Britt. and Rose. Ashingtonia 3: 31–35, 38–62.
——— (1979). Occasional generic review no. 6. *Weingartia* Werd. Ashingtonia 3: 87–139.
Eggli, U. & Taylor, N.P., compilers (1991). IOS Index of names of Cactaceae published 1950–1990. Richmond: Royal Botanic Gardens Kew.
Eggli, U., Muñoz Schick, M. & Leuenberger, B.E. (1995). Cactaceae of South America. The Ritter Collections. Englera 16. Pp. 646. Bot. Gart. u. Mus. Berlin-Dahlem.
Esser, G. (1982). Vegetationsgliederung und Kakteenvegetation von Paraguay. Trop. subtrop. Pflanzenw. 38 [pp. 113].
Foster, R.C. (1958). Catalogue of the ferns and Flowering Plants of Bolivia. Contr. Gray Herb. no. 184 [Cactaceae: 138–142].
Friedrich, H. & Glätzle, W. (1983). Seed-morphology as an aid to classifying the genus *Echinopsis* Zucc. Bradleya 1: 91–104.
Gibson, A.C. (1988–). The systematics and evolution of subtribe Stenocereinae. Cact. Succ. J. (US) 60: 11–16, et seq.
——— & Horak, K. (1979). Systematic anatomy and phylogeny of Mexican columnar cacti. Ann. Missouri Bot. Gard. 65: 995–1057.
Glass, C.E. (1997). Guide to the identification of Threatened Cacti of Mexico / Guía para la identificación de Cactáceas Amenazadas de México, Vol. 1. [108 unpaginated sheets in ring-binder]. Mexico, D.F.: Ediciones Cante, A.C.
——— & Foster, R. (1978). A revision of the genus *Epithelantha*. Cact. Succ. J. (US) 50: 184–187, with figs.
Guzman, L.U. & Arias Montes, S., compilers (1989). Claves para la identificación de las Cactáceas de México. Mexico: Sociedad Mexicana de Cactología.
Heil, K., Armstrong, B. & Schleser, D. (1981). A review of the genus *Pediocactus*. Cact. Succ. J. (US) 53: 17–39.
Herter, G. (1957). Flora Ilustrada del Uruguay. Pt. 13 [Cactaceae: 581–600, figs. 2241–2320].
Hoffmann, A.E. (1989). Cactáceas en la flora silvestre de Chile. Santiago: Fundación Claudio Gay.
——— & Flores, A.R. (1989). The conservation status of Chilean succulent plants: a preliminary assessment. In Benoit, I.C., ed., Red Book on Chilean terrestrial flora (Part One), 107–121. Santiago: Corporación Nacional Forestal (CONAF).
Howard, R. (1989). Flora of the Lesser Antilles 5 [Cactaceae: 398–422].
Hunt, D.R. (1967, repr. 1979). Cactaceae. In Hutchinson, J., The Genera of Flowering Plants 2: 427–467. Oxford: Oxford University Press.
——— (1969). A synopsis of *Schlumbergera* Lem. Kew Bull. 23:
——— (1985b). Plant portraits 43: *Weberocereus tonduzii*. Kew Mag. 2(4): 339–341.
——— (1983–87). A new review of *Mammillaria* names. Bradleya 1: 105–128; ibid. 2: 65–96; ibid. 3: 53–66; ibid. 4: 39–64; ibid. 5: 17–48. [Reissued in collated form]
——— (1989). Notes on *Selenicereus* (A. Berger) Britton & Rose and *Aporocactus* Lemaire (Cactaceae-Hylocereinae). Bradleya 7: 89–96.
——— et al. (1989). In Walters, S.M. et al. eds., European Garden Flora 3 [Cactaceae: 202–301]. Cambridge: Cambridge University Press.
——— (1989–92, 1996–98). Mammillaria Postscripts nos 1–7. [Newsletter].
——— (1992). Cactaceae. In Royal Horticultural Society Dictionary of Gardening. London: Macmillan.
——— (1996–98). Cactaceae Consensus Initiatives nos 1–6. [Newsletter].
——— & Taylor, N.P., eds (1986). The genera of Cactaceae: towards a new consensus. Bradleya 4: 65–78.

——— & ——— (1990). The genera of Cactaceae: progress towards consensus. Bradleya 8: 85–107.

——— & ——— (1991). Notes on miscellaneous genera of Cactaceae. Bradleya 9: 81–92.

Kattermann, F. (1994). *Eriosyce* (Cactaceae): the genus revised and amplified. Succulent Plant Research 1. Richmond, Surrey: David Hunt.

Kiesling, R. (1978). El género *Trichocereus* (Cactaceae) I. Las especies de la Rep. Argentina. Darwiniana 21: 263–330.

——— (1982a). The genus *Pterocactus*. Cact. Succ. J. Gr. Brit. 44: 51–56.

——— (1984a). Estudios en Cactaceae de Argentina: *Maihueniopsis*, *Tephrocactus* y géneros afines (Opuntioideae). Darwiniana 25: 171–215.

——— (1984b). Recopilación, en edición facsimilar, de todos los trabajos o referencias sobre Cactáceas publicadas por el Dr. Carlos Spegazzini. Buenos Aires: Librosur.

——— (1990). Cactus de la Patagonia. Flora Patagónica 5: 218–243.

——— & Ferrari, O. (1990). *Parodia* sensu strictu [sic] in Argentina. Cact. Succ. J. (US) 62: 194–198, 244–250.

Kimnach, M. (1960). A revision of *Borzicactus*. Cact. Succ. J. (US) 32: 8–13, 57–60, 92–96, 109–112.

——— (1993). The genus *Disocactus*. Haseltonia 1: 95–139.

Leuenberger, B.E. (1986). *Pereskia* (Cactaceae). Memoirs of the New York Botanical Garden 41: 1–141.

——— (1987). A preliminary list of Cactaceae from the Guianas and recommendations for future collecting and preparation of specimens. Willdenowia 16: 497–510.

——— (1997). *Maihuenia* – Monograph of a Patagonian genus of Cactaceae. Bot. Jahrb. Syst. 119(1): 1–92.

——— (1997). Cactaceae. In Görts-van-Rijn, A.R.A. & Jansen-Jacobs, M.J. (eds), Flora of the Guianas, ser. A: Phanerogams, fasc. 18. Pp. 63 + map. Richmond, Surrey, Royal Botanic Gardens, Kew.

Liogier, A.H. & Martorell, L.F. (1982). Flora of Puerto Rico and Adjacent Islands, a systematic synopsis [Cactaceae: 116–118].

Lüthy, J.M. (1995). Taxonomische Untersuchung der Gattung *Mammillaria* Haw. (Cactaceae). Place of publ. not stated. Verlag AfM and the author.

Madsen, J.E. (1989). Cactaceae. In Harling & Andersson, eds. Flora of Ecuador, no. 35.

McMillan, A.J.S. & Horobin, J.F. (1995). Christmas cacti. The genus *Schlumbergera* and its hybrids. Succulent Plant Research 4. Milborne Port, Somerset: David Hunt.

Meregalli, M. (1985). Il genere *Gymnocalycium* Pfeiffer. Piante Grasse 51: 5–63.

Metzing, D. (1993). Cactaceae in Paraguay; specie, ecologia e minaccia di estinzione. Piante Grasse 13(4): Suppl. 5–64.

———, Meregalli, M. & Kiesling, R. (1995). An annotated checklist of the genus *Gymnocalycium* Pfeiffer ex Mittler (Cactaceae). Allionia 33: 181–228.

Molina, R.A. (1975). Enumeración de las Plantas de Honduras. Ceiba 19(1) [Cactaceae: 79–80].

Moscoso, R.M. (1941). Las Cactáceas de la Flora de Santo Domingo. An. Univ. Santo Domingo 5: 58–90.

Oldfield, S. (comp.) (1997). Cactus and Succulent Plants – Status Survey and Conservation Action Plan. Pp. x + 212. IUCN/SSC Cactus and Succulent Specilaist Group. Gland, Switzerland, and Cambridge, U.K., International Union for Conservation of Nature and Natural Resources.

Pilbeam, J. (1985). *Sulcorebutia* and *Weingartia*. A Collector's Guide. London: Batsford.

——— (1995). *Gymnocalycium*. A Collector's Guide. Rotterdam: A.A. Balkema.

Preston-Mafham, R. & K. (1991). Cacti. The Illustrated Dictionary. London: Blandford. ISBN 0-7137-2092-1. [A useful photographic reference with nearly 1100 colour pictures of 'globular' cacti.]

Proctor, G.R. (1984). Flora of the Cayman Islands [Cactaceae: 319–326].

Rauh, W. (1958). Beitrag zur Kenntnis der peruanischen Kakteenvegetation. Sitzungsberichte der Heidelberger Akademie der Wissenschaften, Math.-naturw. Kl. 1958: 1–542.

Rausch, W. (1986). *Lobivia* '85 [incl. 'Ergänzung 86']. Vienna: Rudolf Herzig.

Reppenhagen, W. (1991–92). Die Gattung *Mammillaria*. 2 vols. Titisee-Neustadt, Verlag Druckerei Steinhart.

Ritter, F. (1979–81). Kakteen in Südamerika 1–4. Spangenberg: Friedrich Ritter Selbstverlag.

Rowley, G.D. (1982). A Checklist of *Lobivia* names. Cact. Succ. J. Gr. Brit. 44(4): 75–81.
Sánchez-Mejorada, H. (1974). Revisión del género *Peniocereus* (Las Cactáceas). Dir. Agric. Gan. Gob. Est. de Mexico.
Standley, P.C. (1937). Flora of Costa Rica. Publ. Field Mus. Bot. Ser. 18 [Cactaceae: 749–759].
——— & Calderon, S. (1944). Flora Salvadorena. Lista Preliminar, ed. 2 [Cactaceae: 200–201].
——— & Williams, L.O. (1962). Flora of Guatemala. Fieldiana 24(7) [Opuntiales: 187–234].
Taylor, N.P. (1979). A Commentary on the genus *Echinofossulocactus* Lawr. Cact. Succ. J. Gt. Brit. 41: 35–42.
——— (1981a). Reconsolidation of *Discocactus*. Cact. Succ. J. Gt. Brit. 43(2/3): 37–40.
——— (1981b). A commentary on *Copiapoa*. Cact. Succ. J. Gt. Brit. 43(2/3): 49–60.
——— (1984). A review of *Ferocactus* Britton & Rose. Bradleya 2: 19–38.
——— (1985). The Genus *Echinocereus*. Kew Magazine Monograph.
——— (1986). The Identification of Escobarias. Brit. Cact. Succ. J. 4: 36–44.
——— (1988, 1989). Supplementary notes on Mexican *Echinocereus* (1). Bradleya 6: 65–84; ibid. (2). Bradleya 7: 73–77.
——— (1991). The genus *Melocactus* (Cactaceae) in Central and South America. Bradleya 9: 1–80.
——— (1994). Ulteriori studi su *Echinocereus* (Further notes on *Echinocereus*). Piante Grasse 13(4): Suppl. 79–96.
——— & Iliff, J. (1996). Nomenclatural notes on Andean Opuntioideae (Cactaceae). Bradleya 14: 17–19.
——— & Zappi, D.C. (1990). Brief notes on *Leocereus* Britton & Rose. Bradleya 8: 107–108.
Trujillo, B. & Ponce, M. (1988). Lista-inventario de Cactaceae silvestres en Venezuela con sinonimia y otros aspectos relacionados. Ernstia no. 47: 1–20.
Wagenaar-Hummelinck, P. (1938). Notes on the Cactaceae of Curaçao, Aruba, Bonaire and the Venezuelan Islands. Rec. Trav. bot. Néerl. 35: 29–55.
Wallace, R.S. & Forquer, E.D. (1995). Molecular evidence for the systematic placement of *Echinocereus pensilis* (K. Brandegee) J. Purpus (Cactaceae). Haseltonia 3: 71–76.
Woodson, R.E. & Schery, R.W. (1958). Flora of Panama. Ann. Missouri Bot. Gard. 45. [Cactaceae: 68–91].
Zappi, D.C. (1994). The genus *Pilosocereus* (Cactaceae) in Brazil. Succulent Plant Research 3. Milborne Port, Somerset: David Hunt.
Zimmerman, A.D. (1985, ined.). Systematics of the genus *Coryphantha* (Cactaceae). PhD thesis, University of Texas.

4. List of accepted taxa with distribution and synonymy

Acanthocalycium ferrarii Rausch . . . . AR
®*Acanthocalycium variiflorum* Backeberg
Acanthocalycium klimpelianum (Weidlich & Werdermann) Backeberg . . . . AR
Echinopsis peitscheriana (Backeberg) Friedrich & Rowley
Acanthocalycium peitscherianum Backeberg
Acanthocalycium spiniflorum (Schumann) Backeberg . . . . AR
Echinopsis spiniflora (Schumann) Berger
Acanthocalycium violaceum (Werdermann) Backeberg

Acanthocereus baxaniensis (Karwinsky ex Pfeiffer) Borg . . . . CU
Acanthocereus colombianus Britton & Rose . . . . CO
Acanthocereus horridus Britton & Rose . . . . MX, GT, SV
Acanthocereus occidentalis Britton & Rose . . . . MX
Acanthocereus subinermis Britton & Rose . . . . MX
Acanthocereus tetragonus (Linnaeus) Hummelinck
. . . . US, MX, CU, GT, BZ, HN, NI, SV, CR, PA, WI, TT, AN, VE
Acanthocereus floridanus Small ex Britton & Rose
Acanthocereus pentagonus (Linnaeus) Britton & Rose
Cereus pentagonus (Linnaeus) Haworth
Cereus variabilis Engelmann

Ariocarpus agavoides (Castañeda) Anderson . . . . ♦MX
Neogomesia agavoides Castañeda♦
Ariocarpus bravoanus Hernandez & Anderson . . . . ♦MX
Ariocarpus bravoanus ssp. **bravoanus** . . . . ♦MX
Ariocarpus bravoanus ssp. **hintonii** (Stuppy & Taylor) Anderson & Fitz Maurice ♦MX
Ariocarpus fissuratus (Engelmann) Schumann . . . . ♦US, MX
Roseocactus fissuratus (Engelmann) Berger♦
Roseocactus intermedius Backeberg & Kilian♦
Roseocactus lloydii (Rose) Berger♦
Ariocarpus kotschoubeyanus (Lemaire) Schumann . . . . ♦MX
Roseocactus kotschoubeyanus (Lemaire) Berger♦
Ariocarpus kotschoubeyanus ssp. *albiflorus* (Backeberg) Glass♦
Roseocactus kotschoubeyanus ssp. *macdowellii* Backeberg♦
Ariocarpus retusus Scheidweiler . . . . ♦MX
Ariocarpus elongatus (Salm-Dyck) M.H. Lee♦
Ariocarpus retusus ssp. *scapharostroides* Halda & Horacek♦
Ariocarpus retusus ssp. **retusus** . . . . ♦MX
Ariocarpus confusus Halda & Horacek♦
Ariocarpus furfuraceus (Watson) H.C. Thompson♦
Ariocarpus retusus ssp. **trigonus** (Weber) Anderson & Fitz Maurice . . . . ♦MX
Ariocarpus trigonus (Weber) Schumann♦
Ariocarpus scaphirostris Boedeker . . . . ♦MX
®*Ariocarpus scapharostrus* Boedeker♦

Armatocereus arduus Ritter . . . . PE
Armatocereus brevispinus Madsen . . . . EC
Armatocereus cartwrightianus (Britton & Rose) Backeberg ex A.W. Hill . . . . EC, PE
Lemaireocereus cartwrightianus Britton & Rose
Armatocereus riomajensis Rauh & Backeberg . . . . PE
Armatocereus godingianus (Britton & Rose) Backeberg ex E. Salisbury . . . . EC
Lemaireocereus godingianus Britton & Rose
Armatocereus humilis (Britton & Rose) Backeberg . . . . CO
Lemaireocereus humilis Britton & Rose
Armatocereus laetus (Kunth) Backeberg ex A.W. Hill . . . . PE
Lemaireocereus laetus (Kunth) Britton & Rose

Armatocereus mataranus Ritter . . . . PE
Armatocereus mataranus ssp. **mataranus** . . . . PE
Armatocereus mataranus ssp. **ancashensis** (Ritter) Ostolaza . . . . PE
Armatocereus matucanensis Backeberg ex A.W. Hill . . . . EC, PE
Armatocereus arboreus Rauh & Backeberg
Armatocereus churinensis Rauh & Backeberg
Armatocereus oligogonus Rauh & Backeberg . . . . PE
Armatocereus procerus Rauh & Backeberg . . . . PE
Armatocereus rauhii Backeberg . . . . PE
Armatocereus rauhii ssp. **rauhii** . . . . PE
Armatocereus rauhii ssp. **balsasensis** (Ritter) Ostolaza . . . . PE
Armatocereus balsasensis Ritter
Armatocereus rupicola Ritter . . . . PE

Arrojadoa ×albiflora Buining & Brederoo . . . . BR
Arrojadoa bahiensis (Braun & Esteves Pereira) Taylor & Eggli . . . . BR
Floribunda bahiensis Braun & Esteves Pereira
Pierrebraunia bahiensis (Braun & Esteves Pereira) Esteves
Arrojadoa dinae Buining & Brederoo . . . . BR
Arrojadoa dinae ssp. **dinae** . . . . BR
Arrojadoa beateae Braun & Esteves
Arrojadoa dinae ssp. *nana* (Braun & Esteves Pereira) Braun & Esteves Pereira
Arrojadoa multiflora Ritter
Arrojadoa dinae ssp. **eriocaulis** (Buining & Brederoo) Taylor & Zappi . . . . BR
Arrojadoa eriocaulis Buining & Brederoo
Arrojadoa eriocaulis ssp. *albicoronata* (Van Heek et al.) Braun & Esteves Pereira
Arrojadoa penicillata (Guerke) Britton & Rose . . . . BR
Arrojadoa rhodantha (Guerke) Britton & Rose . . . . BR
Arrojadoa aureispina Buining & Brederoo
Arrojadoa canudosensis Buining & Brederoo
Arrojadoa horstiana Braun & Heimen
Arrojadoa rhodantha ssp. *aureispina* (Buining & Brederoo) Braun & Esteves Pereira
Arrojadoa rhodantha ssp. *canudosensis* (Buining & Brederoo) Braun
Arrojadoa rhodantha ssp. *reflexa* Braun
Arrojadoa theunisseniana Buining & Brederoo

Arthrocereus glaziovii (Schumann) Taylor & Zappi . . . . BR
Arthrocereus campos-portoi (Werdermann) Backeberg
Leocereus glaziovii (Schumann) Britton & Rose
Arthrocereus itabiriticola Braun
®*Arthrocereus damazioi* P.V. Heath
Arthrocereus melanurus (Schumann) Diers et al. . . . . BR
Leocereus melanurus (Schumann) Britton & Rose
Arthrocereus melanurus ssp. **melanurus** . . . . BR
Arthrocereus melanurus ssp. *estevesii* (Diers) Braun & Esteves Pereira
Arthrocereus melanurus ssp. *mello-barretoi*
(Backeberg & Voll) Braun & Esteves Pereira
Arthrocereus mello-barretoi Backeberg & Voll
Arthrocereus melanurus ssp. **magnus** Taylor & Zappi . . . . BR
Arthrocereus melanurus ssp. **odorus** (Ritter) Taylor & Zappi . . . . BR
Arthrocereus odorus Ritter
Arthrocereus rondonianus Backeberg & Voll . . . . BR
Arthrocereus spinosissimus (Buining & Brederoo) Ritter . . . . BR
Eriocereus spinosissimus Buining et al.

Astrophytum asterias (Karwinski ex Zuccarini) Lemaire . . . . ♦US, MX
Echinocactus asterias Karwinski ex Zuccarini♦
Astrophytum capricorne (Dietrich) Britton & Rose . . . . MX
Astrophytum senile Fric

Astrophytum myriostigma Lemaire . . . . MX
Astrophytum coahuilense (Moeller) Kayser
Astrophytum columnare (Schumann) Sadovsky & Schuetz
Astrophytum myriostigma ssp. *potosinum* (Moeller) Kayser
Astrophytum myriostigma ssp. *tulense* Kayser
Astrophytum tulense (Kayser) Sadovsky & Schuetz
Astrophytum ornatum (De Candolle) Britton & Rose . . . . MX

Austrocactus bertinii (Cels ex Herincq) Britton & Rose . . . . AR
Cereus bertinii Cels ex Herincq
Austrocactus dusenii (Weber) Spegazzini
Austrocactus gracilis Backeberg
Austrocactus coxii (Schumann) Backeberg . . . . AR
Austrocactus patagonicus (Weber ex Spegazzini) Hosseus . . . . AR
Austrocactus philippii (Regel & Schmidt) Buxbaum & Ritter CL
Austrocactus hibernus Ritter
Erdisia philippii (Regel & Schmidt) Britton & Rose
Austrocactus spiniflorus (Philippi) Ritter CL
Erdisia spiniflora (Philippi) Britton & Rose
Corryocactus spiniflorus (Philippi) Hutchison

Aztekium hintonii Glass & Fitz Maurice . . . . MX
Aztekium ritteri (Boedeker) Boedeker . . . . ♦MX

Bergerocactus emoryi (Engelmann) Britton & Rose . . . . US, MX
Cereus emoryi Engelmann

Blossfeldia liliputana Werdermann . . . . BO, AR
Blossfeldia atroviridis Ritter
Blossfeldia campaniflora Backeberg
Blossfeldia fechseri Backeberg
Parodia liliputana (Werdermann) Taylor
Blossfeldia minima Ritter
Blossfeldia pedicellata Ritter

Brachycereus nesioticus (Schumann ex Robinson) Backeberg . . . . EC

Brasilicereus markgrafii Backeberg & Voll . . . . BR
Cereus markgrafii (Backeberg & Voll) Braun
Brasilicereus phaeacanthus (Guerke) Backeberg . . . . BR
Brasilicereus breviflorus Ritter
Cephalocereus phaeacanthus (Guerke) Britton & Rose
Brasilicereus phaeacanthus ssp. *breviflorus* (Ritter) Braun & Esteves Pereira

Browningia albiceps Ritter . . . . PE
Browningia altissima (Ritter) Buxbaum . . . . PE
Gymnocereus altissimus (Ritter) Backeberg
Browningia amstutziae (Rauh & Backeberg) Hutchison ex Krainz . . . . PE
Gymnocereus amstutziae Rauh & Backeberg
Browningia caineana (Cardenas) Hunt . . . . PY, BO
Castellanosia caineana Cardenas
Browningia candelaris (Meyen) Britton & Rose . . . . PE, CL
Browningia icaensis Ritter
Browningia chlorocarpa (Kunth) Marshall . . . . PE
Seticereus chlorocarpus (Kunth) Backeberg
Browningia columnaris Ritter . . . . PE
Browningia hertlingiana (Backeberg) Buxbaum . . . . PE
Azureocereus hertlingianus (Backeberg) Backeberg
Browningia microsperma (Werdermann & Backeberg) Marshall . . . . PE
Gymnocereus microspermus (Werdermann & Backeberg) Backeberg.

Browningia pilleifera (Ritter) Hutchison . . . PE
Gymnanthocereus macracanthus Ritter
Gymnanthocereus pilleifer Ritter
Browningia viridis (Rauh & Backeberg) Buxbaum . . . PE
Azureocereus viridis Rauh & Backeberg

Calymmanthium substerile Ritter . . . PE
Calymmanthium fertile Ritter

Carnegiea gigantea (Engelmann) Britton & Rose . . . US, MX
Cereus giganteus Engelmann

Cephalocereus apicicephalium Dawson . . . MX
Cephalocereus columna-trajani (Karwinsky ex Pfeiffer) Schumann . . . MX
Haseltonia columna-trajani (Karwinski ex Pfeiffer) Backeberg
Pachycereus columna-trajani (Karwinski ex Pfeiffer) Britton & Rose
Cephalocereus hoppenstedtii (Roezl ex Ruempler) Schumann
Cephalocereus nizandensis (Bravo & MacDougall) Buxbaum . . . MX
Neodawsonia apicicephalium (Dawson) Backeberg
Neodawsonia nizandensis Bravo & MacDougall
Cephalocereus senilis (Haworth) Pfeiffer . . . MX
Cephalocereus totolapensis (Bravo & MacDougall) Buxbaum . . . MX
Neodawsonia totolapensis Bravo & MacDougall

Cephalocleistocactus chrysocephalus Ritter . . . BO

Cereus adelmarii (Rizzini & Mattos) Braun . . . BR
Monvillea adelmarii Rizzini & Mattos
Cereus phatnospermus ssp. *adelmarii* (Rizzini & Mattos) Braun & Esteves Pereira
Cereus aethiops Haworth . . . BR, UY, AR
Piptanthocereus aethiops (Haworth) Ritter
Cereus azureus Parmentier ex Pfeiffer
Cereus chalybaeus Otto ex Walpers
Cereus albicaulis (Britton & Rose) Luetzelburg . . . BR
Acanthocereus albicaulis Britton & Rose
Mirabella albicaulis (Britton & Rose) Ritter
Monvillea albicaulis (Britton & Rose) Kiesling
Cereus argentinensis Britton & Rose . . . AR
Cereus bicolor Rizzini & Mattos . . . BR
Cereus braunii Cardenas . . . BO
Cereus cochabambensis Cardenas . . . BO
Cereus comarapanus Cardenas . . . BO
Piptanthocereus comarapanus (Cardenas) Ritter
Cereus fernambucensis Lemaire . . . BR
Cereus fernambucensis ssp. **fernambucensis** . . . BR
Ⓤ*Cereus pernambucensis* auctt
Ⓤ?*Cereus neotetragonus* Backeberg
Ⓜ*Cereus tetragonus* auctt
Cereus fernambucensis ssp. **sericifer** (Ritter) Taylor & Zappi . . . BR
Cereus sericifer (Ritter) Braun
Piptanthocereus sericifer Ritter
Cereus fricii Backeberg . . . VE
Cephalocereus russel(l)ianus Rose ex Bailey
Subpilocereus russelianus (Salm-Dyck) Backeberg
Ⓤ*Cereus russelianus* Hort. Berol. ex Salm-Dyck
Cereus haageanus (Backeberg) Taylor . . . PY
Monvillea haageana Backeberg
Cereus hankeanus Weber ex Schumann . . . BO, AR
Cereus hexagonus (Linnaeus) Miller . . . TT, GY, GF, SR, BR, VE
Cereus longiflorus Alexander
Cereus perlucens Schumann

Cereus hildmannianus Schumann . . . . BR, PY, UY, AR
Cereus hildmannianus ssp. *xanthocarpus* (Schumann) Braun & Esteves Pereira
Cereus neonesioticus (Ritter) Braun
Piptanthocereus neonesioticus Ritter
Cereus xanthocarpus Schumann
Piptanthocereus xanthocarpus (Schumann) Ritter
®*Cereus peruvianus* auctt non (Linnaeus) Miller
Cereus hildmannianus ssp. **hildmannianus** . . . . BR, PY
Cereus milesimus Rost
Cereus hildmannianus ssp. **uruguayanus** (Kiesling) Taylor . . . . BR, UY, AR
Piptanthocereus bageanus Ritter
Cereus uruguayanus Kiesling
®*Piptanthocereus uruguayanus* Ritter
Cereus horrispinus Backeberg . . . . CO, VE
Subpilocereus horrispinus (Backeberg) Backeberg
Subpilocereus ottonis Backeberg
Cereus huilunchu Cardenas . . . . BO
Piptanthocereus huilunchu (Cardenas) Ritter
Cereus insularis Hemsley . . . . BR
Monvillea insularis (Hemsley) Britton & Rose
Cereus jamacaru De Candolle . . . . BR
Cereus jamacaru ssp. **jamacaru** . . . . BR
Cereus jamacaru ssp. **calcirupicola** (Ritter) Taylor & Zappi . . . . BR
Piptanthocereus cabralensis Ritter
Cereus calcirupicola (Ritter) Rizzini
Piptanthocereus calcirupicola Ritter
Cereus calcirupicola ssp. *cabralensis* (Ritter) Braun & Esteves Pereira
Cereus calcirupicola ssp. *cipoensis* (Ritter) Braun & Esteves Pereira
Piptanthocereus cipoensis Ritter
Cereus jamacaru ssp. ***goiasensis*** (Ritter) Braun & Esteves Pereira . . . . BR
Cereus goiasensis (Ritter) Braun
Piptanthocereus goiasensis Ritter
Cereus kroenleinii Taylor . . . . BR, BO, PY
Cereus phatnospermus ssp. *kroenleinii* (Taylor) Braun & Esteves Pereira
®*Monvillea kroenleinii* Kiesling
Cereus lamprospermus Schumann . . . . PY, ?BO
?*Cereus lamprospermus* ssp. *colosseus* (Ritter) Braun & Esteves Pereira . . . . . . BO
?*Piptanthocereus colosseus* Ritter
Cereus lanosus (Ritter) Braun . . . . PY
Piptanthocereus lanosus Ritter
Cereus mirabella Taylor . . . . BR
Mirabella minensis Ritter
Monvillea minensis (Ritter) Kiesling
Cereus mortensenii (Croizat) Hunt & Taylor . . . . VE
Pilosocereus gruberi Schatzl & Till
Pilosocereus mortensenii (Croizat) Backeberg
Subpilocereus mortensenii (Croizat) Trujillo
Cereus pachyrrhizus Schumann . . . . PY
Piptanthocereus pachyrhizus (Schumann) Ritter
Cereus phatnospermus Schumann . . . . PY
Monvillea phatnosperma (Schumann) Britton & Rose
Cereus repandus (Linnaeus) Miller . . . . AN, VE
Cereus atroviridis Backeberg
Subpilocereus atroviridis (Backeberg) Backeberg
Cereus grenadensis Britton & Rose
Subpilocereus grenadensis (Britton & Rose) Backeberg
Cereus margaritensis J.R. Johnston
Cereus peruvianus (Linnaeus) Miller
Subpilocereus remolinensis (Backeberg) Backeberg
Subpilocereus repandus (Linnaeus) Backeberg
Subpilocereus repandus ssp. *micracanthus* (Hummelinck) Trujillo & Ponce

Cereus ridleii Andrade-Lima ex Backeberg . . . . BR
Cereus roseiflorus Spegazzini . . . . AR
Cereus saddianus (Rizzini & Mattos) Braun . . . . BR
Monvillea saddiana Rizzini & Mattos
Cereus spegazzinii Weber . . . . BR, AR, BO, PY
Cereus anisitsii Schumann
Monvillea ebenacantha Ritter
Monvillea lindenzweigiana (Guerke) Backeberg
Piptanthocereus lindenzweigianus (Guerke) Ritter
Monvillea spegazzinii (Weber) Britton & Rose
Cereus stenogonus Schumann . . . . BO, PY, AR
Cereus dayami Spegazzini
Piptanthocereus dayamii (Spegazzini) Ritter
Piptanthocereus stenogonus (Schumann) Ritter
Cereus tacuaralensis Cardenas . . . . BO
Cereus trigonodendron Schumann ex Vaupel . . . . PE
Cereus validus Haworth . . . . AR
Cereus forbesii Hort. Berol. ex Foerster
Cereus vargasianus Cardenas . . . . PE

Cintia knizei Riha . . . . BO

Cipocereus bradei (Backeberg & Voll) Zappi & Taylor . . . . BR
Pilosocereus bradei (Backeb. & Voll) Byles & Rowley
Pseudopilocereus bradei (Backeberg & Voll) Buxbaum
Cipocereus crassisepalus (Buining & Brederoo) Zappi & Taylor . . . . BR
Cereus crassisepalus Buining & Brederoo
Piptanthocereus crassisepalus (Buining & Brederoo) Ritter
Cipocereus laniflorus Taylor & Zappi . . . . BR
Cipocereus minensis (Werdermann) Ritter . . . . BR
Cipocereus minensis ssp. **minensis** . . . . BR
Pilosocereus minensis (Werdermann) Byles & Rowley
Cipocereus minensis ssp. **pleurocarpus** (Ritter) Taylor & Zappi . . . . BR
Cipocereus pleurocarpus Ritter
Pilosocereus pleurocarpus (Ritter) Braun
Cipocereus pusilliflorus (Ritter) Zappi & Taylor . . . . BR
Floribunda pusilliflora Ritter
Pilosocereus pusilliflorus (Ritter) Braun

Cleistocactus acanthurus (Vaupel) Hunt . . . . PE
Cleistocactus acanthurus ssp. **acanthurus** . . . . PE
Borzicactus acanthurus (Vaupel) Britton & Rose
Loxanthocereus acanthurus (Vaupel) Backeberg
Loxanthocereus bicolor Ritter
Loxanthocereus canetensis Rauh & Backeberg
Loxanthocereus convergens Ritter
Loxanthocereus cullmannianus Backeberg
Loxanthocereus eremiticus Ritter
Loxanthocereus erigens Rauh & Backeberg
Loxanthocereus eriotrichus (Werdermann & Backeberg) Backeberg
Loxanthocereus eulalianus Rauh & Backeberg
Loxanthocereus gracilispinus Rauh & Backeberg
Loxanthocereus keller-badensis Backeberg & Krainz
Loxanthocereus multifloccosus Rauh & Backeberg
Loxanthocereus neglectus Ritter
Loxanthocereus pacaranensis Ritter
Haageocereus paradoxus Rauh & Backeberg
Cleistocactus acanthurus ssp. **faustianus** (Backeberg) Ostolaza . . . . PE
Loxanthocereus faustianus (Backeberg) Backeberg
Cleistocactus acanthurus ssp. **pullatus** (Rauh & Backeberg) Ostolaza . . . . PE
Loxanthocereus pullatus Rauh & Backeberg

Cleistocactus baumannii (Lemaire) Lemaire . . . . BR, PY, BO, AR
Cleistocactus baumannii ssp. **baumannii** . . . . PY, BO, AR
Cleistocactus aureispinus Fric
Cereus baumannii Lemaire
Cleistocactus bruneispinus Backeberg
Cleistocactus flavispinus (Schumann) Backeberg
Cleistocactus baumannii ssp. ***anguinus*** (Guerke) Braun & Esteves Pereira . . . . PY
Cleistocactus anguinus (Guerke) Britton & Rose
Cleistocactus baumannii ssp. ***chacoanus*** (Ritter) Braun & Esteves Pereira . . . . BO
Cleistocactus chacoanus Ritter
Cleistocactus baumannii ssp. ***croceiflorus*** (Ritter) Braun & Esteves Pereira . . . . PY
Cleistocactus croceiflorus Ritter
Cleistocactus baumannii ssp. **horstii** (Braun) Taylor . . . . BR
Cleistocactus horstii Braun
Cleistocactus baumannii ssp. ***santacruzensis*** (Backeberg) Mottram . . . . BO
Cleistocactus santacruzensis Backeberg
Cleistocactus brookeae Cardenas . . . . BO
Cleistocactus wendlandiorum Backeberg
Cleistocactus buchtienii Backeberg . . . . BO
Cleistocactus angosturensis Cardenas
Cleistocactus ayopayanus Cardenas
Cleistocactus ressinianus Cardenas
Cleistocactus sucrensis Cardenas
Cleistocactus candelilla Cardenas . . . . BO
Cleistocactus ianthinus Cardenas
Cleistocactus pojoensis (Cardenas) Backeberg
Cleistocactus vallegrandensis Cardenas
Cleistocactus chotaensis Weber ex Roland-Gosselin . . . . PE
Loxanthocereus trujilloensis Ritter
Cleistocactus ×crassiserpens Rauh & Backeberg . . . . PE
Loxanthocereus crassiserpens (Rauh & Backeberg) Backeberg
Cleistocactus clavispinus (Rauh & Backeberg) Ostolaza . . . . PE
?*Loxanthocereus deserticola* Ritter
?*Loxanthocereus ferrugineus* Rauh & Backeberg
Cleistocactus dependens Cardenas . . . . BO
Seticleistocactus dependens (Cardenas) Backeberg
Cleistocactus ferrarii Kiesling . . . . AR
Cleistocactus fieldianus (Britton & Rose) Hunt . . . . PE
Cleistocactus fieldianus ssp. **fieldianus** . . . . PE
Borzicactus cajamarcensis Ritter
Borzicactus calviflorus Ritter
Clistanthocereus calviflorus (Ritter) Backeberg
Borzicactus fieldianus Britton & Rose
Clistanthocereus fieldianus (Britton & Rose) Backeberg
Cleistocactus fieldianus ssp. **samnensis** (Ritter) Ostolaza . . . . PE
Borzicactus samnensis Ritter
Clistanthocereus samnensis (Ritter) Backeberg
Cleistocactus fieldianus ssp. **tessellatus** (Akers & Buining) Ostolaza . . . . PE
Clistanthocereus tessellatus (Akers & Buining) Backeberg
Cleistocactus grossei (Weingart) Backeberg . . . . PY
Cleistocactus hildegardiae Ritter . . . . BO
Cleistocactus hystrix (Rauh & Backeberg) Ostolaza . . . . PE
Loxanthocereus hystrix Rauh & Backeberg
Loxanthocereus montanus Ritter
Cleistocactus hyalacanthus (Schumann) Roland-Gosselin . . . . AR
Cleistocactus jujuyensis (Backeberg) Backeberg
Cleistocactus icosagonus (Kunth) Weber ex Roland-Gosselin . . . . EC, PE
Borzicactus aurivillus (Schumann) Britton & Rose
Cleistocactus humboldtii (Kunth) Weber ex Roland-Gosselin
Matucana humboldtii (Kunth) Buxbaum

Seticereus humboldtii (Kunth) Backeberg
Borzicactus icosagonus (Kunth) Britton & Rose
Seticereus icosagonus (Kunth) Backeberg
Cleistocactus laniceps (Schumann) Roland-Gosselin . . . . BO
Cleistocactus luribayensis Cardenas . . . . BO
Cleistocactus glaucus Ritter
Cleistocactus granjaensis Ritter
Cleistocactus micropetalus Ritter . . . . BO
Cleistocactus morawetzianus Backeberg . . . . PE
Cleistocactus apurimacensis Johnson ex Backeberg
Cleistocactus pycnacanthus (Rauh & Backeberg) Backeberg
Cleistocactus villaazulensis Ritter
®*Cleistocactus luminosus* Johnson ex Backeberg
Cleistocactus muyurinensis Ritter . . . . BO
Cleistocactus neoroezlii (Ritter) Buxbaum . . . . EC, PE
Borzicactus neoroezlii Ritter
Cleistocactus orthogonus Cardenas . . . . BO
Cleistocactus pachycladus (Rauh & Backeberg) Ostolaza . . . . PE
Loxanthocereus pachycladus Rauh & Backeberg
Loxanthocereus piscoensis Rauh & Backeberg
Oreocereus piscoensis (Rauh & Backeberg) Ritter
Loxanthocereus yauyosensis Ritter
Cleistocactus palhuayensis Ritter & Shahori . . . . BO
Cephalocleistocactus pallidus Backeberg
Cleistocactus viridiflorus Backeberg
Cleistocactus paraguariensis Ritter . . . . PY
Cleistocactus parapetiensis Cardenas . . . . BO
Cleistocactus azerensis Cardenas
Cleistocactus parviflorus (Schumann) Roland-Gosselin . . . . BO
Cleistocactus areolatus (Muehlenpfordt ex Schumann) Riccobono
Cleistocactus fusiflorus Cardenas
Cleistocactus herzogianus Backeberg
Cleistocactus peculiaris (Rauh & Backeberg) Ostolaza . . . . PE
Cleistocactus brevispinus Ritter
Loxanthocereus brevispinus Rauh
Loxanthocereus cantaensis Rauh & Backeberg
Haageocereus peculiaris (Rauh & Backeberg) Ritter
Loxanthocereus peculiaris Rauh & Backeberg
Cleistocactus piraymirensis Cardenas . . . . BO
Seticleistocactus piraymirensis (Cardenas) Backeberg
Cleistocactus plagiostoma (Vaupel) Hunt . . . . PE
Borzicactus plagiostoma (Vaupel) Britton & Rose
Borzicactus purpureus Ritter
Cleistocactus pungens Ritter . . . . PE
Cleistocactus reae Cardenas . . . . BO
Cleistocactus ritteri Backeberg . . . . BO
Cephalocleistocactus ritteri (Backeberg) Backeberg
Cleistocactus roezlii (Haage ex Schumann) Backeberg . . . . PE, BO
Seticereus roezlii (Haage ex Schumann) Backeberg
Cleistocactus samaipatanus (Cardenas) Hunt . . . . BO
Bolivicereus brevicaulis Ritter
Bolivicereus croceus Ritter
Bolivicereus rufus Ritter
Bolivicereus samaipatanus Cardenas
Borzicactus samaipatanus (Cardenas) Kimnach
Cleistocactus sepium (Kunth) Weber ex Roland-Gosselin . . . . EC
Borzicactus aequatorialis Backeberg
Loxanthocereus jajoianus (Backeberg) Backeberg
Borzicactus morleyanus Britton & Rose
Borzicactus sepium (Kunth) Britton & Rose

Borzicactus ventimigliae Riccobono
Borzicactus websterianus Backeberg
Cleistocactus serpens (Kunth) Weber ex Roland-Gosselin .... PE
Loxanthocereus granditessellatus Rauh & Backeberg
Cleistocactus leonensis Madsen
Loxanthocereus otuscensis Ritter
Loxanthocereus parvitesselatus Ritter
Borzicactus pseudothelegonus (Rauh & Backeberg) Rauh & Backeberg
Bolivicereus serpens (Kunth) Backeberg
Borzicactus serpens (Kunth) Kimnach
Borzicactus sulcifer (Rauh & Backeberg) Kimnach
Loxanthocereus sulcifer Rauh & Backeberg
Cleistocactus sextonianus (Backeberg) Hunt .... PE
Loxanthocereus aticensis Rauh & Backeberg
Loxanthocereus camanaensis Rauh & Backeberg
Loxanthocereus gracilis (Akers & Buining) Backeberg
Loxanthocereus nanus Akers ex Backeberg
Loxanthocereus puquiensis Ritter
Loxanthocereus riomajensis Rauh & Backeberg
Borzicactus sextonianus (Backeberg) Kimnach
Loxanthocereus sextonianus (Backeberg) Backeberg
Loxanthocereus splendens Akers ex Backeberg
Loxanthocereus variabilis Ritter
Cleistocactus smaragdiflorus (Weber) Britton & Rose .... BO, AR
Cleistocactus strausii (Heese) Backeberg .... BO
Cleistocactus tarijensis Cardenas .... BO
Cleistocactus compactus Backeberg
Cleistocactus tenuiserpens Rauh & Backeberg .... PE
Bolivicereus tenuiserpens (Rauh & Backeberg) Backeberg
Borzicactus tenuiserpens (Rauh & Backeberg) Kimnach
Cleistocactus tominensis (Weingart) Backeberg .... BO
Cleistocactus capadalensis Ritter
Cleistocactus clavicaulis Cardenas
Cleistocactus crassicaulis Cardenas
Cleistocactus mendozae Cardenas
Cleistocactus viridialabastri Cardenas
Cleistocactus tupizensis (Vaupel) Backeberg .... BO
Cleistocactus varispinus Ritter .... BO
Cephalocleistocactus schattatianus Backeberg
Cleistocactus vulpis-cauda Ritter & Cullmann .... BO
Cleistocactus winteri Hunt .... BO
Hildewintera aureispina (Ritter) Ritter
Cleistocactus xylorhizus (Ritter) Ostolaza .... PE
Loxanthocereus xylorhizus Ritter

×***Cleistocana mirabilis*** (Buining) Hunt .... PE
Arequipa mirabilis (Buining) Backeberg
Borzicactus mirabilis (Buining) Donald
Matucana mirabilis Buining

Coleocephalocereus aureus Ritter .... BR
Buiningia aurea (Ritter) Buxbaum
Coleocephalocereus aureus ssp. *brevicylindricus* (Buining) Braun
Coleocephalocereus aureus ssp. *elongatus* (Buining) Braun
Coleocephalocereus aureus ssp. *longispinus* (Buining) Braun
Buiningia brevicylindrica Buining
Coleocephalocereus brevicylindricus (Buining) Ritter
Coleocephalocereus elongatus (Buining) Braun
Coleocephalocereus buxbaumianus Buining .... BR
Coleocephalocereus buxbaumianus ssp. **buxbaumianus** .... BR
Coleocephalocereus braunii Diers & Esteves Pereira

Coleocephalocereus buxbaumianus ssp. **flavisetus** (Ritter) Taylor & Zappi . . . . . . BR
Coleocephalocereus estevesii Diers & Esteves Pereira
Coleocephalocereus flavisetus Ritter
Coleocephalocereus fluminensis (Miquel) Backeberg . . . . BR
Austrocephalocereus fluminensis (Miquel) Buxbaum
Cephalocereus fluminensis (Miquel) Britton & Rose
Coleocephalocereus fluminensis ssp. **fluminensis** . . . . BR
Coleocephalocereus diersianus Braun & Esteves Pereira
Coleocephalocereus fluminensis ssp. *braamhaarii* (Braun) Braun & Esteves Pereira
Coleocephalocereus fluminensis ssp. *paulensis* (Ritter) Braun & Esteves Pereira
Coleocephalocereus paulensis Ritter
Leocereus paulensis Spegazzini
Coleocephalocereus fluminensis ssp. **decumbens** (Ritter) Taylor & Zappi . . . . BR
Coleocephalocereus decumbens Ritter
Coleocephalocereus goebelianus (Vaupel) Buining . . . . BR
Coleocephalocereus pachystele Ritter
Coleocephalocereus pluricostatus Buining & Brederoo . . . . BR
Coleocephalocereus pluricostatus ssp. *uebelmanniorum* Braun & Esteves Pereira
Coleocephalocereus purpureus (Buining & Brederoo) Ritter . . . . BR
Buiningia purpurea Buining & Brederoo

Copiapoa bridgesii (Pfeiffer) Backeberg CL
Copiapoa calderana Ritter CL
Copiapoa calderana ssp. **calderana** CL
Copiapoa atacamensis Middleditch
?*Copiapoa boliviana* (Pfeiffer) Ritter
®*Copiapoa lembckei* Backeberg
Copiapoa calderana ssp. **longistaminea** (Ritter) Taylor CL
Copiapoa longistaminea Ritter
®*Copiapoa cinerea* ssp. *longistaminea* (Ritter) Meregalli
Copiapoa chanaralensis Ritter CL
®*Copiapoa chaniaralensis* Ritter
Copiapoa cinerascens (Salm-Dyck) Britton & Rose CL
®*Copiapoa applanata* Backeberg
Copiapoa cinerea (Philippi) Britton & Rose CL
Copiapoa cinerea ssp. **cinerea** CL
®?*Copiapoa cinerea* ssp. *columna-alba* (Ritter) Meregalli
Copiapoa cinerea ssp. **haseltoniana** (Backeberg) Taylor CL
Copiapoa cinerea ssp. *gigantea* (Backeberg) Slaba
Copiapoa eremophila Ritter
Copiapoa gigantea Backeberg
Copiapoa haseltoniana Backeberg
Copiapoa melanohystrix Ritter
Copiapoa tenebrosa Ritter
Copiapoa cinerea ssp. **krainziana** (Ritter) Slaba CL
Copiapoa krainziana Ritter
®***Copiapoa copiapensis*** (Pfeiffer) Meregalli CL
Copiapoa coquimbana (Karwinski ex Ruempler) Britton & Rose CL
Copiapoa alticostata Ritter
Copiapoa pendulina Ritter
Copiapoa pepiniana (Lemaire ex Salm-Dyck) Backeberg
Copiapoa serenana Voldan
Copiapoa vallenarensis Ritter
®*Copiapoa wagenknechtii* Ritter ex Backeberg
Copiapoa desertorum Ritter CL
Copiapoa echinoides (Salm-Dyck) Britton & Rose CL
Copiapoa cuprea Ritter
Copiapoa cupreata (Poselger ex Ruempler) Backeberg
Copiapoa dura Ritter

Copiapoa fiedleriana (Schumann) Backeberg . . . CL

Copiapoa echinata Ritter

Copiapoa intermedia Ritter ex Backeberg

Copiapoa pseudocoquimbana Ritter

Copiapoa totoralensis Ritter

Copiapoa hornilloensis Ritter . . . CL

Copiapoa humilis (Philippi) Hutchison . . . CL

Copiapoa esmeraldana Ritter

Copiapoa longispina Ritter

Copiapoa paposoensis Ritter

Copiapoa taltalensis (Werdermann) Looser

Copiapoa hypogaea Ritter . . . CL

⊚*Copiapoa barquitensis* Ritter

Copiapoa laui Diers & Esteves Pereira . . . CL

⊚***Copiapoa macracantha*** (Salm-Dyck) Meregalli . . . CL

Copiapoa malletiana (Lemaire ex Salm-Dyck) Backeberg . . . CL

Copiapoa carrizalensis Ritter

Copiapoa dealbata Ritter

Copiapoa cinerea ssp. *dealbata* (Ritter) Slaba

Copiapoa marginata (Salm-Dyck) Britton & Rose . . . CL

Copiapoa streptocaulon (Hooker) Ritter

Copiapoa megarhiza Britton & Rose . . . CL

Copiapoa brunnescens Backeberg

Copiapoa montana Ritter . . . CL

Copiapoa montana ssp. **montana** . . . CL

Copiapoa mollicula Ritter

Copiapoa olivana Ritter

Copiapoa rarissima Ritter

Copiapoa montana ssp. **grandiflora** (Ritter) Taylor . . . CL

Copiapoa grandiflora Ritter

Copiapoa rupestris Ritter . . . CL

Copiapoa rubriflora Ritter

Copiapoa serpentisulcata Ritter . . . CL

Copiapoa castanea Ritter

Copiapoa solaris (Ritter) Ritter . . . CL

Pilocopiapoa solaris Ritter

⊚*Copiapoa ferox* Lembcke & Backeberg

⊗*Copiapoa conglomerata* (Philippi) Lembcke

⊚***Copiapoa tenuissima*** Ritter . . . CL

Copiapoa tocopillana Ritter . . . CL

Copiapoa varispinata Ritter . . . CL

Corryocactus acervatus Ritter . . . PE

Corryocactus apiciflorus (Vaupel) Hutchison ex Buxbaum . . . PE

Erdisia apiciflora (Vaupel) Werdermann

Erdisia maxima Backeberg

Corryocactus maximus (Backeberg) Hutchison ex Buxbaum

Corryocactus aureus (Meyen) Hutchison ex Buxbaum . . . PE

Erdisia meyenii Britton & Rose

Corryocactus ayacuchoensis Rauh & Backeberg . . . PE

Corryocactus ayopayanus Cardenas . . . BO

Corryocactus brachycladus Ritter . . . PE

Corryocactus brachypetalus (Vaupel) Britton & Rose . . . PE

Corryocactus brevispinus Rauh & Backeberg . . . PE

Corryocactus brevistylus (Schumann) Britton & Rose . . . PE, CL

Corryocactus brevistylus ssp. **brevistylus** . . . PE, CL

Corryocactus krausii Backeberg

Corryocactus brevistylus ssp. **puquiensis** (Rauh & Backeberg) Ostolaza . . . PE

Corryocactus puquiensis Rauh & Backeberg

Corryocactus chachapoyensis Ochoa & Backeberg . . . PE

Corryocactus charazanensis Cardenas . . . . BO
Corryocactus chavinilloensis Ritter . . . . PE
Corryocactus cuajonesensis Ritter . . . . PE
Corryocactus erectus (Backeberg) Ritter . . . . PE
Erdisia aureispina Backeberg & Jacobsen
Erdisia erecta Backeberg
Bolivicereus pisacensis Knize
Borzicactus pisacensis (Knize) Rowley
Erdisia ruthae Johnson ex Backeberg
Bolivicereus soukupii Knize
Borzicactus soukupii (Knize) Rowley
Corryocactus gracilis Ritter . . . . PE
Corryocactus heteracanthus Backeberg . . . . PE
Corryocactus huincoensis Ritter . . . . PE
Corryocactus matucanensis Ritter
Corryocactus megarhizus Ritter . . . . PE
Corryocactus melaleucus Ritter . . . . PE
Corryocactus melanotrichus (Schumann) Britton & Rose . . . . BO
Corryocactus odoratus Ritter . . . . PE
Corryocactus otuyensis Cardenas . . . . BO
Corryocactus perezianus Cardenas . . . . BO
Corryocactus pilispinus Ritter . . . . PE
Corryocactus prostratus Ritter . . . . PE
Corryocactus pulquinensis Cardenas . . . . BO
Corryocactus pachycladus Rauh & Backeberg
Corryocactus pyroporphyranthus Ritter . . . . PE
Corryocactus quadrangularis (Rauh & Backeberg) Ritter . . . . PE
Erdisia quadrangularis Rauh & Backeberg
Corryocactus quivillanus Ritter . . . . PE
Corryocactus serpens Ritter . . . . PE
Corryocactus solitarius Ritter . . . . PE
Corryocactus squarrosus (Vaupel) Hutchison ex Buxbaum . . . . PE
Erdisia squarrosa (Vaupel) Britton & Rose
Corryocactus tarijensis Cardenas . . . . BO
Corryocactus tenuiculus (Backeberg) Hutchison ex Buxbaum . . . . PE
Erdisia fortalezensis Ritter
Erdisia tenuicula Backeberg

Coryphantha calipensis Bravo . . . . MX
Coryphantha calochlora Boedeker . . . . MX
Coryphantha clavata (Scheidweiler) Backeberg . . . . MX
Neolloydia clavata (Scheidweiler) Britton & Rose
Coryphantha compacta (Engelmann) Britton & Rose . . . . MX
¶*Coryphantha palmeri* Britton & Rose
Coryphantha cornifera (De Candolle) Lemaire . . . . MX
Coryphantha cornuta (Hildmann) Berger . . . . MX
Coryphantha delaetiana (Quehl) Berger . . . . MX
Coryphantha cuencamensis Bremer
Coryphantha gladiispina (Boedeker) Berger
Coryphantha laui Bremer
Coryphantha pseudonickelsiae Backeberg
®*Coryphantha salm-dyckiana* (Scheer ex Salm-Dyck) Britton & Rose
Coryphantha difficilis (Quehl) Orcutt . . . . MX
Coryphantha durangensis (Schumann) Britton & Rose . . . . MX
Coryphantha echinus (Engelmann) Orcutt . . . . US, MX
Coryphantha pectinata (Engelmann) Britton & Rose
Coryphantha echinoidea (Quehl) Britton & Rose . . . . MX
Coryphantha schwarziana Boedeker
Coryphantha elephantidens (Lemaire) Lemaire . . . . MX
Coryphantha bumamma (Ehrenberg) Britton & Rose
Coryphantha garessii Bremer

Coryphantha greenwoodii Bravo
Coryphantha recurvispina (Engelmann) Bremer
Coryphantha erecta (Pfeiffer) Lemaire . . . . MX
Coryphantha georgii Boedeker . . . . MX
Coryphantha villarensis Backeberg
Coryphantha glanduligera (Otto ex Dietrich) Lemaire . . . . MX
Coryphantha bergeriana Boedeker
Coryphantha gracilis Bremer & Lau . . . . MX
Coryphantha grata Bremer . . . . MX
Coryphantha guerkeana (Boedeker) Britton & Rose . . . . MX
Coryphantha indensis Bremer . . . . MX
Coryphantha jalpanensis F. Buchenau . . . . MX
®***Coryphantha jaumavei*** Fric . . . . MX
®*Coryphantha palmeri* auctt
Coryphantha longicornis Boedeker . . . . MX
Coryphantha grandis Bremer
Coryphantha macromeris (Engelmann) Lemaire . . . . US, MX
Lepidocoryphantha macromeris (Engelmann) Backeberg
Coryphantha pirtlei Werdermann
Coryphantha macromeris ssp. **macromeris** . . . . US, MX
Coryphantha macromeris ssp. **runyonii** (Britton & Rose) Taylor . . . . US
Coryphantha runyonii Britton & Rose
Lepidocoryphantha runyonii (Britton & Rose) Backeberg
Coryphantha maiz-tablasensis Schwarz ex Backeberg . . . . MX
Coryphantha maliterrarum Bremer . . . . MX
Coryphantha melleospina Bravo . . . . MX
Coryphantha nickelsiae (K. Brandegee) Britton & Rose . . . . US, MX
Coryphantha neglecta Bremer . . . . MX
Coryphantha octacantha (De Candolle) Britton & Rose . . . . MX
Coryphantha clava (Pfeiffer) Lemaire
Coryphantha odorata Boedeker . . . . MX
Cumarinia odorata (Boedeker) Buxbaum
Neolloydia odorata (Boedeker) Backeberg
Coryphantha ottonis (Pfeiffer) Lemaire . . . . MX
Coryphantha asterias (Cels ex Salm-Dyck) Hubner
Coryphantha bussleri (Mundt) Scheinvar
Coryphantha exsudans (Zuccarini) Lemaire
Coryphantha pallida Britton & Rose . . . . MX
Coryphantha poselgeriana (Dietrich) Britton & Rose . . . . MX
Coryphantha valida (Purpus) Bremer
Coryphantha potosiana (Jacobi) Glass & Foster . . . . MX
Coryphantha pseudoechinus Boedeker . . . . MX
Coryphantha pseudoradians Bravo . . . . MX
Coryphantha pulleineana (Backeberg) Glass . . . . MX
Neolloydia pulleineana Backeberg.
Coryphantha pusilliflora Bremer . . . . MX
Coryphantha pycnacantha (Martius) Lemaire . . . . MX
Coryphantha andreae (J. Purpus & Boedeker) Berger
Coryphantha connivens Britton & Rose
Coryphantha radians (De Candolle) Britton & Rose . . . . MX
Coryphantha bernalensis Bremer
Coryphantha delicata Bremer
Coryphantha ramillosa Cutak . . . . US
Coryphantha recurvata (Engelmann) Britton & Rose . . . . US, MX
Coryphantha reduncispina Boedeker . . . . MX
Coryphantha retusa (Pfeiffer) Britton & Rose . . . . MX
Coryphantha robustispina (Schott ex Engelmann) Britton & Rose . . . . US, MX
Coryphantha robustispina ssp. **robustispina** . . . . US, MX
®*Coryphantha muehlenpfordtii* ssp. *robustispina* (Schott ex Engelmann) Dicht

Coryphantha robustispina ssp. **scheeri** (Lemaire) Taylor .... US, MX
Coryphantha neoscheeri Backeberg
Coryphantha scheeri Lemaire
®*Coryphantha muehlenpfordtii* Britton & Rose
Coryphantha robustispina ssp. **uncinata** (Benson) Taylor .... US
®*Coryphantha muehlenpfordtii* ssp. *uncinata* (Benson) Dicht
Coryphantha salinensis (Poselger) A. Zimmerman ex Dicht & A. Luethy ...... MX
Coryphantha borwigii (J. Purpus) Berger
Coryphantha roederiana Boedeker
Coryphantha sulcata (Engelmann) Britton & Rose .... US
Coryphantha obscura Boedeker
Coryphantha speciosa Boedeker
Coryphantha sulcolanata (Lemaire) Lemaire .... MX
Coryphantha conimamma (Linke) Berger
Coryphantha tripugionacantha Lau .... MX
Coryphantha unicornis Bodeker .... MX
Coryphantha vaupeliana Boedeker .... MX
Coryphantha vogtherriana Werdermann & Boedeker .... MX
Coryphantha werdermannii Boedeker .... ♦MX
Coryphantha densispina Werdermann♦
Coryphantha wohlschlageri Holzeis .... MX

Dendrocereus nudiflorus (Engelmann ex Sauvalle) Britton & Rose .... CU
Cereus nudiflorus Engelmann ex Sauvalle
Dendrocereus undulosus (De Candolle) Britton & Rose .... HT
Acanthocereus undulosus (De Candolle) Croizat

Denmoza rhodacantha (Salm-Dyck) Britton & Rose .... AR
Denmoza erythrocephala (Schumann) Berger

Discocactus bahiensis Britton & Rose .... ♦BR
Discocactus bahiensis ssp. *subviridigriseus*
(Buining & Brederoo) Braun & Esteves Pereira♦
Discocactus subviridigriseus Buining et al.♦
Discocactus ferricola Buining & Brederoo .... ♦BR, BO
Discocactus heptacanthus (Rodrigues) Britton & Rose .... ♦BR, BO
Discocactus paranaensis Backeberg♦
Discocactus heptacanthus ssp. **heptacanthus** .... ♦BR, BO, PY
Discocactus boliviensis Buining et al.♦
Discocactus cangaensis Diers & Esteves Pereira♦
Discocactus catingicola ssp. *griseus*
(Buining & Brederoo) Braun & Esteves Pereira♦
Discocactus catingicola ssp. *rapirhizus*
(Buining & Brederoo) Braun & Esteves Pereira♦
Discocactus cephaliaciculosus Buining & Brederoo♦
Discocactus crassispinus Braun & Esteves Pereira♦
Discocactus crassispinus ssp. *araguaiensis* Braun & Esteves Pereira♦
Discocactus diersianus Esteves♦
Discocactus diersianus ssp. *goianus*
(Diers & Esteves Pereira) Braun & Esteves Pereira♦
Discocactus estevesii Diers & Esteves Pereira♦
Discocactus flavispinus Buining et al.♦
Discocactus goianus Diers & Esteves Pereira♦
Discocactus griseus Buining & Brederoo♦
Discocactus hartmannii ssp. *setosiflorus* Braun & Esteves Pereira♦
Discocactus heptacanthus ssp. *melanochlorus*
(Buining et al.) Braun & Esteves Pereira♦
Discocactus lindanus Diers & Esteves Pereira♦
Discocactus melanochlorus Buining et al.♦
Discocactus prominentigibbus Diers & Esteves Pereira♦

Discocactus rapirhizus Buining & Brederoo♦
Discocactus semicampaniflorus Buining & Brederoo♦
Discocactus silicicola Buining & Brederoo♦
Discocactus silvaticus Buining et al.♦
Discocactus squamibaccatus Buining et al.♦
Discocactus subterraneo-proliferans Diers & Esteves Pereira♦
®*Discocactus cephaliaciculosus* ssp. *nudicephalus* Braun & Esteves Pereira♦
®*Discocactus lindaianus* Diers & Esteves Pereira
Discocactus heptacanthus ssp. **catingicola** (Buining & Brederoo) Taylor & Zappi ♦BR
Discocactus catingicola Buining & Brederoo♦
Discocactus nigrisaetosus Buining et al.♦
Discocactus piauiensis Braun & Esteves Pereira♦
Discocactus spinosior Buining et al.♦
Discocactus heptacanthus ssp. **magnimammus** (Buining & Brederoo) Taylor & Zappi . . . . ♦BR, PY
Discocactus hartmannii (Schumann) Britton & Rose♦
Discocactus hartmannii ssp. *giganteus* Braun & Esteves Pereira♦
Discocactus hartmannii ssp. *magnimammus* (Buining & Brederoo) Braun & Esteves Pereira♦
Discocactus hartmannii ssp. *patulifolius* (Buining & Brederoo) Braun & Esteves Pereira♦
Discocactus magnimammus Buining & Brederoo♦
Discocactus magnimammus ssp. *bonitoensis* Buining et al.♦
Discocactus mamillosus Buining & Brederoo♦
Discocactus pachythele Buining & Brederoo♦
Discocactus patulifolius Buining & Brederoo♦
Discocactus horstii Buining & Brederoo . . . . ♦BR
Discocactus woutersianus Brederoo & [van den] Broek♦
Discocactus placentiformis (Lehmann) Schumann . . . . ♦BR
Discocactus alteolens Lemaire ex Dietrich♦
Discocactus crystallophilus Diers & Esteves Pereira♦
Discocactus insignis Pfeiffer♦
Discocactus latispinus Buining et al.♦
Discocactus latispinus ssp. *pseudolatispinus* (Diers & Esteves Pereira) Braun & Esteves Pereira♦
Discocactus latispinus ssp. *pulvinicapitatus* (Buining & Brederoo) Braun & Esteves Pereira♦
Discocactus multicolorispinus Braun & Brederoo♦
Discocactus placentiformis ssp. *alteolens* (Lemaire) Braun & Esteves Pereira♦
Discocactus placentiformis ssp. *multicolorispinus* (Braun & Brederoo) Braun & Esteves Pereira♦
Discocactus placentiformis ssp. *pugionacanthus* (Buining et al.) Braun & Esteves Pereira♦
Discocactus pseudolatispinus Diers & Esteves Pereira♦
Discocactus pugionacanthus Buining et al.♦
Discocactus pulvinicapitatus Buining & Brederoo♦
Discocactus tricornis Monville ex Pfeiffer♦
Discocactus pseudoinsignis Taylor & Zappi . . . . ♦BR
Discocactus zehntneri Britton & Rose . . . . ♦BR
Discocactus zehntneri ssp. **zehntneri** . . . . ♦BR
Discocactus albispinus Buining & Brederoo♦
Discocactus zehntneri ssp. *albispinus* (Buining & Brederoo) Braun & Esteves Pereira♦
Discocactus zehntneri ssp. **boomianus** (Buining & Brederoo) Taylor & Zappi . . ♦BR
Discocactus araneispinus Buining et al.♦
Discocactus boomianus Buining & Brederoo♦
Discocactus buenekeri Abraham♦
Discocactus zehntneri ssp. *araneispinus* (Buining et al.) Braun & Esteves Pereira♦
Discocactus zehntneri ssp. *buenekeri* (Abraham) Braun & Esteves Pereira♦
Discocactus zehntneri ssp. *horstiorum* (Braun) Braun & Esteves Pereira♦

Disocactus ackermannii (Lindley) Barthlott . . . . MX
Epiphyllum ackermannii Haworth
Nopalxochia ackermannii (Haworth) Knuth
Nopalxochia conzattiana MacDougall
Pseudonopalxochia conzattiana (MacDougall) Backeberg
Disocactus amazonicus (Schumann) Hunt . . . . CR, PA, BR, CO, VE, EC, PE
Wittia amazonica Schumann
Wittiocactus amazonicus (Schumann) Rauschert
Wittia panamensis Britton & Rose
Wittiocactus panamensis (Britton & Rose) Rauschert
Disocactus aurantiacus (Kimnach) Barthlott . . . . MX, HN, NI
Heliocereus aurantiacus Kimnach
Disocactus biformis (Lindley) Lindley . . . . GT, HN
Epiphyllum biforme (Lindley) Don
Disocactus cinnabarinus (Eichlam) Barthlott . . . . MX, GT, HN, SV
Heliocereus cinnabarinus (Eichlam ex Weingart) Britton & Rose
Heliocereus heterodoxus Standley & Steyermark
Disocactus eichlamii (Weingart) Britton & Rose . . . . GT
Epiphyllum eichlamii (Weingart) L.O. Williams
Disocactus flagelliformis (Linnaeus) Barthlott . . . . MX
Aporocactus flagelliformis (Linnaeus) Lemaire
Aporocactus flagriformis (Pfeiffer) Lemaire
Aporocactus leptophis (De Candolle) Britton & Rose
Disocactus kimnachii Rowley . . . . CR
Nopalxochia horichii Kimnach
Disocactus macdougallii (Alexander) Barthlott . . . . ♦MX
Lobeira macdougallii Alexander♦
Nopalxochia macdougallii (Alexander) W.T. Marshall♦
Disocactus macranthus (Alexander) Kimnach & Hutchison . . . . MX
Pseudorhipsalis macrantha Alexander
Disocactus martianus (Zuccarini) Barthlott . . . . MX
Aporocactus conzattii Britton & Rose
Aporocactus martianus (Zuccarini) Britton & Rose
Disocactus nelsonii (Britton & Rose) Lindinger . . . . MX, GT, HN
Chiapasia nelsonii (Britton & Rose) Britton & Rose
Epiphyllum nelsonii Britton & Rose
Disocactus phyllanthoides (De Candolle) Barthlott . . . . MX
Nopalxochia phyllanthoides (De Candolle) Britton & Rose
Disocactus quezaltecus (Standley & Steyermark) Kimnach . . . . GT
Epiphyllum quezaltecum (Standley & Steyermark) L.O. Williams
Disocactus schrankii (Zuccarini ex Seitz) Barthlott . . . . MX
Heliocereus elegantissimus Britton & Rose
Heliocereus luzmariae Scheinvar
Heliocereus schrankii (Zuccarini ex Seitz) Britton & Rose
Disocactus speciosus (Cavanilles) Barthlott . . . . MX
Heliocereus speciosissimus (De Candolle) Y. Ito
Heliocereus speciosus (Cavanilles) Britton & Rose

Echinocactus grusonii Hildmann . . . . MX
Echinocactus horizonthalonius Lemaire . . . . US, MX
Meyerocactus horizonthalonius (Lemaire) Doweld
Echinocactus parryi Engelmann . . . . MX
Emorycactus parryi (Engelmann) Doweld
Echinocactus platyacanthus Link & Otto . . . . MX
Echinocactus grandis Rose
Echinocactus ingens Zuccarini ex Pfeiffer
Echinocactus palmeri Rose
Echinocactus visnaga Hooker
Echinocactus polycephalus Engelmann & Bigelow . . . . US, MX
Emorycactus polycephalus (Engelmann & Bigelow) Doweld

Echinocactus polycephalus ssp. **polycephalus** .... US, MX
Echinocactus polycephalus ssp. **xeranthemoides** (J.M. Coulter) Taylor .... US
Emorycactus xeranthemoides (Engelmann ex J. Coulter) Doweld
Echinocactus texensis Hopffer .... US, MX
Homalocephala texensis (Hopffer) Britton & Rose

Echinocereus adustus Engelmann .... MX
Echinocereus adustus ssp. **adustus** .... MX
Echinocereus radians Engelmann
Echinocereus rufispinus Engelmann
Echinocereus adustus ssp. **bonatzii** (R.C. Roemer) Taylor .... MX
Echinocereus bonatzii R.C. Roemer
Echinocereus pamanesiorum ssp. *bonatzii* (R.C. Roemer) R.C. Roemer
Echinocereus adustus ssp. **schwarzii** (Lau) Taylor .... MX
Echinocereus schwarzii Lau
Echinocereus apachensis Blum & Rutow .... US
Echinocereus barthelowanus Britton & Rose .... MX
Echinocereus berlandieri (Engelmann) Hort. F.A. Haage .... US, MX
⊗*Echinocereus blanckii* Palmer
Echinocereus bonkerae Thornber & Bonker .... US
Echinocereus fasciculatus ssp. *bonkerae* (Thornber & Bonker) Taylor
Echinocereus boyce-thompsonii Orcutt .... US
Echinocereus fasciculatus ssp. *boyce-thompsonii* (Orcutt) Taylor
Echinocereus brandegeei (J. Coulter) Schumann .... MX
Echinocereus mamillatus (Engelmann) Britton & Rose
Echinocereus bristolii W.T. Marshall .... MX
Echinocereus chisoensis W.T. Marshall .... US, MX
Echinocereus fobeanus Oehme
Echinocereus fobeanus ssp. *metornii* (G.R.W Frank) Blum & Lange
Echinocereus metornii G.R.W. Frank
Echinocereus cinerascens (De Candolle) Lemaire .... MX
Echinocereus cinerascens ssp. **cinerascens** .... MX
Echinocereus chlorophthalmus (Hooker) Britton & Rose
Echinocereus cinerascens ssp. *ehrenbergii* (Pfeiffer) Blum & Rutow
Echinocereus ehrenbergii (Pfeiffer) Ruempler
Echinocereus glycimorphus Ruempler
Echinocereus cinerascens ssp. **septentrionalis** (Taylor) Taylor .... MX
Echinocereus cinerascens ssp. **tulensis** (Bravo) Taylor .... MX
Echinocereus tulensis Bravo
Echinocereus coccineus Engelmann .... US, MX
Echinocereus coccineus ssp. **coccineus** .... US, MX
Echinocereus arizonicus Rose ex Orcutt
Echinocereus arizonicus ssp. *matudae* (Bravo) Rutow
Echinocereus arizonicus ssp. *nigrihorridispinus* Blum & Rutow
Echinocereus canyonensis Clover & Jotter
Echinocereus coccineus ssp. *aggregatus* (Engelmann ex B.D. Jackson) Blum et al.
Echinocereus coccineus ssp. *rosei* (Wooton & Standley) Blum & Rutow
Echinocereus decumbens Clover & Jotter
Echinocereus hexaedrus (Engelmann) Ruempler
Echinocereus krausei Smet ex Ruempler
Echinocereus kunzei Guerke
Echinocereus matudae Bravo
Echinocereus roemeri Hort. F.A. Haage
Echinocereus rosei Wooton & Standley
Echinocereus toroweapensis (Fischer) Fuersch
?*Echinocereus neomexicanus* Standley
Echinocereus coccineus ssp. *paucispinus* (Engelmann) Blum et al.
Echinocereus coccineus ssp. *roemeri* (Muehlenpfordt) Blum et al.
Echinocereus dasyacanthus Engelmann .... US, MX
Echinocereus ctenoides (Engelmann) Lemaire

Echinocereus hildmannii Arendt
Echinocereus pectinatus ssp. *ctenoides* (Engelmann) G.R.W. Frank
Echinocereus steereae Clover
Echinocereus dasyacanthus ssp. ***rectispinus*** (Trocha & Fethke) Blum et al. ..... MX
Echinocereus engelmannii (Parry ex Engelmann) Lemaire .... US, MX
Echinocereus munzii (Parish) L. Benson
Echinocereus engelmannii ssp. ***decumbens*** (Clover & Jotter) Blum & Lange ..... US
Echinocereus enneacanthus Engelmann .... US, MX
Echinocereus enneacanthus ssp. **enneacanthus** .... US, MX
Echinocereus dubius (Engelmann) Ruempler
Echinocereus merkeri Hildmann ex Schumann
Echinocereus sarissophorus Britton & Rose
Echinocereus uspenskii Hort. A. Blanc ex Haage
Echinocereus enneacanthus ssp. **brevispinus** (W.O. Moore) Taylor .... US, MX
Echinocereus fasciculatus (Engelmann ex B.D. Jackson) L. Benson .... US, MX
Echinocereus abbeae Parsons
Echinocereus engelmannii ssp. *fasciculatus* (Engelmann ex Jackson) Blum et al.
Echinocereus fendleri (Engelmann) Seitz .... US, MX
Echinocereus fendleri ssp. **fendleri** .... US, MX
Echinocereus kuenzleri Castetter, Pierce & Schwerin
Echinocereus fendleri ssp. ***hempelii*** (Fobe) Blum .... MX
Echinocereus hempelii Fobe
Echinocereus fendleri ssp. **rectispinus** (Peebles) Taylor .... US, MX
Echinocereus rectispinus Peebles
Echinocereus ferreirianus H.E. Gates .... MX
Echinocereus ferreirianus ssp. **ferreirianus** .... MX
Echinocereus ferreirianus ssp. **lindsayi** (Meyran) Taylor .... ♦MX
Echinocereus lindsayi Meyran♦
Echinocereus freudenbergeri G.R.W. Frank .... MX
Echinocereus longisetus ssp. *freudenbergeri* (G.R.W. Frank) Blum
Echinocereus grandis Britton & Rose .... MX
Echinocereus klapperi Blum .... MX
Echinocereus knippelianus Liebner .... MX
Echinocereus knippelianus ssp. *kruegeri* (Glass & Foster) Glass
Echinocereus knippelianus ssp. *reyesii* (Lau) Blum & Lange
Echinocereus laui G.R.W. Frank .... MX
Echinocereus ledingii Peebles .... US
Echinocereus leucanthus Taylor .... MX
Wilcoxia albiflora Backeberg
Echinocereus longisetus (Engelmann) Lemaire .... MX
Echinocereus longisetus ssp. **longisetus** .... MX
Echinocereus longisetus ssp. **delaetii** (Guerke) Taylor .... MX
Echinocereus delaetii (Guerke) Guerke
Echinocereus mapimiensis E.F. Anderson et al. .... MX
Echinocereus maritimus (M.E. Jones) Schumann .... MX
Echinocereus maritimus ssp. ***hancockii*** (Dawson) Blum & Rutow .... MX
Echinocereus hancockii E. Dawson
Echinocereus mojavensis (Engelmann & Bigelow) Ruempler .... US, MX
Echinocereus coccineus ssp. *mojavensis* (Engelmann & Bigelow) Taylor
Echinocereus triglochidiatus ssp. *mojavensis* (Engelmann & Bigelow) Blum & Lange
Echinocereus nicholii (L. Benson) Parfitt .... US, MX
Echinocereus nicholii ssp. ***llanuraensis*** Rutow .... MX
Echinocereus nivosus Glass & Foster .... MX
®*Echinocereus albatus* Backeberg
Echinocereus ortegae Rose ex J.G. Ortega .... MX
Echinocereus ortegae ssp. *koehresianus* (G.R.W. Frank) W. Rischer & G.R.W. Frank
Echinocereus palmeri Britton & Rose .... MX
Echinocereus pamanesiorum Lau .... MX
Echinocereus papillosus Linke ex Ruempler .... US, MX
Echinocereus angusticeps Clover

Echinocereus parkeri Taylor . . . . MX
Echinocereus parkeri ssp. **parkeri** . . . . MX
Echinocereus parkeri ssp. ***arteagensis*** Blum & Lange . . . . MX
Echinocereus parkeri ssp. **gonzalezii** (Taylor) Taylor . . . . MX
Echinocereus parkeri ssp. **mazapilensis** Blum & Lange . . . . MX
Echinocereus pectinatus (Scheidweiler) Engelmann . . . . US, MX
Echinocereus pectinatus ssp. ***wenigeri*** (Benson) Blum & Rutow . . . . US
Echinocereus pensilis (K. Brandegee) J. Purpus . . . . MX
  Morangaya pensilis (K. Brandegee) Rowley
Echinocereus pentalophus (De Candolle) Lemaire . . . . US, MX
Echinocereus pentalophus ssp. **pentalophus** . . . . US, MX
  Echinocereus leptacanthus (Salm-Dyck) Ruempler
Echinocereus pentalophus ssp. **leonensis** (Mathsson) Taylor . . . . MX
  Echinocereus leonensis Mathsson
Echinocereus pentalophus ssp. **procumbens** (Engelmann) Blum & Lange . . . US, MX
Echinocereus polyacanthus Engelmann . . . . US, MX
Echinocereus polyacanthus ssp. **polyacanthus** . . . . US, MX
  Echinocereus durangensis Poselger ex Ruempler
  Echinocereus leeanus (Hooker) Lemaire
  Echinocereus santaritensis Blum & Rutow
Echinocereus polyacanthus ssp. **acifer** (Otto ex Salm-Dyck) Taylor . . . . MX
  Echinocereus acifer (Otto ex Salm-Dyck) Hort. F.A. Haage
  Echinocereus acifer ssp. *tubiflorus* Rischer
  Echinocereus marksianus F. Schwarz ex Backeberg
Echinocereus polyacanthus ssp. **huitcholensis** (Weber) Taylor . . . . MX
  Echinocereus acifer ssp. *huitcholensis* (Weber) Lange
  Echinocereus huitcholensis (Weber) Guerke
  ®*Echinocereus matthesianus* Backeberg
Echinocereus polyacanthus ssp. **pacificus** (Engelmann) Breckwoldt . . . . MX
  Echinocereus mombergerianus G.R.W. Frank
  Echinocereus pacificus (Engelmann) Hort. F.A. Haage
  Echinocereus pacificus ssp. *mombergerianus* (G.R.W. Frank) Blum & Rutow
  Echinocereus polyacanthus ssp. *mombergerianus* (G.R.W. Frank) Breckwoldt
Echinocereus poselgeri Lemaire . . . . US, MX
  Wilcoxia kroenleinii A. Cartier
  Cereus poselgeri (Lemaire) Coulter
  Wilcoxia poselgeri (Lemaire) Britton & Rose
  Echinocereus poselgeri ssp. *kroenleinii* (Cartier) Lange
  Wilcoxia tamaulipensis Werdermann
  Echinocereus tamaulipensis ssp. *deherdtii* Lange
  Echinocereus tamaulipensis ssp. *waldeisii* (Haugg) Lange
  Wilcoxia tuberosa Kreuzinger
  Echinocereus waldeisii Haugg
Echinocereus primolanatus F. Schwarz ex Taylor . . . . MX
Echinocereus pseudopectinatus (Taylor) Taylor . . . . US, MX
  Echinocereus scopulorum ssp. *pseudopectinatus* (Taylor) Blum & Lange
Echinocereus pulchellus (Martius) Seitz . . . . MX
Echinocereus pulchellus ssp. **pulchellus** . . . . MX
  Echinocereus amoenus (Dietrich) Schumann
Echinocereus pulchellus ssp. **acanthosetus** (Arias & Guzman) Blum . . . . MX
Echinocereus pulchellus ssp. **sharpii** (Taylor) Taylor . . . . MX
Echinocereus pulchellus ssp. **weinbergii** (Weingart) Taylor . . . . MX
  Echinocereus pulchellus ssp. *venustus* Blum & Rischer
  Echinocereus weinbergii Weingart
Echinocereus rayonesensis Taylor . . . . MX
Echinocereus reichenbachii (Terscheck ex Walpers) Hort. F.A. Haage . . . . . US, MX
  Echinocereus caespitosus (Engelmann) Engelmann
  Echinocereus mariae Backeberg
  Echinocereus purpureus Lahman
Echinocereus reichenbachii ssp. **reichenbachii** . . . . US, MX

Echinocereus reichenbachii ssp. **armatus** (Poselger) Taylor .... MX
Echinocereus armatus (Poselger) Berger
Echinocereus fitchii ssp. *armatus* (Poselger) Blum et al.
Echinocereus reichenbachii ssp. **baileyi** (Rose) Taylor .... US
Echinocereus albispinus Lahman
Echinocereus baileyi Rose
Echinocereus reichenbachii ssp. **fitchii** (Britton & Rose) Taylor .... US, MX
Echinocereus fitchii Britton & Rose
Echinocereus fitchii ssp. *albertii* (Benson) Blum & Lange
®*Echinocereus melanocentrus* Lowry
Echinocereus reichenbachii ssp. **perbellus** (Britton & Rose) Taylor .... US
Echinocereus perbellus Britton & Rose
®*Echinocereus reichenbachii* ssp. *caespitosus* (Engelmann) Blum & Lange
Echinocereus rigidissimus (Engelmann) Hort. F.A. Haage .... US, MX
Echinocereus rigidissimus ssp. **rigidissimus** .... US, MX
Echinocereus rigidissimus ssp. **rubispinus** (G.R.W. Frank & Lau) Taylor ...... MX
Echinocereus ×roetteri (Engelmann) Ruempler .... US
Echinocereus lloydii Britton & Rose
Echinocereus russanthus Weniger .... US, MX
Echinocereus russanthus ssp. *fiehnii* (Trocha) Blum & Lange
?*Echinocereus russanthus* ssp. *weedinii* Leuck ex Blum & Lange
Echinocereus scheeri (Salm-Dyck) Scheer .... MX
Echinocereus scheeri ssp. **scheeri** .... MX
?*Echinocereus salm-dyckianus* Scheer
Echinocereus salm-dyckianus ssp. *obscuriensis* (Lau) Blum
?*Echinocereus salmianus* Ruempler
Echinocereus sanpedroensis Raudonat & Rischer
Echinocereus scheeri ssp. **gentryi** (Clover) Taylor .... MX
Echinocereus gentryi Clover
Echinocereus schereri G.R.W. Frank .... MX
Echinocereus schmollii (Weingart) Taylor .... ♦MX
Wilcoxia nerispina [Schmoll ex] A. Cartier♦
Wilcoxia schmollii (Weingart) Backeberg♦
Echinocereus sciurus (K. Brandegee) Dams .... MX
®?*Echinocereus subterraneus* Backeberg
Echinocereus sciurus ssp. **sciurus** .... MX
Echinocereus sciurus ssp. **floresii** (Backeberg) Taylor .... MX
Echinocereus bristolii ssp. *floresii* Blum & Lange
Echinocereus floresii Schwarz ex Backeberg
Echinocereus scopulorum Britton & Rose .... MX
Echinocereus spinigemmatus Lau .... MX
Echinocereus stoloniferus W.T. Marshall .... MX
Echinocereus stoloniferus ssp. **stoloniferus** .... MX
Echinocereus stoloniferus ssp. **tayopensis** (W.T. Marshall) Taylor .... MX
Echinocereus tayopensis W.T. Marshall
Echinocereus stramineus (Engelmann) Seitz .... US, MX
Echinocereus stramineus ssp. **stramineus** .... US, MX
Echinocereus conglomeratus Foerster ex Schumann
Echinocereus stramineus ssp. **occidentalis** (Taylor) Taylor .... MX
Echinocereus subinermis Salm-Dyck ex Scheer .... MX
Echinocereus subinermis ssp. **subinermis** .... MX
Echinocereus luteus Britton & Rose
Echinocereus subinermis ssp. **ochoterenae** (J.G. Ortega) Taylor .... MX
Echinocereus ochoterenae J.G. Ortega
Echinocereus triglochidiatus Engelmann .... US, MX
Echinocereus gonacanthus (Engelmann & Bigelow) Lemaire
Echinocereus viereckii Werdermann .... MX
Echinocereus viereckii ssp. **viereckii** .... MX
Echinocereus viereckii ssp. **morricalii** (Riha) Taylor .... MX
Echinocereus morricalii Riha
Echinocereus viereckii ssp. *huastecensis* Blum et al.

Echinocereus viridiflorus Engelmann . . . . US, MX
Echinocereus viridiflorus ssp. **viridiflorus** . . . . US
Echinocereus standleyi Britton & Rose
Echinocereus viridiflorus ssp. **chloranthus** (Engelmann) Taylor . . . . US, MX
Echinocereus chloranthus (Engelmann) Hort. F.A. Haage
?*Echinocereus carmenensis* Blum et al.
?*Echinocereus chloranthus* ssp. *neocapillus* (Weniger) Fuersch
?*Echinocereus chloranthus* ssp. *rhyolithensis* Blum & Lange
?*Echinocereus neocapillus* (Weniger) Blum & Lange
Echinocereus viridiflorus ssp. ***correllii*** (Benson) Blum & Lange . . . . US
Echinocereus viridiflorus ssp. **cylindricus** (Engelmann) Taylor . . . . US
Echinocereus chloranthus ssp. *cylindricus* (Engelmann) Fuersch
Echinocereus viridiflorus ssp. **davisii** (Houghton) Taylor . . . . US
Echinocereus davisii Houghton
Echinocereus websterianus G. Lindsay . . . . MX

Echinopsis adolfofriedrichii Moser . . . . PY
Echinopsis ancistrophora Spegazzini . . . . BO, AR
Pseudolobivia ancistrophora (Spegazzini) Backeberg
Echinopsis hamatacantha Backeberg
Pseudolobivia hamatacantha (Backeberg) Backeberg
Pseudolobivia leucorhodantha (Backeberg) Backeberg ex Krainz
Pseudolobivia pelecyrhachis (Backeberg) Backeberg ex Krainz
Echinopsis polyancistra Backeberg
Pseudolobivia polyancistra (Backeberg) Backeberg
Echinopsis ancistrophora ssp. ***arachnacantha*** (Buining & Ritter) Rausch . . . . BO
Echinopsis arachnacantha (Buining & Ritter) Friedrich
Lobivia arachnacantha Buining & Ritter
Echinopsis kratochviliana Backeberg
Pseudolobivia kratochviliana (Backeberg) Backeberg
Echinopsis torrecillasensis Cardenas
Pseudolobivia torrecillasensis (Cardenas) Backeberg
Echinopsis ancistrophora ssp. ***cardenasiana*** (Rausch) Rausch . . . . BO
Echinopsis cardenasiana (Rausch) Friedrich
Lobivia cardenasiana Rausch
Echinopsis ancistrophora ssp. ***pojoensis*** (Rausch) Rausch . . . . BO
Lobivia pojoensis Rausch
Echinopsis angelesiae (Kiesling) Rowley . . . . AR
Trichocereus angelesii Kiesling
Echinopsis antezanae (Cardenas) Friedrich & Rowley . . . . BO
Helianthocereus antezanae (Cardenas) Backeberg
Trichocereus antezanae Cardenas
Echinopsis arboricola (Kimnach) Mottram . . . . AR
Trichocereus arboricola Kimnach
Echinopsis arebaloi Cardenas . . . . BO
Echinopsis atacamensis (Philippi) Friedrich & Rowley . . . . BO, AR, CL
Helianthocereus atacamensis (Philippi) Backeberg
Echinopsis atacamensis ssp. **atacamensis** CL
Echinopsis atacamensis ssp. **pasacana** (Weber) G. Navarro . . . . AR, BO
Trichocereus eremophilus Ritter
Echinopsis pasacana (Weber ex Ruempler) Friedrich & Rowley
Helianthocereus pasacana (Weber) Backeberg
Trichocereus pasacana (Weber) Britton & Rose
Echinopsis rivierei (Backeberg) Friedrich & Rowley
Leucostele rivierei Backeberg
Trichocereus rivierei (Backeberg) Krainz
?*Echinopsis formosissima* Labouret
Echinopsis aurea Britton & Rose . . . . AR
Hymenorebutia aurea (Britton & Rose) Ritter
Lobivia aurea (Britton & Rose) Backeberg

Pseudolobivia aurea (Britton & Rose) Backeberg
Echinopsis cylindracea (Backeberg) Friedrich
Lobivia cylindracea Backeberg
Lobivia cylindrica Backeberg
Echinopsis fallax (Oehme) Friedrich
Lobivia fallax ssp. *aurea* (Britton & Rose) Herzog
Echinopsis leucomalla (Wessner) Friedrich
Echinopsis quinesensis (Rausch) H. Friedrich & Glaetzle
Hymenorebutia quinesensis (Rausch) Ritter
Lobivia shaferi Britton & Rose
Lobivia shaferi ssp. *fallax* (Oehme) Herzog
Lobivia shaferi ssp. *leucomalla* (Wessner) Herzog
Lobivia shaferi ssp. *rubriflora* Herzog
®*Pseudolobivia luteiflora* Backeberg
Echinopsis backebergii Werdermann ex Backeberg .... PE, BO
Lobivia backebergii (Werdermann) Backeberg
Lobivia backebergii ssp. *wrightiana* (Backeb.) Rausch ex Rowley
Lobivia backebergii ssp. *zecheri* (Rausch) Rausch ex Rowley
Echinopsis boedekeriana Harden
Lobivia oxyalabastra Cardenas & Rausch
Lobivia winteriana Ritter
Neolobivia winteriana (Ritter) Ritter
Lobivia wrightiana Backeberg
Lobivia zecheri Rausch
Echinopsis baldiana Spegazzini .... AR
Echinopsis bertramiana (Backeberg) Friedrich & Rowley .... BO
Helianthocereus bertramianus (Backeberg) Backeberg
Trichocereus bertramianus Backeberg
Echinopsis boyuibensis Ritter .... BO
Pseudolobivia boyuibensis (Ritter) Backeberg
Echinopsis brasiliensis Fric ex Pazout .... BR
Echinopsis bridgesii Salm-Dyck .... BO
Trichocereus tenuispinus Ritter
Echinopsis bridgesii ssp. ***yungasensis*** (Ritter) Braun & Esteves Pereira .... BO
Echinopsis yungasensis Ritter
Echinopsis bruchii (Britton & Rose) Castellanos & Lelong .... AR
Lobivia bruchii Britton & Rose
Soehrensia bruchii (Britton & Rose) Backeberg
Trichocereus bruchii (Britton & Rose) Ritter
Lobivia formosa ssp. *bruchii* (Britton & Rose) Rausch
Lobivia formosa ssp. *grandis* (Britton & Rose) Rausch
Echinopsis grandis (Britton & Rose) Friedrich & Rowley
Lobivia grandis Britton & Rose
Soehrensia grandis (Britton & Rose) Backeberg
Trichocereus grandis (Britton & Rose) Ritter
Echinopsis ingens (Backeberg) Friedrich & Rowley
Soehrensia ingens Backeberg
Trichocereus ingens (Backeberg) Ritter
Echinopsis ×cabrerae (Kiesling) G. Rowley .... AR
Trichocereus cabrerae Kiesling
Echinopsis caineana (Cardenas) Hunt .... BO
Lobivia caineana Cardenas
Echinopsis cajasensis Ritter .... BO
Echinopsis calliantholilacina Cardenas .... BO
Echinopsis callichroma Cardenas .... BO
Pseudolobivia callichroma (Cardenas) Backeberg
Echinopsis calochlora Schumann .... BR
?*Echinopsis grandiflora* Linke
Echinopsis calochlora ssp. ***glaetzleana*** Braun & Esteves Pereira .... BR
Echinopsis camarguensis (Cardenas) Friedrich & Rowley .... BO
Trichocereus camarguensis Cardenas

Echinopsis candicans (Gillies ex Salm-Dyck) Hunt ... AR
Trichocereus candicans (Salm-Dyck) Britton & Rose
Echinopsis courantii (Schumann) Friedrich & Rowley
Trichocereus courantii (Schumann) Backeberg
Trichocereus neolamprochlorus Backeberg
®*Echinopsis pseudocandicans* (Kiesling) H. Friedrich & Glaetzle
®*Helianthocereus pseudocandicans* Backeberg
®*Trichocereus pseudocandicans* Backeberg ex Kiesling
Echinopsis cephalomacrostibas (Werdermann & Backeb.) Friedrich & Rowle ... PE
Haageocereus cephalomacrostibas (Backeberg) P.V. Heath
Trichocereus cephalomacrostibas (Werdermann & Backeberg) Backeberg
Weberbauerocereus cephalomacrostibas (Werdermann & Backeberg) Ritter
Echinopsis cerdana Cardenas ... BO
Echinopsis chalaensis (Rauh & Backeberg) Friedrich & Rowley ... PE
Trichocereus chalaensis Rauh & Backeberg
Echinopsis chamaecereus H. Friedrich & Glaetzle ... AR
Chamaecereus silvestrii (Spegazzini) Britton & Rose
Lobivia silvestrii (Spegazzini) G. Rowley
Echinopsis chiloensis (Colla) Friedrich & Rowley ... CL
Trichocereus chiloensis (Colla) Britton & Rose
Echinopsis chrysantha Werdermann ... AR
Hymenorebutia chrysantha (Werdermann) Ritter
Lobivia chrysantha (Werdermann) Backeberg
Lobivia jujuiensis W. Haage
Lobivia polaskiana Backeberg
Echinopsis chrysochete Werdermann ... AR
Lobivia chrysochete (Werdermann) Wessner
Lobivia hystrix Ritter
Lobivia markusii Rausch
Echinopsis cinnabarina (Hooker) Labouret ... BO
Lobivia charcasina Cardenas
Lobivia cinnabarina (Hooker) Britton & Rose
Lobivia cinnabarina ssp. *prestoana* (Cardenas) Rausch ex Rowley
Lobivia cinnabarina ssp. *taratensis* (Cardenas) Rowley
Lobivia draxleriana Rausch
Lobivia oligotricha Cardenas
Lobivia prestoana Cardenas
Lobivia taratensis Cardenas
Lobivia walterspielii Boedeker
Lobivia zudanensis Cardenas
®*Lobivia acanthoplegma* (Backeberg) Backeberg
®*Pseudolobivia acanthoplegma* Backeberg
®*Lobivia cinnabarina* ssp. *acanthoplegma* (Backeberg) Rausch
®*Lobivia neocinnabarina* Backeberg
®*Lobivia pseudocinnabarina* Backeberg
Echinopsis clavatus (Ritter) Hunt ... BO
Trichocereus clavatus Ritter
Echinopsis cochabambensis Backeberg ... BO
Echinopsis comarapana Cardenas ... BO
Echinopsis ayopayana Ritter & Rausch
Echinopsis pereziensis Cardenas
Echinopsis conaconensis (Cardenas) Friedrich & Rowley ... BO
Helianthocereus conaconensis (Cardenas) Backeberg
Trichocereus conaconensis Cardenas
Echinopsis coquimbana (Molina) Friedrich & Rowley ... CL
Trichocereus coquimbanus (Molina) Britton & Rose
Trichocereus serenanus Ritter
Echinopsis coronata Cardenas ... BO
Echinopsis cotacajesii Cardenas ... BO

Echinopsis crassicaulis (Kiesling) H. Friedrich & Glaetzle .... AR
Lobivia crassicaulis Kiesling
®*Helianthocereus crassicaulis* Backeberg
Echinopsis cuzcoensis (Britton & Rose) Friedrich & Rowley .... PE
Trichocereus cuzcoensis Britton & Rose
Echinopsis densispina Werdermann .... AR
Hymenorebutia chlorogona (Wessner) Ritter
Lobivia haematantha ssp. *densispina* (Werderm.) Rausch ex Rowley
Lobivia napina Pazout
Echinopsis rebutioides (Backeberg) Friedrich
Lobivia rebutioides Backeberg
Lobivia scoparia (Werderm.) Werderm. ex Backeb. & F. Knut
Echinopsis derenbergii Fric .... PY
Echinopsis deserticola (Werdermann) Friedrich & Rowley CL
Trichocereus deserticola (Werdermann) Looser
Echinopsis fulvilana (Ritter) Friedrich & Rowley
Trichocereus fulvilanus Ritter
Echinopsis escayachensis (Cardenas) Friedrich & Rowley .... BO
Helianthocereus escayachensis (Cardenas) Backeberg
Trichocereus escayachensis Cardenas
Echinopsis eyriesii (Turpin) Pfeiffer & Otto .... BR, UY, AR
Echinopsis pudantii Pfersd.
Echinopsis turbinata (Pfeiffer) Pfeiffer & Otto
Echinopsis fabrisii (Kiesling) G. Rowley .... AR
Trichocereus fabrisii Kiesling
Echinopsis famatinensis (Spegazzini) Werdermann .... AR
Lobivia famatimensis (Spegazzini) Britton & Rose
Rebutia famatinensis (Spegazzini) Spegazzini ex Hosseus
Reicheocactus pseudoreicheanus Backeberg
Echinopsis ferox (Britton & Rose) Backeberg .... BO, AR
Lobivia aureolilacina Cardenas
Lobivia backebergiana Ito
Lobivia claeysiana Backeberg
Lobivia ferox Britton & Rose
Pseudolobivia ferox (Britton & Rose) Backeberg
Lobivia hastifera Werdermann
Lobivia horrida Ritter
Echinopsis lecoriensis Cardenas
Pseudolobivia lecoriensis (Cardenas) Backeberg
Echinopsis longispina (Britton & Rose) Werdermann ex Backeberg
Lobivia longispina Britton & Rose
Pseudolobivia longispina (Britton & Rose) Backeberg ex Krainz
Lobivia pachyacantha Ito
Lobivia pictiflora Ritter
Echinopsis potosina Werdermann
Lobivia potosina (Werdermann) Friedrich
Pseudolobivia potosina (Werdermann) Backeberg ex Krainz
Lobivia varispina Ritter
?*Echinopsis orurensis* (Cardenas) Friedrich & Rowley
?*Helianthocereus orurensis* (Cardenas) Backeberg
?*Trichocereus orurensis* Cardenas
®*Lobivia wilkeae* (Backeberg) Friedrich
®*Pseudolobivia wilkeae* Backeberg
Echinopsis formosa (Pfeiffer) Jacobi ex Salm-Dyck .... AR, CL
Lobivia formosa (Pfeiffer) Dodds
Soehrensia formosa (Pfeiffer) Backeberg
Trichocereus formosus (Pfeiffer) Ritter
Lobivia kieslingii Rausch
Soehrensia oreopepon (Spegazzini) Backeberg

Echinopsis randallii (Cardenas) Friedrich & Rowley
Helianthocereus randallii (Cardenas) Backeberg
Trichocereus randallii Cardenas
Lobivia rosarioana Rausch
®?*Echinopsis uebelmanniana* (Lembcke & Backeberg) A.E. Hoffmann
®?*Soehrensia uebelmanniana* Lembcke & Backeberg
Echinopsis friedrichii Rowley . . . . AR
Echinopsis glauca (Ritter) Friedrich & Rowley . . . . PE, CL
Trichocereus glaucus Ritter
Echinopsis glaucina Friedrich & Rowley . . . . AR
Echinopsis aurantiaca (Rausch) Friedrich & Rowley
Acanthocalycium aurantiacum Rausch
Acanthocalycium glaucum Ritter
Echinopsis graciliflora Cardenas . . . . BO
Echinopsis haematantha (Spegazzini) Hunt . . . . BO, AR
Echinopsis amblayensis (Rausch) Friedrich
Lobivia amblayensis Rausch
Lobivia cabradae Fric ex Wilhelm
Lobivia cerasiflora Fric ex Wilhelm
Lobivia chorrillosensis Rausch
Hymenorebutia drijveriana (Backeberg) Ritter
Lobivia drijveriana Backeberg
Echinopsis elongata (Backeberg) Friedrich
Lobivia elongata Backeberg
Lobivia haematantha (Spegazzini) Britton & Rose
Lobivia haematantha ssp. *chorrillosensis* (Rausch) Rausch ex Rowley
Lobivia haematantha ssp. *kuehnrichii* (Fric) Rausch ex Rowley
Echinopsis hualfinensis (Rausch) H. Friedrich & Glaetzle
Lobivia hualfinensis Rausch
Echinopsis kuehnrichii (Fric) H. Friedrich & Glaetzle
Hymenorebutia kuehnrichii Ritter
Lobivia kuehnrichii Fric
Lobivia mirabunda Backeberg
Echinopsis hahniana (Backeberg) R.S. Wallace . . . . PY
Harrisia hahniana (Backeberg) Kimnach & Hutchison
Mediocactus hahnianus Backeberg
Echinopsis hammerschmidii Cardenas . . . . BO, BR
Echinopsis hertrichiana (Backeberg) Hunt . . . . PE, BO
Lobivia allegraiana Backeberg
Lobivia backebergii ssp. *hertrichiana* (Backeberg) Rausch ex Rowley
Lobivia binghamiana Backeberg
Lobivia echinata Rausch
Neolobivia echinata (Rausch) Ritter
Lobivia hertrichiana Backeberg
Neolobivia hertrichiana (Backeberg) Ritter
Lobivia huilcanota Rauh & Backeberg
Lobivia incaica Backeberg
Neolobivia incaica (Backeberg) Ritter
Lobivia laui Donald
Lobivia minuta Ritter
Neolobivia minuta (Ritter) Ritter
Lobivia planiceps Backeberg
Lobivia simplex Rausch
Lobivia vilcabambae Ritter
Neolobivia vilcabambae (Ritter) Ritter
®*Lobivia larabei* Johnson ex Backeberg
Echinopsis huascha (Weber) Friedrich & Rowley . . . . AR
Lobivia andalgalensis (Weber ex Schumann) Britton & Rose
Trichocereus catamarcensis Ritter

Lobivia grandiflora Britton & Rose
Helianthocereus grandiflorus (Britton & Rose) Backeberg
Trichocereus grandiflorus Backeberg
Helianthocereus huascha (Weber) Backeberg
Lobivia huascha (Weber) W.T. Marshall
Trichocereus huascha (Weber) Britton & Rose
Helianthocereus hyalacanthus (Spegazzini) Backeberg
Trichocereus lobivioides Graeser & Ritter ex Ritter
Echinopsis pecheretiana (Backeberg) Friedrich & Rowley
Helianthocereus pecheretianus Backeberg
Lobivia purpureominiata Ritter
Echinopsis rowleyi Friedrich
Trichocereus rowleyi (Friedrich) Kiesling
Echinopsis huotii (Cels) Labouret . . . . BO
Echinopsis pamparuizii Cardenas
®*Echinopsis semidenudata* Cardenas ex W. Haage
Echinopsis huotii ssp. ***vallegrandensis*** (Cardenas) G. Navarro . . . . BO
Echinopsis vallegrandensis Cardenas
Echinopsis hystrichoides Ritter . . . . BO
Echinopsis korethroides Werdermann ex Backeberg . . . . AR
Soehrensia korethroides (Werdermann) Backeberg
Trichocereus korethroides (Werdermann) Ritter
Echinopsis ibicuatensis Cardenas . . . . BO
Echinopsis kladiwana Rausch . . . . BO
Echinopsis klingleriana Cardenas . . . . BO
Echinopsis knuthiana (Backeberg) Friedrich & Rowley . . . . PE
Azureocereus deflexispinus Backeberg ex Rauh
Trichocereus knuthianus Backeberg
Echinopsis lageniformis (Foerster) Friedrich & Rowley . . . . BO
Trichocereus bridgesii (Salm-Dyck) Britton & Rose
Echinopsis lamprochlora (Lemaire) Weber ex Friedrich & Glaetzle . . . . AR
Trichocereus lamprochlorus (Lemaire) Britton & Rose
Echinopsis purpureopilosa (Weingart) Friedrich & Rowley
Echinopsis lateritia Guerke . . . . BO
Lobivia camataquiensis Cardenas
Lobivia carminantha Backeb.
Hymenorebutia cintiensis (Cardenas) Ritter
Lobivia cintiensis Cardenas
Lobivia kupperiana Backeberg
Lobivia lateritia (Guerke) Britton & Rose
Lobivia scopulina Backeberg
Hymenorebutia torataensis Ritter
Hymenorebutia torreana Ritter
Echinopsis leucantha (Gillies ex Salm-Dyck) Walpers . . . . AR
Echinopsis cordobensis Spegazzini
Echinopsis intricatissima Spegazzini
Echinopsis melanopotamica Spegazzini
Echinopsis shaferi Britton & Rose
Echinopsis spegazziniana Britton & Rose
Echinopsis litoralis (Johow) Friedrich & Rowley CL
Trichocereus litoralis (Johow) Looser
Echinopsis macrogona (Salm-Dyck) Friedrich & Rowley . . . . BO
Trichocereus macrogonus Riccobono
Echinopsis mamillosa Guerke . . . . BO, AR
Echinopsis kermesina (Krainz) Krainz
Pseudolobivia kermesina Krainz
Echinopsis ritteri Boedeker
?*Echinopsis herbasii* Cardenas
?*Echinopsis roseolilacina* Cardenas
Echinopsis mamillosa ssp. ***silvatica*** (Ritter) Braun & Esteves Pereira . . . . BO
Echinopsis silvatica Ritter

Echinopsis marsoneri Werdermann . . . . BO, AR

Lobivia buiningiana Ritter

Lobivia chrysantha ssp. *jajoiana* (Backeberg) Rausch ex Rowley

Lobivia chrysantha ssp. *marsoneri* (Werdermann) Rausch ex Rowley

Lobivia glauca Rausch

Lobivia haageana Backeberg

Lobivia iridescens Backeberg

Lobivia jajoiana Backeberg

Lobivia marsoneri (Werdermann) Backeberg

Lobivia nigrostoma Kreuzinger & Buining

Lobivia rubescens (Backeberg) Backeberg

Lobivia tuberculosa Ritter

Lobivia uitewaaleana Buining

®*Lobivia muhriae* Backeberg

Echinopsis mataranensis Cardenas . . . . BO

Echinopsis maximiliana Heyder ex A. Dietrich . . . . PE, BO

Lobivia caespitosa (J. Purpus) Britton & Rose

Lobivia cariquinensis Cardenas

Lobivia charazanensis Cardenas

Lobivia corbula (Herrera) Britton & Rose

Lobivia coriquinensis Wilhelm ex W. Haage

Lobivia cruciaureispina Knize

Lobivia hermanniana Backeberg

Lobivia intermedia Rausch

Lobivia maximiliana (Heyder) Backeberg

Lobivia maximiliana ssp. *caespitosa* (J. Purpus) Rausch ex Rowley

Lobivia maximiliana ssp. *westii* (Hutchison) Rausch ex Rowley

Lobivia miniatiflora Ritter

Lobivia pseudocariquinensis Cardenas

Lobivia sicuaniensis Rausch

Lobivia westii Hutchison

Echinopsis meyeri Heese . . . . PY

Echinopsis mieckleyi R. Meyer . . . . BO

Echinopsis minuana Spegazzini . . . . AR

Echinopsis mirabilis Spegazzini . . . . AR

Setiechinopsis mirabilis (Spegazzini) Backeberg ex de Haas

Echinopsis molesta Spegazzini . . . . AR

Echinopsis nigra Backeberg . . . . AR

Echinopsis obrepanda (Salm-Dyck) Schumann . . . . BO

Echinopsis fiebrigii Guerke

Pseudolobivia obrepanda (Salm-Dyck) Backeberg ex Krainz

?*Pseudolobivia orozasana* (Ritter) Backeberg

?*Echinopsis rojasii* Cardenas

?*Pseudolobivia rojasii* (Cardenas) Backeberg

?*Echinopsis toralapana* Cardenas

?*Pseudolobivia toralapana* (Cardenas) Backeberg

®*Echinopsis carmineiflora* (Hoffman & Backeberg) Friedrich

®*Pseudolobivia carmineoflora* Hoffmann & Backeberg

Echinopsis obrepanda ssp. ***calorubra*** (Cardenas) G. Navarro . . . . BO

Lobivia aguilarii Vasquez

Echinopsis calorubra Cardenas

Lobivia calorubra (Cardenas) Rausch

Pseudolobivia calorubra (Cardenas) Backeberg

Lobivia mizquensis Rausch

Echinopsis rauschii Friedrich

Echinopsis obrepanda ssp. ***tapecuana*** (Ritter) G. Navarro . . . . BO

Echinopsis tapecuana Ritter

Echinopsis oxygona (Link) Zuccarini ex Pfeiffer & Otto . . . . BR, PY, UY, AR

Echinopsis multiplex (Pfeiffer) Pfeiffer & Otto

Echinopsis paraguayensis Mundt ex Ritter

Echinopsis schwantesii Fric

Echinopsis pachanoi (Britton & Rose) Friedrich & Rowley . . . . EC, PE
Trichocereus pachanoi Britton & Rose
Echinopsis pampana (Britton & Rose) Hunt . . . . PE
Lobivia aureosenilis Knize
Lobivia glaucescens Ritter
Lobivia mistiensis (Werdermann & Backeberg) Backeberg
Lobivia pampana Britton & Rose
Echinopsis pentlandii (Hooker) Salm-Dyck ex A. Dietrich . . . . PE, BO
Lobivia aculeata Buining
Lobivia argentea Backeberg
Lobivia boliviensis Britton & Rose
Lobivia brunneo-rosea Backeberg
Echinopsis hardeniana Boedeker
Lobivia higginsiana Backeberg
Lobivia johnsoniana Backeberg
Lobivia larae Cardenas
Lobivia lauramarca Rauh & Backeberg
Lobivia leucorhodon Backeberg
Lobivia leucoviolacea Backeberg
Lobivia omasuyana Cardenas
Lobivia pentlandii (Hooker) Britton & Rose
Echinopsis pentlandii ssp. *hardeniana* (Boedeker) G. Navarro
Echinopsis pentlandii ssp. *larae* (Cardenas) G. Navarro
Lobivia raphidacantha Backeberg
Lobivia scheeri (Salm-Dyck) Backeberg
Lobivia schneideriana Backeberg
Lobivia titicacensis Cardenas
Lobivia varians Backeberg
Lobivia wegheiana Backeberg
ⓤ*Lobivia aurantiaca* Backeberg
ⓤ*Lobivia multicostata* Backeberg
Echinopsis peruviana (Britton & Rose) Friedrich & Rowley . . . . PE
Echinopsis peruviana ssp. **peruviana** . . . . PE
Trichocereus peruvianus Britton & Rose
Trichocereus tacnaensis Ritter
Trichocereus torataensis Ritter
Echinopsis peruviana ssp. **puquiensis** (Rauh & Backberg) Ostolaza . . . . PE
Echinopsis puquiensis (Rauh & Backeberg) Friedrich & Rowley
Trichocereus puquiensis Rauh & Backeberg
Echinopsis pojoensis Cardenas . . . . BO
Echinopsis pseudomamillosa Cardenas . . . . BO
Echinopsis pugionacantha Rose & Boedeker . . . . BO, AR
Lobivia adpressispina Ritter
Lobivia campicola Ritter
Lobivia cornuta Rausch
Lobivia culpinensis Ritter
Lobivia pugionacantha (Rose & Boedeker) Backeberg
Lobivia salitrensis Rausch
Echinopsis stollenwerkiana Boedeker
Lobivia versicolor Rausch
Echinopsis pugionacantha ssp. ***rossii*** (Boedeker) G. Navarro . . . . BO
Lobivia rossii (Boedeker) Boedeker ex Backeberg & F. Knuth
Echinopsis quadratiumbonatus (Ritter) Hunt . . . . BO
Trichocereus quadratiumbonatus Ritter
Echinopsis rhodotricha Schumann . . . . BR, PY, UY, AR
?*Echinopsis forbesii* Hort. Angl. ex A. Dietrich
Echinopsis rhodotricha ssp. ***chacoana*** (Schuetz) Braun & Esteves Pereira . . . . PY
Echinopsis chacoana Schuetz
Echinopsis riviere-de-caraltii Cardenas . . . . BO
Echinopsis saltensis Spegazzini . . . . AR

Lobivia cachensis (Spegazzini) Britton & Rose
Lobivia emmae Backeberg
Echinopsis nealeana (Backeberg) Friedrich
Lobivia nealeana Backeberg
Echinopsis pseudocachensis (Backeberg) Friedrich
Lobivia pseudocachensis Backeberg
Lobivia saltensis (Spegazzini) Britton & Rose
Echinopsis sanguiniflora (Backeberg) Hunt . . . . AR
Lobivia breviflora Backeberg
Lobivia duursmaiana Backeberg
Lobivia polycephala Backeberg
Lobivia sanguiniflora Backeberg
Echinopsis santaensis (Rauh & Backeberg) Friedrich & Rowley . . . . PE
Trichocereus santaensis Rauh & Backeberg
Echinopsis schickendantzii Weber . . . . AR
Echinopsis manguinii (Backeberg) Friedrich & Rowley
Trichocereus manguinii Backeberg
Trichocereus schickendantzii (Weber) Britton & Rose
Trichocereus shaferi Britton & Rose
Trichocereus volcanensis Ritter
Echinopsis schieliana (Backeberg) Hunt . . . . PE, BO
Lobivia backebergii ssp. *schieliana* (Backeberg) Rausch ex Rowley
Lobivia leptacantha Rausch
Lobivia maximiliana ssp. *quiabayensis* (Rausch) Rausch ex Rowley
Lobivia quiabayensis Rausch
Lobivia schieliana Backeberg
Echinopsis schoenii (Rauh & Backeberg) Friedrich & Rowley . . . . PE
Trichocereus schoenii Rauh & Backeberg
Echinopsis schreiteri (Castellanos) Werdermann ex Backeberg . . . . AR
Lobivia schreiteri Castellanos
Lobivia stilowiana Backeberg
Echinopsis scopulicola (Ritter) Mottram . . . . BO
Trichocereus scopulicola Ritter
Echinopsis silvestrii Spegazzini . . . . AR
Echinopsis skottsbergii (Backeberg ex Skottsberg) Friedrich & Rowley CL
Trichocereus skottsbergii Backeberg ex Skottsberg
Echinopsis smrziana Backeberg . . . . AR
Soehrensia smrziana (Backeberg) Backeberg
Trichocereus smrzianus (Backeberg) Backeberg
Echinopsis spachiana (Lemaire) Friedrich & Rowley . . . . BO, AR
Trichocereus santiaguensis (Spegazzini) Backeberg
Cereus spachianus Lemaire
Trichocereus spachianus (Lemaire) Riccobono
Echinopsis spinibarbis (Pfeiffer) A.E. Hoffmann CL
Eulychnia spinibarbis (Pfeiffer) Britton & Rose
Trichocereus spinibarbis (Otto ex Pfeiffer) Ritter
Echinopsis strigosa (Salm-Dyck) Friedrich & Rowley . . . . AR
Trichocereus strigosus (Salm-Dyck) Britton & Rose
Echinopsis subdenudata Cardenas . . . . BO
Echinopsis sucrensis Cardenas . . . . BO
Echinopsis tacaquirensis (Vaupel) Friedrich & Rowley . . . . BO
Cereus tacaquirensis Vaupel
Trichocereus tacaquirensis (Vaupel) Cardenas ex Backeberg
Echinopsis tacaquirensis ssp. ***taquimbalensis*** (Cardenas) G. Navarro . . . . BO
Echinopsis taquimbalensis (Cardenas) Friedrich & Rowley
Trichocereus taquimbalensis Cardenas
Echinopsis taratensis (Cardenas) Friedrich & Rowley . . . . BO
Trichocereus taratensis Cardenas
Echinopsis tarijensis (Vaupel) Friedrich & Rowley . . . . BO, AR
Lobivia formosa ssp. *tarijensis* (Vaupel) Rausch

Echinopsis poco (Backeberg) Friedrich & Rowley
Helianthocereus poco (Backeberg) Backeberg
Trichocereus poco Backeberg
Trichocereus tarijensis (Vaupel) Werdermann
?*Echinopsis narvaecensis* (Cardenas) Friedrich & Rowley
?*Helianthocereus narvaecensis* (Cardenas) Backeberg
?*Trichocereus narvaecensis* Cardenas
®*Helianthocereus tarijensis* (Vaupel) Backeberg
Echinopsis tarijensis ssp. ***herzogianus*** (Cardenas) G. Navarro . . . . BO
Echinopsis herzogiana (Cardenas) Friedrich & Rowley
Helianthocereus herzogianus (Cardenas) Backeberg
Trichocereus herzogianus Cardenas
Echinopsis tarijensis ssp. ***totorensis*** (Cardenas) G. Navarro . . . . BO
Trichocereus totorensis (Cardenas) Ritter
Echinopsis tarmaensis (Rauh & Backeberg) Friedrich & Rowley . . . . PE
Trichocereus tarmaensis Rauh & Backeberg
Echinopsis tegeleriana (Backeberg) Hunt . . . . PE
Lobivia akersii Rausch
Acantholobivia incuiensis (Rauh & Backeberg) Rauh & Backeberg
Lobivia incuiensis Rauh & Backeberg
Acantholobivia tegeleriana (Backeberg) Backeberg
Echinopsis terscheckii (Parm. ex Pfeiffer) Friedrich & Rowley . . . . BO, AR
Trichocereus terscheckii (Parmentier ex Pfeiffer) Britton & Rose
Echinopsis werdermanniana (Backeberg) Friedrich & Rowley
Trichocereus werdermannianus Backeberg
Echinopsis thelegona (Weber ex Schumann) Friedrich & Rowley . . . . AR
Trichocereus thelegonus (Weber ex Schumann) Britton & Rose
Echinopsis thelegonoides (Spegazzini) Friedrich & Rowley . . . . AR
Trichocereus thelegonoides (Spegazzini) Britton & Rose
®*Echinopsis rubinghiana* (Backeberg) Friedrich & Rowley
®*Trichocereus rubinghianus* Backeberg
Echinopsis thionantha (Spegazzini) Werdermann . . . . AR
Echinopsis brevispina (Ritter) Friedrich & Rowley
Acanthocalycium brevispinum Ritter
Acanthocalycium catamarcense Ritter
Lobivia chionantha (Spegazzini) Britton & Rose
Acanthocalycium chionanthum (Spegazzini) Backeberg
Lobivia thionantha (Spegazzini) Britton & Rose
Acanthocalycium thionanthum (Spegazzini) Backeberg
®*Acanthocalycium griseum* Backeberg
Echinopsis tiegeliana (Wessner) Hunt . . . . BO, AR
Echinopsis fricii (Rausch) Friedrich
Lobivia fricii Rausch
Mediolobivia hirsutissima Cardenas
Lobivia peclardiana Krainz
Hymenorebutia pusilla (Ritter) Ritter
Lobivia pusilla Ritter
Hymenorebutia tiegeliana (Wessner) Ritter
Lobivia tiegeliana Wessner
Echinopsis trichosa (Cardenas) Friedrich & Rowley . . . . BO
Trichocereus trichosus Cardenas
Echinopsis tubiflora (Pfeiffer) Zuccarini ex A. Dietrich . . . . AR
Echinopsis albispinosa Schumann
Echinopsis tulhuayacensis (Ochoa ex Backeberg) Friedrich & Rowley . . . . PE
Trichocereus tulhuayacensis Ochoa ex Backeberg
Echinopsis tunariensis (Cardenas) Friedrich & Rowley . . . . BO
Trichocereus tunariensis Cardenas
Echinopsis uyupampensis (Backeberg) Friedrich & Rowley . . . . PE, BO
Trichocereus uyupampensis Backeberg
?*Trichocereus validus* (Monville ex Salm-Dyck) Backeberg

Echinopsis vasquezii (Rausch) Rowley ... BO
Trichocereus vasquezii Rausch
Echinopsis vatteri (Kiesling) G. Rowley ... AR
Trichocereus vatteri Kiesling
Echinopsis volliana (Backeberg) Friedrich & Rowley ... BO
Trichocereus vollianus Backeberg
Echinopsis walteri (Kiesling) Friedrich & Glaetzle ... AR
Lobivia walteri Kiesling
Trichocereus walteri (Kiesling) J.G. Lambert
Echinopsis werdermannii Fric ex Fleischer ... PY
Echinopsis yuquina Hunt ... BO
Lobivia rauschii Zecher

Epiphyllum anguliger (Lemaire) Don ... MX
Epiphyllum darrahii (Schumann) Britton & Rose
Epiphyllum cartagense (Weber) Britton & Rose ... CR, PA
Epiphyllum caudatum Britton & Rose ... MX
Epiphyllum columbiense (Weber) Dodson & Gentry ... CR, PA, CO, VE, EC
Epiphyllum costaricense (Weber) Britton & Rose ... CR, PA
Epiphyllum macrocarpum (Weber) Backeberg
Epiphyllum crenatum (Lindley) Don ... MX, GT, BZ, HN
Epiphyllum caulorhizum (Lemaire) G. Don ex Loudon
Epiphyllum floribundum Kimnach ... PE
Epiphyllum grandilobum (Weber) Britton & Rose ... CR, PA
Epiphyllum gigas R.E. Woodson & Cutak
Epiphyllum guatemalense Britton & Rose ... MX, GT, HN
Epiphyllum hookeri Haworth ... MX, GT, BZ, HN, NI, SV, CR, CU, TT, VE
Epiphyllum stenopetalum (Foerster) Britton & Rose
Epiphyllum strictum (Lemaire) Britton & Rose
Epiphyllum laui Kimnach ... MX
Epiphyllum lepidocarpum (Weber) Britton & Rose ... CR, PA
Epiphyllum oxypetalum (De Candolle) Haworth ... MX, GT, HN, NI, SV, CR, XS
Epiphyllum phyllanthus (Linnaeus) Haworth
... PA, GY, GF, SR, BR, PY, CO, VE, EC, PE, BO, UY, AR
Epiphyllum pittieri (Weber) Britton & Rose ... NI, CR, PA
Epiphyllum pumilum (Vaupel) Britton & Rose ... MX, GT, BZ
Epiphyllum rubrocoronatum (Kimnach) C. Dodson & A. Gentry ... PA, CO, EC
Epiphyllum thomasianum (Schumann) Britton & Rose ... MX, GT, HN, NI, CR, EC
Epiphyllum trimetrale Croizat ... CO

Epithelantha bokei Benson ... US, MX
Epithelantha micromeris (Engelmann) Weber ... US, MX
Epithelantha micromeris ssp. **micromeris** ... US, MX
Epithelantha micromeris ssp. **greggii** (Engelmann) Taylor ... MX
Epithelantha densispina Bravo
Epithelantha greggii (Engelmann) Orcutt
Epithelantha rufispina Bravo
Epithelantha micromeris ssp. **pachyrhiza** (W.T. Marshall) Taylor ... MX
Epithelantha pachyrhiza (W.T. Marshall) Backeberg
Epithelantha micromeris ssp. **polycephala** (Backeberg) Glass ... MX
Epithelantha polycephala Backeberg
Epithelantha micromeris ssp. **unguispina** (Boedeker) Taylor ... MX

Eriosyce aerocarpa (Ritter) Kattermann ... CL
Neochilenia aerocarpa (Ritter) Backeberg
Thelocephala aerocarpa (Ritter) Ritter
Thelocephala nuda Ritter
Eriosyce andreaeana Kattermann ... AR
®*Neochilenia andreaeana* Backeberg
®*Neoporteria andreaeana* (Backeberg) Donald & Rowley
®*Acanthocalycium andreaeanum* (Backeberg) Donald

Eriosyce aspillagae (Soehrens) Kattermann CL
Neochilenia aspillagae (Soehrens) Backeberg
Neoporteria aspillagae (Soehrens) Backeberg
Pyrrhocactus aspillagae (Soehrens) Ritter
Eriosyce aurata (Pfeiffer) Backeberg CL
Eriosyce algarrobensis Ritter
Eriosyce ihotzkyanae Ritter
Eriosyce lapampaensis Ritter
Eriosyce sandillon (Remy ex Gay) Philippi
Eriosyce spinibarbis Ritter
Ⓜ*Eriosyce ceratistes* Britton & Rose
Eriosyce bulbocalyx (Werdermann) Kattermann . . . . AR
Neoporteria bulbocalyx (Werdermann) Donald & Rowley
Pyrrhocactus bulbocalyx (Werdermann) Backeberg
Neoporteria dubia (Backeberg) Donald & Rowley
Pyrrhocactus dubius Backeberg
Neoporteria megliolii (Rausch) Donald
Pyrrhocactus megliolii Rausch
Ⓤ*Pyrrhocactus marayesensis* (Backeb.) J.G. Lambert
Eriosyce chilensis (Hildmann ex Schumann) Kattermann CL
Neochilenia chilensis (Hildmann ex Schumann) Backeberg
Neoporteria chilensis (Hildmann ex Schumann) Britton & Rose
Pyrrhocactus chilensis (Hildmann ex Schumann) Ritter
Pyrrhocactus krausii Ritter
Eriosyce confinis (Ritter) Kattermann CL
Neochilenia confinis (Ritter) Backeberg
Neoporteria confinis (Ritter) Donald & Rowley
Pyrrhocactus confinis Ritter
Eriosyce crispa (Ritter) Kattermann CL
Eriosyce crispa ssp. **crispa** CL
Neoporteria crispa (Ritter) Donald & Rowley
Horridocactus crispus (Ritter) Backeberg
Pyrrhocactus crispus Ritter
Neoporteria huascensis (Ritter) Donald & Rowley
Pyrrhocactus huascensis Ritter
Neochilenia nigriscoparia Backeberg
Eriosyce crispa ssp. **atroviridis** (Ritter) Kattermann CL
Horridocactus atroviridis (Ritter) Backeberg
Neoporteria atroviridis (Ritter) Ferryman ex Preston-Mafham
Pyrrhocactus atroviridis Ritter
Horridocactus carrizalensis (Ritter) Backeberg
Pyrrhocactus carrizalensis Ritter
Neochilenia huascensis (Ritter) Backeberg
Neochilenia totoralensis (Ritter) Backeberg
Neoporteria totoralensis (Ritter) Donald & Rowley
Pyrrhocactus totoralensis Ritter
Ⓤ*Neoporteria carrizalensis* (Ritter) A.E. Hoffmann
Eriosyce curvispina (Bertero ex Colla) Kattermann CL
Horridocactus aconcaguensis (Ritter) Backeberg
Horridocactus andicola Ritter
Pyrrhocactus andicola (Ritter) Ritter
Neoporteria armata (Ritter) Krainz
Horridocactus armatus (Ritter) Backeberg
Pyrrhocactus armatus Ritter
Horridocactus choapensis (Ritter) Backeberg
Neoporteria choapensis (Ritter) Donald & Rowley
Pyrrhocactus choapensis Ritter
Pyrrhocactus coliguayensis Ritter
Neoporteria curvispina (Bertero ex Colla) Donald & Rowley

Horridocactus curvispinus (Bertero ex Colla) Backeberg
Pyrrhocactus curvispinus (Bertero ex Colla) Backeberg
Horridocactus grandiflorus (Ritter) Backeberg
Pyrrhocactus grandiflorus Ritter
Neoporteria horrida (Remy ex Gay) Hunt
Pyrrhocactus horridus (Remy ex Gay) Backeberg
Neochilenia odoriflora (Ritter) Backeberg
Pyrrhocactus odoriflorus Ritter
Pyrrhocactus pamaensis Ritter
Neochilenia robusta (Ritter) Backeberg
Pyrrhocactus truncatipetalus Ritter
Neoporteria tuberisulcata (Jacobi) Donald & Rowley
Horridocactus tuberisulcatus (Jacobi) Ito
?*Horridocactus froehlichianus* (Schumann) Backeberg
?*Horridocactus kesselringianus* Doelz
®*Horridocactus robustus* (Ritter) Backeberg

Eriosyce engleri (Ritter) Kattermann CL
Horridocactus engleri Ritter
Neoporteria engleri (Ritter) Donald & Rowley
Pyrrhocactus engleri (Ritter) Ritter

Eriosyce esmeraldana (Ritter) Kattermann CL
Chileorebutia esmeraldana Ritter
Neochilenia esmeraldana (Ritter) Backeberg
Neoporteria esmeraldana (Ritter) Donald & Rowley
Thelocephala esmeraldana (Ritter) Ritter

Eriosyce garaventae (Ritter) Kattermann CL
Horridocactus garaventae Ritter
Neoporteria garaventae (Ritter) Ferryman ex Preston-Mafham
Pyrrhocactus garaventae (Ritter) Ritter
Neoporteria subaiana (Backeberg) Donald & Rowley
Pyrrhocactus subaianus Backeberg

Eriosyce heinrichiana (Backeberg) Kattermann CL
Pyrrhocactus chaniarensis Ritter
Neochilenia chorosensis (Ritter) Backeberg
Neoporteria chorosensis (Ritter) Donald & Rowley
Pyrrhocactus chorosensis Ritter
Neochilenia deherdtiana Backeberg
Neoporteria deherdtiana (Backeberg) Donald & Rowley
Pyrrhocactus deherdtianus (Backeberg) Kattermann
Neoporteria heinrichiana (Backeberg) Ferryman ex Preston-Mafham
Horridocactus heinrichianus Backeberg
Pyrrhocactus trapichensis Ritter
?*Neochilenia jussieui* (Monville ex Salm-Dyck) Backeberg
?*Neoporteria jussieui* (Monville ex Salm-Dyck) Britton & Rose
?*Pyrrhocactus jussieui* (Monville ex Salm-Dyck) Ritter
?*Neochilenia occulta* (Philippi) Backeberg
?*Neoporteria occulta* (Philippi) Britton & Rose
®*Neochilenia trapichensis* Backeberg

Eriosyce heinrichiana ssp. **heinrichiana** CL

Eriosyce heinrichiana ssp. **intermedia** (Ritter) Kattermann CL
Neochilenia dimorpha (Ritter) Backeberg
Neoporteria dimorpha (Ritter) Donald & Rowley
Pyrrhocactus dimorphus Ritter
Neoporteria ritteri Donald & Rowley
Neochilenia setosiflora (Ritter) Backeberg
Neoporteria setosiflora (Ritter) Donald & Rowley
Pyrrhocactus setosiflorus Ritter
Neochilenia wagenknechtii (Ritter) Backeberg
Pyrrhocactus wagenknechtii Ritter

Eriosyce heinrichiana ssp. **simulans** (Ritter) Kattermann CL
Neochilenia simulans (Ritter) Backeberg
Neoporteria simulans (Ritter) Donald & Rowley
Pyrrhocactus simulans Ritter
Eriosyce islayensis (Foerster) Kattermann .... PE, CL
Islaya bicolor Akers & Buining
Neoporteria bicolor (Akers & Buining) Donald & Rowley
Islaya brevicylindrica Rauh & Backeberg
Islaya copiapoides Rauh & Backeberg
Islaya divaricatiflora Ritter
Islaya flavida Ritter
Islaya grandiflorens Rauh & Backeberg
Islaya grandis Rauh & Backeberg
Islaya islayensis (Foerster) Backeberg
Neoporteria islayensis (Foerster) Donald & Rowley
Islaya krainziana Ritter
Neoporteria krainziana (Ritter) Donald & Rowley
Islaya maritima Ritter
Islaya minor Backeberg
Islaya minuscula Ritter
Islaya mol(l)endensis (Vaupel) Backeberg
Islaya mollendensis (Vaupel) Backeberg
Islaya paucispina Rauh & Backeberg
Islaya paucispinosa Rauh & Backeberg
Islaya unguispina Ritter
Eriosyce krausii (Ritter) Kattermann CL
Neochilenia krausii (Ritter) Backeberg
Thelocephala krausii (Ritter) Ritter
Thelocephala longirapa Ritter
Chileorebutia malleolata Ritter
Neochilenia malleolata (Ritter) Backeberg
Thelocephala malleolata (Ritter) Ritter
Eriosyce kunzei (Foerster) Kattermann CL
Neochilenia eriosyzoides (Ritter) Backeberg
Neoporteria eriosyzoides (Ritter) Donald & Rowley
Pyrrhocactus eriosyzoides (Ritter) Ritter
Neochilenia kunzei (Foerster) Backeberg
Neoporteria kunzei (Foerster) Backeberg
Neochilenia transitensis (Ritter) Backeberg
Neoporteria transitensis (Ritter) Ferryman ex Preston-Mafham
Pyrrhocactus transitensis Ritter
Horridocactus vallenarensis (Ritter) Backeberg
®*Neoporteria vallenarensis* (Ritter) A.E. Hoffmann
Pyrrhocactus vallenarensis Ritter
?*Horridocactus geissei* (Poselger ex Schumann) Doelz
?*Neoporteria nidus* (Soehrens ex Schumann) Britton & Rose
Eriosyce laui Luethy CL
Eriosyce limariensis (Ritter) Kattermann CL
Neoporteria limariensis (Ritter) Ferryman ex Preston-Mafham
Pyrrhocactus limariensis Ritter
?*Horridocactus nigricans* (Dietrich ex Schumann) Backeberg & Doelz
?*Neoporteria nigricans* (Dietrich ex Schumann) Britton & Rose
Eriosyce marksiana (Ritter) Kattermann CL
Horridocactus lissocarpus (Ritter) Backeberg
Pyrrhocactus lissocarpus Ritter
Neoporteria marksiana (Ritter) Donald & Rowley
Horridocactus marksianus (Ritter) Backeberg
Pyrrhocactus marksianus Ritter

Eriosyce napina (Philippi) Kattermann CL
Neochilenia mitis (Philippi) Backeberg
Neochilenia napina (Philippi) Backeberg
Neoporteria napina (Philippi) Backeberg
Thelocephala napina (Philippi) Ito
Eriosyce napina ssp. **napina** CL
Eriosyce napina ssp. **lembckei** Kattermann CL
Chileorebutia duripulpa Ritter
Neochilenia duripulpa (Ritter) Backeberg
Thelocephala duripulpa (Ritter) Ritter
®*Neochilenia imitans* Backeberg
®*Neochilenia lembckei* Backeberg
®*Thelocephala lembckei* (Backeberg) Ritter
®*Neochilenia neoreichei* Backeberg
®*Reicheocactus neoreichei* (Backeberg) Backeberg
Eriosyce occulta Kattermann CL
®*Pyrrhocactus occultus* (Schumann) Ritter
Eriosyce odieri (Lemaire ex Salm-Dyck) Kattermann CL
?*Neochilenia reichei* (Schumann) Backeberg
?*Thelocephala reichei* (Schumann) Ritter
?*Neoporteria reichii* (Schumann) Backeberg
®?*Neochilenia atra* Backeberg
®?*Neochilenia pseudoreichei* Lembcke & Backeberg
Eriosyce odieri ssp. **odieri** CL
Neochilenia odieri (Lemaire) Backeberg
Neoporteria odieri (Lemaire ex Salm-Dyck) Backeberg
Thelocephala odieri (Salm-Dyck) Ritter
®*Neochilenia monte-amargensis* Backeberg
®*Neoporteria monte-amargensis* (Backeberg) Donald & Rowley
Eriosyce odieri ssp. **fulva** (Ritter) Kattermann CL
Thelocephala fulva (Ritter) Ritter
Eriosyce odieri ssp. **glabrescens** (Ritter) Kattermann CL
Neochilenia glabrescens (Ritter) Backeberg
Thelocephala glabrescens (Ritter) Ritter
®*Neochilenia carneoflora* Kilian ex Backeberg
Eriosyce omasensis (Ostolaza & Mischler) Ostolaza . . . . PE
Islaya omasensis Ostolaza & Mischler
®*Neoporteria omasensis* (Ostolaza & Mischler) Ferryman ex Preston-Mafham
Eriosyce recondita (Ritter) Kattermann CL
Neochilenia recondita (Ritter) Backeberg
Neoporteria recondita (Ritter) Donald & Rowley
Pyrrhocactus reconditus Ritter
Eriosyce recondita ssp. **recondita** CL
Pyrrhocactus vexatus Ritter
Eriosyce recondita ssp. **iquiquensis** (Ritter) Kattermann CL
Neochilenia aricensis (Ritter) Backeberg
Neoporteria aricensis (Ritter) Donald & Rowley
Pyrrhocactus aricensis Ritter
Neochilenia iquiquensis (Ritter) Backeberg
Neoporteria iquiquensis (Ritter) Donald & Rowley
Pyrrhocactus iquiquensis Ritter
Neochilenia residua (Ritter & Buining) Backeberg
Neoporteria residua (Ritter) Donald & Rowley
Pyrrhocactus residuus Ritter & Buining
Pyrrhocactus saxifragus Ritter
®*Pyrrhocactus floribundus* (Backeberg) Ritter
®*Reicheocactus floribundus* Backeberg
Eriosyce rodentiophila Ritter CL
Eriosyce megacarpa Ritter
Rodentiophila megacarpa (Ritter) Y. Ito
®*Rodentiophila atacamensis* Ritter ex Y. Ito

Eriosyce senilis (Backeberg) Kattermann CL
Eriosyce senilis ssp. **senilis** CL
Neoporteria multicolor Ritter
®*Neoporteria gerocephala* Y. Ito
Eriosyce senilis ssp. **coimasensis** (Ritter) Kattermann CL
Neoporteria coimasensis Ritter
Neoporteria robusta Ritter
Eriosyce senilis ssp. **elquiensis** Kattermann CL
Eriosyce sociabilis (Ritter) Kattermann CL
Neoporteria sociabilis Ritter
Eriosyce strausiana (Schumann) Kattermann . . . . AR
Pyrrhocactus atrospinosus Backeberg
Neoporteria backebergii Donald & Rowley
Pyrrhocactus pachacoensis Rausch
Pyrrhocactus platyacanthus Ritter
Neoporteria sanjuanensis (Spegazzini) Donald & Rowley
Pyrrhocactus sanjuanensis (Spegazzini) Backeberg
Neoporteria setiflora (Backeberg) Donald & Rowley
Pyrrhocactus setiflorus Backeberg
Neoporteria strausiana (Schumann) Donald & Rowley
Pyrrhocactus strausianus (Schumann) Backeberg
Neoporteria volliana (Backeberg) Donald & Rowley
Pyrrhocactus vollianus Backeberg
Eriosyce subgibbosa (Haworth) Kattermann CL
Neoporteria castaneoides (Cels) Werdermann
Neoporteria subgibbosa (Haworth) Britton & Rose
Eriosyce subgibbosa ssp. **subgibbosa** CL
Neoporteria castanea Ritter
Neoporteria heteracantha (Backeberg) W.T. Marshall
Neoporteria litoralis Ritter
Neoporteria subcylindrica (Backeberg) Backeberg
Eriosyce subgibbosa ssp. **clavata** (Soehrens ex Schumann) Kattermann CL
Neoporteria clavata (Soehrens ex Schumann) Werdermann
Neoporteria microsperma Ritter
Neoporteria nigrihorrida (Backeberg) Backeberg
Neoporteria vallenarensis Ritter
Neoporteria wagenknechtii Ritter
Eriosyce taltalensis (Hutchison) Kattermann CL
Neochilenia taltalensis (Hutchison) Backeberg
Neoporteria taltalensis Hutchison
Pyrrhocactus taltalensis (Hutchison) Ritter
®?*Neochilenia neofusca* Backeberg
Eriosyce taltalensis ssp. **taltalensis** CL
Neochilenia calderana (Ritter) Backeberg
Neoporteria calderana (Ritter) Donald & Rowley
Pyrrhocactus calderanus Ritter
Neochilenia gracilis (Ritter) Backeberg
Pyrrhocactus gracilis Ritter
Neochilenia intermedia (Ritter) Backeberg
Neoporteria intermedia (Ritter) Donald & Rowley
Pyrrhocactus intermedius Ritter
Neochilenia pulchella Backeberg
Neoporteria pulchella (Ritter) Ferryman ex Preston-Mafham
Pyrrhocactus pulchellus Ritter
Neochilenia pygmaea (Ritter) Backeberg
Pyrrhocactus pygmaeus Ritter
Neochilenia rupicola (Ritter) Backeberg
Neoporteria rupicola (Ritter) Donald & Rowley
Pyrrhocactus rupicola Ritter

Neochilenia scoparia (Ritter) Backeberg
Neoporteria scoparia (Ritter) Donald & Rowley
Pyrrhocactus scoparius Ritter
Pyrrhocactus tenuis Ritter
Neoporteria transiens (Ritter) Ferryman ex Preston-Mafham
Pyrrhocactus transiens Ritter

Eriosyce taltalensis ssp. **echinus** (Ritter) Kattermann CL
Horridocactus echinus (Ritter) Backeberg
Neoporteria echinus (Ritter) Ferryman ex Preston-Mafham
Pyrrhocactus echinus Ritter
Neochilenia eriocephala Backeberg
Neoporteria eriocephala (Backeberg) Donald & Rowley
Neochilenia floccosa (Ritter) Backeberg
Neoporteria floccosa (Ritter) J. Lode
Pyrrhocactus floccosus Ritter
Neochilenia glaucescens (Ritter) Backeberg
Pyrrhocactus glaucescens Ritter

Eriosyce taltalensis ssp. **paucicostata** (Ritter) Kattermann CL
Pyrrhocactus neohankeanus Ritter
Neochilenia paucicostata (Ritter) Backeberg
Neoporteria paucicostata (Ritter) Donald & Rowley
Pyrrhocactus paucicostatus (Ritter) Ritter
®*Delaetia woutersiana* Backeberg
®*Neoporteria woutersiana* (Backeberg) Donald & Rowley
?*Neochilenia fobeana* (Mieckley) Backeberg
?*Neochilenia fusca* (Muehlenpfordt) Backeberg
?*Neoporteria fusca* (Muehlenpfordt) Britton & Rose
?*Neochilenia hankeana* (Foerster) Doelz ex Backeberg
?*Neoporteria hankeana* (Foerster) Donald & Rowley

Eriosyce taltalensis ssp. **pilispina** (Ritter) Kattermann CL
Neochilenia pilispina (Ritter) Backeberg
Neoporteria pilispina (Ritter) Donald & Rowley
Pyrrhocactus pilispinus Ritter

Eriosyce tenebrica (Ritter) Kattermann CL
Thelocephala fankhauseri Ritter
Neochilenia tenebrica Ritter ex W. Haage
Thelocephala tenebrica Ritter

Eriosyce umadeave (Fric ex Werdermann) Kattermann . . . . AR
Neoporteria umadeave (Fric ex Werdermann) Donald & Rowley
Pyrrhocactus umadeave (Werdermann) Backeberg

Eriosyce vertongenii (J.G. Lambert) Hunt . . . . AR
Pyrrhocactus vertongenii J.G. Lambert

Eriosyce villicumensis (Rausch) Kattermann . . . . AR
Neoporteria villicumensis (Rausch) Donald
Pyrrhocactus villicumensis Rausch
®*Neoporteria melanacantha* (Backeberg) Donald & Rowley
®*Pyrrhocactus melanacanthus* Backeberg

Eriosyce villosa (Monville) Kattermann CL
Neoporteria atrispinosa Backeberg
Neoporteria cephalophora Backeberg
Neoporteria laniceps Ritter
Neoporteria polyraphis (Pfeiffer ex Salm-Dyck) Backeberg
Neoporteria villosa (Monville) Berger

Escobaria aguirreana (Glass & Foster) Taylor . . . . MX
Acharagma aguirreana (Glass & Foster) Glass
Gymnocactus aguirreanus Glass & Foster
Thelocactus aguirreanus (Glass & Foster) Bravo

Escobaria albicolumnaria Hester . . . . US
Coryphantha albicolumnaria (Hester) D. Zimmerman

Escobaria alversonii (J. Coulter) Taylor . . . . US
Coryphantha alversonii (J. Coulter) Orcutt
Escobaria chihuahuensis Britton & Rose . . . . MX
¶?*Escobaria strobiliformis* (Poselger) Boedeker
Escobaria chihuahuensis ssp. **chihuahuensis** . . . . MX
Escobaria chihuahuensis ssp. **henricksonii** (Glass & Foster) Taylor . . . . MX
Coryphantha henricksonii (Glass & Foster) Glass & Foster
Escobaria henricksonii Glass & Foster
Escobaria cubensis (Britton & Rose) Hunt . . . . CU
Coryphantha cubensis Britton & Rose
Escobaria dasyacantha (Engelmann) Britton & Rose . . . . US, MX
Mammillaria dasyacantha Engelmann
Escobaria dasyacantha ssp. **dasyacantha** . . . . US, MX
Escobaria dasyacantha ssp. **chaffeyi** (Britton & Rose) Taylor . . . . MX
Escobaria chaffeyi Britton & Rose
Escobaria deserti (Engelmann) Buxbaum . . . . US
Coryphantha chlorantha (Engelmann) Britton & Rose
Coryphantha deserti (Engelmann) Britton & Rose
Escobaria duncanii (Hester) Buxbaum . . . . US
Coryphantha duncanii (Hester) L. Benson
Escobaria emskoetteriana (Quehl) Borg . . . . US, MX
Escobaria bella Britton & Rose
Escobaria muehlbaueriana (Boedeker) F. Knuth
Escobaria runyonii Britton & Rose
Escobaria guadalupensis Brack & Heil . . . . US
Escobaria hesteri (Y. Wright) Buxbaum . . . . US
Escobaria laredoi (Glass & Foster) Taylor . . . . MX
Coryphantha laredoi Glass & Foster
®*Escobaria rigida* Backeberg
Escobaria lloydii Britton & Rose . . . . MX
Escobaria minima (Baird) Hunt . . . . ♦US
Coryphantha minima Baird♦
Escobaria nellieae (Croizat) Backeberg♦
Escobaria missouriensis (Sweet) Hunt . . . . US, MX
Escobaria missouriensis ssp. **missouriensis** . . . . US
Coryphantha missouriensis (Sweet) Britton & Rose
Neobesseya missouriensis (Sweet) Britton & Rose
Escobaria missouriensis ssp. *navajoensis* Hochstaetter
Neobesseya notesteinii (Britton) Britton & Rose
Neobesseya rosiflora Lahman ex G. Turner
Neobesseya similis (Engelmann) Britton & Rose
Neobesseya wissmannii (Hildmann ex Schumann) Britton & Rose
Escobaria missouriensis ssp. **asperispina** (Boedeker) Taylor . . . . MX
Escobaria asperispina (Boedeker) Hunt
Neobesseya asperispina (Boedeker) Boedeker
Escobaria orcuttii Boedeker . . . . US
Coryphantha orcuttii (Rose ex Orcutt) D. Zimmerman
Escobaria organensis (D. Zimmerman) Castetter et al. . . . . US
Coryphantha organensis D. Zimmerman
Escobaria robbinsorum (Earle) Hunt . . . . US, MX
Cochiseia robbinsorum Earle
Coryphantha robbinsorum (Earle) A. Zimmerman
Escobaria roseana (Boedeker) Buxbaum . . . . MX
Gymnocactus roseanus (Boedeker) Glass & Foster
Thelocactus roseanus (Boedeker) Borg
Escobaria roseana ssp. ***galeanensis*** Haugg . . . . MX
Escobaria sandbergii Castetter et al. . . . . US
Escobaria sneedii Britton & Rose . . . . ♦US
Coryphantha sneedii (Britton & Rose) Berger♦

Escobaria sneedii ssp. **sneedii** . . . . ♦US
Escobaria sneedii ssp. **leei** (Rose ex Boedeker) Hunt . . . . ♦US
Escobaria leei (Rose) Boedeker♦
Escobaria tuberculosa (Engelmann) Britton & Rose . . . . US, MX
Escobaria varicolor Tiegel
⊗*Escobaria strobiliformis* auctt
Escobaria villardii Castetter et al. . . . . US
Escobaria vivipara (Nuttall) Buxbaum . . . . CA, US, MX
Coryphantha arizonica (Engelmann) Britton & Rose
Escobaria bisbeeana (Orcutt) Borg
Coryphantha columnaris Lahman
Escobaria neomexicana (Engelmann) Buxbaum
Escobaria radiosa (Engelmann) Frank
Coryphantha vivipara (Nuttall) Britton & Rose
⊗*Coryphantha aggregata* Britton & Rose
Escobaria zilziana (Boedeker) Backeberg . . . . MX

Escontria chiotilla (Weber ex Schumann) Rose . . . . MX
Myrtillocactus chiotilla (Schumann) P.V. Heath

Espostoa baumannii Knize . . . . PE
Espostoa blossfeldiorum (Werdermann) Buxbaum . . . . PE
Thrixanthocereus blossfeldiorum (Werdermann) Backeberg
Espostoa calva Ritter . . . . PE
Espostoa frutescens Madsen . . . . EC
Espostoa guentheri (Kupper) Buxbaum . . . . BO
Cephalocereus guentheri Kupper
Vatricania guentheri (Kupper) Backeberg
Espostoa huanucoensis Johnson ex Ritter . . . . PE
Espostoa hylaea Ritter . . . . PE
Espostoa lanata (Kunth) Britton & Rose . . . . EC, PE
Espostoa dautwitzii (Haage f.) Borg
Espostoa laticornua Rauh & Backeberg
Espostoa procera Rauh & Backeberg
Espostoa lanianuligera Ritter . . . . PE
Espostoa melanostele (Vaupel) Borg . . . . PE
Pseudoespostoa melanostele (Vaupel) Backeberg
⊗*Espostoa haagei* (Poselger ex Ruempler) Borg
Espostoa mirabilis Ritter . . . . PE
Espostoa nana Ritter . . . . PE
Pseudoespostoa nana (Ritter) Backeberg
Espostoa ritteri Buining . . . . PE
Espostoa ruficeps Ritter . . . . PE
Espostoa senilis (Ritter) Taylor . . . . PE
Thrixanthocereus cullmannianus Ritter
Thrixanthocereus longispinus Ritter
Thrixanthocereus senilis Ritter
Espostoa superba Ritter . . . . PE

×***Espostocactus mirabilis*** (Rauh & Backeberg ex Backeberg) Rowley . . . . PE
Neobinghamia mirabilis Rauh & Backeberg ex Backeberg

Espostoopsis dybowskii (Roland-Gosselin) Buxbaum . . . . BR
Austrocephalocereus dybowskii (Roland-Gosselin) Backeberg
Cephalocereus dybowskii (Roland-Gosselin) Britton & Rose
Coleocephalocereus dybowskii (Roland-Gosselin) Brandt
Gerocephalus dybowskii (Roland-Gosselin) Ritter

Eulychnia acida Philippi CL
Eulychnia aricensis Ritter CL

Eulychnia breviflora Philippi . . . CL
Eulychnia barquitensis Ritter
Eulychnia saint-pieana Ritter
Eulychnia castanea Philippi . . . CL
Philippicereus castaneus (Philippi) Backeberg
Eulychnia iquiquensis (Schumann) Britton & Rose . . . CL
Eulychnia morromorenoensis Ritter
Ⓤ***Eulychnia procumbens*** Backeberg . . . CL
Eulychnia ritteri Cullmann . . . PE

Facheiroa cephaliomelana Buining & Brederoo . . . BR
Facheiroa cephaliomelana ssp. **cephaliomelana** . . . BR
Facheiroa pilosa Ritter
Facheiroa tenebrosa Braun & Esteves
Facheiroa cephaliomelana ssp. **estevesii** (Braun) Taylor & Zappi . . . BR
Facheiroa estevesii Braun
Facheiroa squamosa (Guerke) Braun . . . BR
Facheiroa chaetacantha (Ritter) Braun
Zehntnerella chaetacantha Ritter
Zehntnerella polygona Ritter
Zehntnerella squamulosa Britton & Rose
Facheiroa squamosa ssp. *polygona* (Ritter) Braun & Esteves Pereira
Facheiroa ulei (Guerke) Werdermann . . . BR
Facheiroa pubiflora Britton & Rose

Ferocactus alamosanus (Britton & Rose) Britton & Rose . . . MX
Ferocactus alamosanus ssp. **alamosanus** . . . MX
Ferocactus alamosanus ssp. **reppenhagenii** (Unger) Taylor . . . MX
Ferocactus reppenhagenii Unger
Ferocactus chrysacanthus (Orcutt) Britton & Rose . . . MX
Ferocactus chrysacanthus ssp. **chrysacanthus** . . . MX
Ferocactus chrysacanthus ssp. **grandiflorus** (Lindsay) Taylor . . . MX
Ferocactus cylindraceus (Engelmann) Orcutt . . . US, MX
Ferocactus rostii Britton & Rose
Ⓜ*Ferocactus acanthodes* (Lemaire) Britton & Rose
Ferocactus cylindraceus ssp. **cylindraceus** . . . US, MX
Ferocactus cylindraceus ssp. **lecontei** (Engelmann) Taylor . . . US, MX
Ferocactus hertrichii Kreuzinger
Ferocactus lecontei (Engelmann) Britton & Rose
Ferocactus cylindraceus ssp. **tortulispinus** (Gates) Taylor . . . MX
Ferocactus tortulispinus Gates
Ⓤ*Ferocactus tortulospinus* Gates
Ferocactus diguetii (Weber) Britton & Rose . . . MX
Ferocactus eastwoodiae (Benson) Benson . . . US
Ferocactus echidne (De Candolle) Britton & Rose . . . MX
Ferocactus victoriensis (Rose) Backeberg
Ⓤ*Ferocactus rhodanthus* Schwarz ex Backeberg
Ferocactus emoryi (Engelmann) Orcutt . . . US, MX
Ferocactus emoryi ssp. **emoryi** . . . US, MX
Ferocactus covillei Britton & Rose
Ferocactus emoryi ssp. **rectispinus** (Engelmann) Taylor . . . MX
Ferocactus rectispinus (Engelmann) Britton & Rose
Ferocactus flavovirens (Scheidweiler) Britton & Rose . . . MX
Ferocactus fordii (Orcutt) Britton & Rose . . . MX
Ferocactus glaucescens (De Candolle) Britton & Rose . . . MX
Ferocactus pfeifferi (Zuccarini) Backeberg
Ferocactus gracilis Gates . . . MX
Ferocactus gracilis ssp. **gracilis** . . . MX
Ferocactus gracilis ssp. **coloratus** (Gates) Taylor . . . MX
Ferocactus coloratus Gates
Ferocactus viscainensis Gates

Ferocactus gracilis ssp. **gatesii** (Lindsay) Taylor . . . . MX
Ferocactus gatesii Lindsay
Ferocactus haematacanthus (Salm-Dyck) Bravo ex Backeberg & F. Knuth . . . . . MX
Ferocactus hamatacanthus (Muehlenpfordt) Britton & Rose . . . . US, MX
Ferocactus hamatacanthus ssp. **hamatacanthus** . . . . US, MX
Hamatocactus hamatacanthus (Muehlenpfordt) F. Knuth
Ferocactus hamatacanthus ssp. **sinuatus** (A. Dietrich) Taylor . . . . US, MX
Hamatocactus sinuatus (Dietrich) Orcutt
Ferocactus herrerae J.G. Ortega . . . . MX
Ferocactus histrix (De Candolle) Lindsay . . . . MX
Ferocactus melocactiformis Britton & Rose
Ferocactus johnstonianus Britton & Rose . . . . MX
Ferocactus latispinus (Haworth) Britton & Rose . . . . MX
®?*Ferocactus nobilis* (Linnaeus) Britton & Rose
?*Ferocactus recurvus* (Miller) Borg
Ferocactus latispinus ssp. **latispinus** . . . . MX
Ferocactus latispinus ssp. **spiralis** (Karwinsky ex Pfeiffer) Taylor . . . . MX
Ferocactus lindsayi Bravo . . . . MX
Ferocactus macrodiscus (Martius) Britton & Rose . . . . MX
Ferocactus macrodiscus ssp. **macrodiscus** . . . . MX
Ferocactus macrodiscus ssp. **septentrionalis** (Meyran) Taylor . . . . MX
Ferocactus peninsulae (Weber) Britton & Rose . . . . MX
Ferocactus horridus Britton & Rose
Ferocactus pilosus (Galeotti ex Salm-Dyck) Werdermann . . . . MX
Ferocactus pringlei (Coulter) Britton & Rose
Ferocactus stainesii (Salm-Dyck) Britton & Rose
?*Ferocactus piliferus* (Lemaire ex Ehrenberg) Unger
Ferocactus pottsii (Salm-Dyck) Backeberg . . . . MX
Ferocactus robustus (Pfeiffer) Britton & Rose . . . . MX
Ferocactus santa-maria Britton & Rose . . . . MX
Ferocactus schwarzii Lindsay . . . . MX
Ferocactus tiburonensis (Lindsay) Backeberg . . . . MX
Ferocactus townsendianus Britton & Rose . . . . MX
Ferocactus viridescens (Torrey & Gray) Britton & Rose . . . . US, MX
Ferocactus orcuttii (Engelmann) Britton & Rose
Ferocactus wislizeni (Engelmann) Britton & Rose . . . . US, MX

Frailea buenekeri W.R. Abraham . . . . BR
Frailea buenekeri ssp. ***densispina*** Hofacker & Herm . . . . BR
Frailea buiningiana Prestle . . . . [?]None given]
Frailea castanea Backeberg . . . . BR, UY, AR
Frailea castanea ssp. ***harmoniana*** (Ritter) Braun & Esteves Pereira . . . . BR
Frailea cataphracta (Dams) Britton & Rose . . . . BR, PY, BO
Frailea cataphracta ssp. **cataphracta** . . . . BR, PY
Frailea cataphracta ssp. *tuyensis* (Buining & Moser) Braun & Esteves Pereira
Frailea cataphracta ssp. **duchii** (Moser) Braun & Esteves Pereira . . . . . . BR, PY, BO
Frailea cataphracta ssp. *melitae* (Buining & Brederoo) Braun & Esteves Pereira
Frailea cataphractoides Backeberg
Frailea matoana Buining & Brederoo
Frailea melitae Buining & Brederoo
Frailea uhligiana Backeberg
Frailea chiquitana Cardenas . . . . BO
Frailea larae Vasquez
Frailea pullispina Backeberg
Frailea curvispina Buining & Brederoo . . . . BR
Frailea friedrichii Buining & Moser . . . . PY
Frailea gracillima (Lemaire) Britton & Rose . . . . BR, PY, UY
Frailea gracillima ssp. **gracillima** . . . . BR, PY, UY
Frailea alacriportana Backeberg & Voll
Frailea albifusca Ritter

Frailea gracillima ssp. *alacriportana* (Backeberg) Gerloff
Frailea gracillima ssp. *albifusca* (Ritter) Braun & Esteves Pereira
Frailea lepida Buining & Brederoo
Frailea pseudogracillima Ritter
Frailea gracillima ssp. **horstii** (Ritter) Braun & Esteves Pereira . . . . BR
Frailea horstii Ritter
Frailea horstii ssp. *fecotrigensis* Prestle
Frailea grahliana (Schumann) Britton & Rose . . . . PY, AR
Frailea grahliana ssp. ***moseriana*** (Buining & Brederoo) Prestle . . . . PY
Frailea moseriana Buining & Brederoo
Frailea knippeliana (Quehl) Britton & Rose . . . . PY
Frailea mammifera Buining & Brederoo . . . . BR, AR
®*Frailea magnifica* Buining ex Prestle
Frailea perumbilicata Ritter . . . . BR
Frailea phaeodisca (Spegazzini) Spegazzini . . . . BR, UY
Frailea perbella Prestle
Frailea pseudopulcherrima Ito . . . . UY
Frailea pumila (Lemaire) Britton & Rose . . . . BR, PY, AR, UY
Frailea albiareolata Buining & Brederoo
Frailea carminifilamentosa Kilian ex Backeberg
Frailea colombiana (Werdermann) Backeberg
Frailea hlineckyana Cervinka
Frailea jajoiana Cervinka
Frailea pumila ssp. *albiareolata* (Buining & Brederoo) Braun & Esteves Pereira
Frailea pumila ssp. *colombiana* (Werdermann) Prestle
Frailea pumila ssp. *hlineckyana* (Cervinka) Prestle
Frailea pumila ssp. *jajoiana* (Cervinka) Prestle
Frailea pumila ssp. *maior* (Ritter) Braun & Esteves Pereira
®*Frailea chrysacantha* Hrabe
Frailea pumila ssp. ***deminuta*** (Buining & Brederoo) Prestle . . . . BR
Frailea deminuta Buining & Brederoo
Frailea pygmaea (Spegazzini) Britton & Rose . . . . BR, UY, AR
Frailea pulcherrima (Arechavaleta) Spegazzini
Frailea pygmaea ssp. **pygmaea** . . . . BR, UY, AR
Frailea asperispina Ritter
Frailea aureinitens Buining & Brederoo
Frailea aureispina Ritter
Frailea pygmaea ssp. *altigibbera* (Ritter) Braun & Esteves Pereira
Frailea pygmaea ssp. *asperispina* (Ritter) Braun & Esteves Pereira
Frailea pygmaea ssp. *aureinitens* (Buining & Brederoo) Braun & Esteves Pereira
Frailea pygmaea ssp. *aureispina* (Ritter) Braun & Esteves Pereira
Frailea pygmaea ssp. *lilalunula* (Ritter) Braun & Esteves Pereira
Frailea pygmaea ssp. **albicolumnaris** (Ritter) Hofacker . . . . BR
Frailea albicolumnaris Ritter
Frailea pygmaea ssp. ***fulviseta*** (Buining & Brederoo) Braun & Esteves Pereira . . . . BR
Frailea fulviseta Buining & Brederoo
Frailea schilinzkyana (Schumann) Britton & Rose . . . . PY, AR
Frailea concepcionensis Buining & Moser
Frailea grahliana ssp. *concepcionensis* (Buining & Moser) Prestle
Frailea grahliana ssp. *ybatensis* (Buining & Moser) Prestle
Frailea ignacionensis Buining & Moser
Frailea schilinzkyana ssp. *concepcionensis*
(Buining & Moser) Braun & Esteves Pereira
Frailea ybatensis Buining & Moser

Geohintonia mexicana Glass & Fitz Maurice . . . . MX

Gymnocalycium ambatoense Piltz . . . . AR
Gymnocalycium amerhauseri H. Till . . . . AR
Gymnocalycium andreae (Boedeker) Backeberg . . . . AR

Gymnocalycium andreae ssp. ***carolinense*** Neuhuber . . . . AR
Gymnocalycium angelae Meregalli . . . . AR
Gymnocalycium anisitsii (Schumann) Britton & Rose . . . . BR, PY, BO
Gymnocalycium damsii (Schumann) Britton & Rose
Gymnocalycium griseopallidum Backeberg
Gymnocalycium joossensianum (Boedeker) Britton & Rose
Gymnocalycium anisitsii ssp. ***multiproliferum*** (Braun) Braun & Esteves Pereira . . . BR
Gymnocalycium baldianum (Spegazzini) Spegazzini . . . . AR
Gymnocalycium sanguiniflorum (Werdermann) Werdermann
Gymnocalycium bayrianum H. Till . . . . AR
Gymnocalycium berchtii G.J.A. Neuhuber . . . . AR
Gymnocalycium bodenbenderianum (Hosseus ex Berger) A.W. Hill . . . . AR
Gymnocalycium bodenbenderianum ssp. ***intertextum*** (Backeberg ex H. Till) H. Till AR
Gymnocalycium intertextum Backeberg ex H. Till
Gymnocalycium moserianum Schuetz
Gymnocalycium borthii Koop ex H. Till . . . . AR
Gymnocalycium bruchii (Spegazzini) Hosseus . . . . AR
Gymnocalycium albispinum Backeberg
Gymnocalycium bueneckeri Swales . . . . BR
Gymnocalycium calochlorum (Boedeker) Ito . . . . AR
Gymnocalycium proliferum (Backeberg) Backeberg
Gymnocalycium capillaense (Schick) Hosseus . . . . AR
Gymnocalycium sigelianum (Schick) Hosseus
Gymnocalycium sutterianum (Schick) Hosseus
Gymnocalycium carminanthum Borth & Koop . . . . AR
Gymnocalycium castellanosii Backeberg . . . . AR
Gymnocalycium acorrugatum J.G. Lambert
Gymnocalycium bozsingianum Schuetz
®*Gymnocalycium ferox* (Backeberg) Slaba
Gymnocalycium catamarcense H. Till & W. Till . . . . AR
Gymnocalycium catamarcense ssp. ***acinacispinum*** H. Till & W. Till . . . . AR
Gymnocalycium catamarcense ssp. ***schmidianum*** H. Till & W. Till . . . . AR
Gymnocalycium chiquitanum Cardenas . . . . BO
Gymnocalycium hammerschmidii Backeberg
Gymnocalycium deeszianum Doelz . . . . AR
Gymnocalycium delaetii (Schumann) Hosseus . . . . AR
Gymnocalycium denudatum (Link & Otto) Pfeiffer ex Mittler . . . . BR, UY
Gymnocalycium erinaceum J.G. Lambert . . . . AR
Gymnocalycium eurypleurum Plesnik ex Ritter . . . . PY
Gymnocalycium eytianum Cardenas . . . . BO
®***Gymnocalycium fleischerianum*** Backeberg . . . . PY
Gymnocalycium gibbosum (Haworth) Pfeiffer ex Mittler . . . . AR
Gymnocalycium brachypetalum Spegazzini
Gymnocalycium chubutense (Spegazzini) Spegazzini
Gymnocalycium gerardii (Boedeker) Ito
Gymnocalycium reductum (Link) Pfeiffer ex Mittler
Gymnocalycium gibbosum ssp. ***ferox*** (Labouret ex Ruempler) Papsch . . . . AR
Gymnocalycium horstii Buining . . . . BR
Gymnocalycium hossei (F. Haage) Berger . . . . AR
Gymnocalycium guanchinense Schuetz
Gymnocalycium mazanense (Backeberg) Backeberg
Gymnocalycium nidulans Backeberg
Gymnocalycium weissianum Backeberg
®*Gymnocalycium rhodantherum* (Boedeker) hort.
Gymnocalycium hybopleurum (Schumann) Backeberg . . . . AR
Gymnocalycium nigriareolatum Backeberg
Gymnocalycium hyptiacanthum (Lemaire) Britton & Rose . . . . UY
Gymnocalycium kieslingii Ferrari . . . . AR
Gymnocalycium leeanum (Hooker) Britton & Rose . . . . AR
Gymnocalycium leptanthum (Spegazzini) Spegazzini . . . . AR

Gymnocalycium mackieanum (Hooker) Metzing et al. . . . . AR
Gymnocalycium schatzlianum Strigl & W. Till
Gymnocalycium marsoneri Fric ex Ito . . . . BR, PY, BO, AR
Gymnocalycium fricianum Plesnik
Gymnocalycium hamatum Ritter
Gymnocalycium onychacanthum Ito
Gymnocalycium marsoneri ssp. ***matoense***
(Buining & Brederoo) Braun & Esteves Pereira . . . . BR, BO
Gymnocalycium matoense Buining & Brederoo
Gymnocalycium megatae Ito
Gymnocalycium pseudomalacocarpus Backeberg
Gymnocalycium tortuga hort. ex Backeberg
Gymnocalycium tudae Ito
®*Gymnocalycium brevistylum* Ritter
Gymnocalycium mesopotamicum Kiesling . . . . AR
Gymnocalycium mihanovichii (Fric & Guerke) Britton & Rose . . . . PY, AR
Gymnocalycium monvillei (Lemaire) Britton & Rose . . . . AR
Gymnocalycium multiflorum (Hooker) Britton & Rose
Gymnocalycium ourselianum (Cels ex Salm-Dyck) Y. Ito
Gymnocalycium schuetzianum H. Till & Schatzl
?*Gymnocalycium megalothelon* (Sencke ex Schumann) Britton & Rose
Gymnocalycium monvillei ssp. ***achirasense*** (H. Till & Schatzl ex H. Till) H. Till . . AR
Gymnocalycium achirasense H. Till & Schatzl ex H. Till
Gymnocalycium monvillei ssp. ***brachyanthum*** (Guerke) H. Till . . . . AR
Gymnocalycium brachyanthum (Guerke) Britton & Rose
Gymnocalycium monvillei ssp. ***horridispinum*** (G. Frank ex H. Till) H. Till . . . . AR
Gymnocalycium horridispinum G. Frank ex H. Till
Gymnocalycium mostii (Guerke) Britton & Rose . . . . AR
Gymnocalycium bicolor Schuetz
Gymnocalycium grandiflorum Backeberg
Gymnocalycium immemoratum Castellanos & Lelong
Gymnocalycium kurtzianum (Guerke) Britton & Rose
Gymnocalycium tobuschianum Schick
Gymnocalycium valnicekianum Jajo
Gymnocalycium mucidum Oehme . . . . AR
Gymnocalycium ferrarii Rausch
Gymnocalycium glaucum Ritter
Gymnocalycium netrelianum (Monville ex Labouret) Britton & Rose . . . . UY
Gymnocalycium neuhuberi H. Till & W. Till . . . . AR
Gymnocalycium curvispinum Fric
Gymnocalycium obductum J. Piltz . . . . AR
Gymnocalycium ochoterenae Backeberg . . . . AR
Gymnocalycium ochoterenae ssp. **ochoternae** . . . . AR
Gymnocalycium ochoterenae ssp. ***herbsthoferianum*** H. Till & Neuhuber . . . . AR
Gymnocalycium ochoterenae ssp. **vatteri** (Buining) Papsch . . . . AR
Gymnocalycium vatteri Buining
Gymnocalycium oenanthemum Backeberg . . . . AR
Gymnocalycium paraguayense (Schumann) Schuetz . . . . PY
Gymnocalycium parvulum (Spegazzini) Spegazzini . . . . AR
Gymnocalycium pediophilum Ritter ex Schuetz . . . . PY
®*Gymnocalycium paediophilum* Ritter ex Schuetz
Gymnocalycium pflanzii (Vaupel) Werdermann . . . . PY, BO, AR
Gymnocalycium chuquisacanum Cardenas
Gymnocalycium comarapense Backeberg
Gymnocalycium izozogsii Cardenas
Gymnocalycium lagunillasense Cardenas
Gymnocalycium marquezii Cardenas
Gymnocalycium millaresii Cardenas
Gymnocalycium pflanzii ssp. *argentinense* H. Till & W. Till
Gymnocalycium riograndense Cardenas
Gymnocalycium zegarrae Cardenas

Gymnocalycium platense (Spegazzini) Britton & Rose . . . . AR
Gymnocalycium pugionacanthum Backeberg . . . . AR
Gymnocalycium quehlianum (F. Haage ex Quehl) Vaupel ex Hosseus . . . . AR
Gymnocalycium ragonesei Castellanos . . . . AR
Gymnocalycium rauschii H. Till & W. Till . . . . UY
Gymnocalycium riojense Fric ex H. Till & W. Till . . . . AR
Gymnocalycium triacanthum Backeberg
®*Gymnocalycium platygonum* (H. Till & W. Till) Pilbeam
Gymnocalycium riojense ssp. ***kozelskyanum*** Schuetz ex Till & Till . . . . AR
Gymnocalycium riojense ssp. ***paucispinum*** Backeberg ex H. Till & W. Till . . . . . . AR
Gymnocalycium riojense ssp. ***piltziorum*** Schuetz ex H. Till & W. Till . . . . AR
Gymnocalycium piltziorum Schuetz
Gymnocalycium ritterianum Rausch . . . . AR
Gymnocalycium rosae H. Till . . . . AR
Gymnocalycium saglionis (Cels) Britton & Rose . . . . AR
®*Gymnocalycium saglione* (Cels) Britton & Rose
Gymnocalycium saglionis ssp. ***tilcarense*** (Backeberg) H. Till & W. Till . . . . AR
Brachycalycium tilcarense (Backeberg) Backeberg
®*Gymnocalycium tilcarense* (Backeberg) Schuetz
Gymnocalycium schickendantzii (Weber) Britton & Rose . . . . AR
Gymnocalycium antherostele Ritter
Gymnocalycium michoga Fric ex Ito
Gymnocalycium schroederianum Osten . . . . UY, AR
Gymnocalycium schroederianum ssp. ***bayense*** Kiesling . . . . AR
Gymnocalycium schroederianum ssp. ***paucicostatum*** Kiesling . . . . AR
Gymnocalycium spegazzinii Britton & Rose . . . . AR
Gymnocalycium horizonthalonium Fric ex Kreuzinger
®*Gymnocalycium loricatum* (Spegazzini) Spegazzini
Gymnocalycium spegazzinii ssp. **cardenasianum** (Ritter) Kiesling & Metzing . . . . BO
Gymnocalycium armatum Ritter
Gymnocalycium cardenasianum Ritter
Gymnocalycium stellatum Spegazzini . . . . AR
Gymnocalycium asterium Y. Ito ex Castellanos
Gymnocalycium stellatum ssp. ***occultum*** Fric ex H. & W. Till . . . . AR
®*Gymnocalycium occultum* Fric ex Schuetz
Gymnocalycium stenopleurum Ritter . . . . PY, BO
®*Gymnocalycium friedrichii* (Werdermann) Pazout
Gymnocalycium striglianum Jeggle ex H. Till . . . . AR
Gymnocalycium stuckertii (Spegazzini) Britton & Rose . . . . AR
Gymnocalycium taningaense J. Piltz . . . . AR
Gymnocalycium terweemeanum (Teucq ex Duursma) Borgmann & Piltz . . . . AR
Gymnocalycium tillianum Rausch . . . . AR
Gymnocalycium uebelmannianum Rausch . . . . AR
Gymnocalycium uruguayense (Arechavaleta) Britton & Rose . . . . BR, UY
Gymnocalycium artigas Herter
Gymnocalycium guerkeanum (Heese) Britton & Rose
Gymnocalycium melanocarpum (Arechavaleta) Britton & Rose

Haageocereus albispinus (Akers) Backeberg . . . . PE
Haageocereus peniculatus Rauh & Backeberg
Haageocereus australis Backeberg . . . . PE, CL
Haageocereus multicolorispinus Buining
Haageocereus chalaensis Ritter . . . . PE
Haageocereus chryseus Ritter . . . . PE
Haageocereus decumbens (Vaupel) Backeberg . . . . PE
Haageocereus ambiguus Rauh & Backeberg
Borzicactus decumbens (Vaupel) Britton & Rose
Haageocereus litoralis Rauh & Backeberg
Haageocereus mamillatus Rauh & Backeberg
Haageocereus fascicularis (Meyen) Ritter CL

Haageocereus fulvus Ritter . . . . PE
Haageocereus icensis Backeberg ex Ritter . . . . PE
Haageocereus icosagonoides Rauh & Backeberg . . . . PE
Haageocereus lanugispinus Ritter . . . . PE
Haageocereus acranthus (Vaupel) Backeberg . . . . PE
Haageocereus acranthus ssp. **acranthus** . . . . PE
Haageocereus achaetus Rauh & Backeberg
Haageocereus clavispinus Rauh & Backeberg
Haageocereus deflexispinus Rauh & Backeberg
Haageocereus lachayensis Rauh & Backeberg
Haageocereus pseudoacranthus Rauh & Backeberg
Haageocereus zonatus Rauh & Backeberg
®?*Haageocereus limensis* (Salm-Dyck) Ritter
Haageocereus acranthus ssp. **olowinskianus** (Backeberg) Ostolaza . . . . PE
Haageocereus olowinskianus Backeberg
Haageocereus pacalaensis Backeberg . . . . PE
Haageocereus horrens Rauh & Backeberg
Haageocereus laredensis Backeberg
Haageocereus repens Rauh & Backeberg
Haageocereus tenuispinus Rauh & Backeberg
Haageocereus platinospinus (Werdermann & Backeberg) Backeberg . . . . PE
Haageocereus pluriflorus Rauh & Backeberg . . . . PE
Haageocereus pseudomelanostele (Werdermann & Backeberg) Backeberg . . . . PE
Haageocereus acanthocladus Rauh & Backeberg
Haageocereus akersii Backeberg
Haageocereus chosicensis (Werdermann & Backeberg) Backeberg
Haageocereus chrysacanthus (Akers) Backeberg
Haageocereus clavatus (Akers) Cullmann
Haageocereus crassiareolatus Rauh & Backeberg
Haageocereus dichromus Rauh & Backeberg
Haageocereus divaricatispinus Rauh & Backeberg
Haageocereus longiareolatus Rauh & Backeberg
Haageocereus pachystele Rauh & Backeberg
Haageocereus piliger Rauh & Backeberg
Haageocereus setosus (Akers) Backeberg
Haageocereus symmetros Rauh & Backeberg
Haageocereus viridiflorus (Akers) Backeberg
Haageocereus zehnderi Rauh & Backeberg
?*Haageocereus multangularis* (Willdenow) Ritter
Haageocereus pseudomelanostele ssp. **pseudomelanostele** . . . . PE
Haageocereus pseudomelanostele ssp. **carminiflorus** (Rauh & Backeberg) Ostolaza PE
Haageocereus pseudomelanostele ssp. **aureispinus** (Rauh & Backeberg) Ostolaza . . PE
Haageocereus aureispinus Rauh & Backeberg
Haageocereus pseudomelanostele ssp. **turbidus** (Rauh & Backeberg) Ostolaza . . . . PE
Haageocereus turbidus Rauh & Backeberg
Haageocereus pseudoversicolor Rauh & Backeberg . . . . PE
Haageocereus subtilispinus Ritter . . . . PE
Haageocereus tenuis Ritter . . . . PE
Haageocereus versicolor (Werdermann & Backeberg) Backeberg . . . . PE
Haageocereus vulpes Ritter . . . . PE
Haageocereus zangalensis Ritter . . . . PE

×***Haagespostoa albisetata*** (Akers) Rowley . . . . PE
Haageocereus albisetatus (Akers) Backeberg
Neobinghamia multiareolata Rauh & Backeberg
Neobinghamia villigera Rauh & Backeberg
×***Haagespostoa climaxantha*** (Werdermann) Rowley . . . . PE
Neobinghamia climaxantha (Werdermann) Backeberg
Haageocereus climaxanthus (Werdermann) Croizat

Harrisia aboriginum Small ex Britton & Rose .... US
Harrisia adscendens (Guerke) Britton & Rose .... BR
Eriocereus adscendens (Guerke) Berger
Harrisia balansae (Schumann) Taylor & Zappi .... BR, PY, AR
Eriocereus guelichii (Spegazzini) Berger
Harrisia guelichii (Spegazzini) Britton & Rose
Harrisia brookii Britton .... BS
Harrisia divaricata (Lamarck) Backeberg .... HT, DO
Harrisia serruliflora (Haworth) Lourteig
Harrisia earlei Britton & Rose .... CU
Harrisia eriophora (Pfeiffer) Britton .... CU
Cereus eriophorus Pfeiffer
Harrisia fernowii Britton .... CU
Harrisia fragrans Small ex Britton & Rose .... US
Harrisia gracilis (Miller) Britton .... JM, KY, US
Harrisia donae-antoniae M.L. Hooten
Cereus gracilis Miller
Harrisia hurstii W.T. Marshall .... DO
Harrisia martinii (Labouret) Britton .... PY, AR
Cereus martinii Labouret
Eriocereus martinii (Labouret) Riccobono
Harrisia nashii Britton .... HT, DO
Harrisia pomanensis (Weber) Britton & Rose .... PY, BO, AR
Eriocereus polyacanthus Ritter
Eriocereus pomanensis (Weber) Berger
Harrisia pomanensis ssp. *tarijensis* (Ritter) Braun & Esteves Pereira
Eriocereus tarijensis Ritter
Ⓤ*Harrisia pomanensis* ssp. *bonplandii* (Parmentier) Braun & Esteves Pereira
Ⓜ*Eriocereus bonplandii* (Parmentier ex Pfeiffer) Riccobono
Ⓜ*Harrisia bonplandii* (Parmentier ex Pfeiffer) Britton & Rose
Harrisia pomanensis ssp. ***regelii*** (Weingart) Kiesling .... AR
Eriocereus regelii (Weingart) Backeberg
Harrisia regelii (Weingart) Borg
Harrisia portoricensis Britton .... PR
Harrisia simpsonii Small ex Britton & Rose .... US
Harrisia deeringii Backeberg
Harrisia taetra Areces .... CU
Harrisia taylori Britton .... CU
Harrisia tetracantha (Labouret) Hunt .... BO
Eriocereus tephracanthus (Labouret) Riccobono
Roseocereus tephracanthus (Labouret) Backeberg
Harrisia tortuosa (Forbes) Britton & Rose .... PY, UY, AR
Eriocereus arendtii (Schumann) Ritter
Cereus tortuosus Forbes ex Otto & Dietrich
Eriocereus tortuosus (Forbes) Riccobono

Hatiora epiphylloides (Campos-Porto & Werdermann) F. Buxbaum .... BR
Pseudozygocactus epiphylloides (Campos-Porto & Werdermann) Backeberg
Hatiora epiphylloides ssp. **epiphylloides** .... BR
Hatiora epiphylloides ssp. **bradei** (Campos-Porto & Castellanos) Barthlott & Taylor BR
Hatiora gaertneri (Regel) Barthlott .... BR
Epiphyllopsis gaertneri (Regel) Berger
Rhipsalidopsis gaertneri (Regel) Moran
Schlumbergera gaertneri (Regel) Britton & Rose
Hatiora ×graeseri (Werdermann) Barthlott XC
Hatiora herminiae (Campos-Porto & Castellanos) Backeberg ex Barthlott .... BR
Rhipsalis herminiae (Campos-Porto & Castellanos) Kimnach
Hatiora rosea (Lagerheim) Barthlott .... BR
Rhipsalidopsis rosea (Lagerheim) Britton & Rose

Hatiora salicornioides (Haworth) Britton & Rose ex Bailey . . . . BR
Hatiora bambusoides (Weber) Britton & Rose
Hatiora cylindrica Britton & Rose

Hylocereus calcaratus (Weber) Britton & Rose . . . . CR
Hylocereus costaricensis (Weber) Britton & Rose . . . . NI, CR, PA
Hylocereus escuintlensis Kimnach . . . . GT
Hylocereus estebanensis Backeberg . . . . VE
Hylocereus guatemalensis (Eichlam) Britton & Rose . . . . GT, SV
Hylocereus lemairei (Hooker) Britton & Rose . . . . TT, AN, VE, [GF, SR]
Hylocereus trinitatensis (Lemaire & Herment) Berger
Hylocereus venezuelensis Britton & Rose
Wilmattea venezuelensis (Britton & Rose) Croizat
®?*Cereus extensus* Hooker
®?*Hylocereus extensus* Britton & Rose ex Bailey
Hylocereus microcladus Backeberg . . . . CO, PE
Hylocereus minutiflorus Britton & Rose . . . . GT, BZ, HN
Wilmattea minutiflora (Britton & Rose) Britton & Rose
Hylocereus monacanthus (Lemaire) Britton & Rose . . . . CR, PA, CO
Hylocereus ocamponis (Salm-Dyck) Britton & Rose . . . . MX
Hylocereus peruvianus Backeberg . . . . PE
Hylocereus polyrhizus (Weber) Britton & Rose . . . . NI, CR, PA, CO, EC
Hylocereus purpusii (Weingart) Britton & Rose . . . . MX
Hylocereus scandens (Salm-Dyck) Backeberg . . . . GY, SR
Hylocereus stenopterus (Weber) Britton & Rose . . . . CR, PA
Hylocereus triangularis (Linnaeus) Britton & Rose . . . . CU, JM, DO
Hylocereus compressus Y. Ito
Hylocereus cubensis Britton & Rose
Hylocereus trigonus (Haworth) Safford . . . . PR, VI, WI
Hylocereus antiguensis Britton & Rose
Hylocereus napoleonis (Graham) Britton & Rose
Hylocereus plumieri (Roland-Gosselin) Lourteig
Mediocactus pomifer (Weingart) Backeberg
Hylocereus undatus (Haworth) Britton & Rose XC
Cereus undatus Haworth

Jasminocereus thouarsii (Weber) Backeberg . . . . EC
Jasminocereus galapagensis (Weber) Britton & Rose
Jasminocereus howellii Dawson
Jasminocereus sclerocarpus (Schumann) Backeberg

Lasiocereus fulvus Ritter . . . . PE
Lasiocereus rupicola Ritter . . . . PE

Leocereus bahiensis Britton & Rose . . . . BR
Leocereus bahiensis
ssp. *barreirensis* (Braun & Esteves Pereira) Braun & Esteves Pereira
ssp. *exiguospinus* (Braun & Esteves Pereira) Braun & Esteves Pereira
ssp. *robustispinus* (Braun & Esteves Pereira) Braun & Esteves Pereira
ssp. *urandianus* (Ritter) Braun & Esteves Pereira
Leocereus estevesii Braun
Leocereus urandianus Ritter

Lepismium aculeatum (Weber) Barthlott . . . . AR
Rhipsalis aculeata Weber
Lepismium bolivianum (Britton) Barthlott . . . . BO
Rhipsalis boliviana (Britton) Lauterbach ex Buchtien
Lepismium brevispinum Barthlott . . . . PE
Rhipsalis brevispina (Barthlott) Kimnach
®*Acanthorhipsalis brevispina* Ritter

Ⓤ*Pfeiffera brevispina* P.V. Heath
Ⓤ*Rhipsalis brevispina* Kimnach
Lepismium crenatum (Britton) Barthlott . . . . BO
Acanthorhipsalis crenata Britton & Rose
Pfeiffera crenata (Britton) P.V. Heath
Rhipsalis crenata (Britton) Vaupel
Lepismium cruciforme (Vellozo) Miquel . . . . BR, PY, AR
Rhipsalis myosurus (De Candolle) Foerster
Rhipsalis squamulosa (Salm-Dyck) Schumann
Lepismium houlletianum (Lemaire) Barthlott . . . . BR, AR
Acanthorhipsalis houlletiana (Lemaire) Volgin
Rhipsalis houlletiana Lemaire
Lepismium ianthothele (Monville) Barthlott . . . . BO, AR
Pfeiffera erecta Ritter
Lepismium erectum (Ritter) Suepplie
Pfeiffera gracilis Ritter
Pfeiffera ianthothele (Monville) Weber
Lepismium mataralense (Ritter) Suepplie
Pfeiffera mataralensis Ritter
Pfeiffera multigona Cardenas
Lepismium incachacanum (Cardenas) Barthlott . . . . BO
Acanthorhipsalis incachacana (Cardenas) Volgin
Rhipsalis incachacana Cardenas
Lepismium lorentzianum (Grisebach) Barthlott . . . . BO, AR
Rhipsalis lorentziana Grisebach
Lepismium lumbricoides (Lemaire) Barthlott . . . . BR, PY, BO, UY, AR
Rhipsalis leucorhaphis Schumann
Rhipsalis loefgrenii Britton & Rose
Rhipsalis lumbricoides (Lemaire) Lemaire ex Salm-Dyck
Lepismium micranthum (Vaupel) Barthlott . . . . PE
Rhipsalis asperula Vaupel
Acanthorhipsalis micrantha (Vaupel) Britton & Rose
Lymanbensonia micrantha (Vaupel) Kimnach
Pfeiffera micrantha (Vaupel) P.V. Heath
Lepismium miyagawae (Barthlott & Rauh) Barthlott . . . . BO
Pfeiffera miyagawae Barthlott & Rauh
Rhipsalis miyagawae (Barthlott & Rauh) Kimnach
Lepismium monacanthum (Grisebach) Barthlott . . . . BO, AR
Acanthorhipsalis incahuasina Cardenas
Pfeiffera incahuasina (Cardenas) P.V. Heath
Acanthorhipsalis monacantha (Grisebach) Britton & Rose
Pfeiffera monacantha (Grisebach) P.V. Heath
Acanthorhipsalis samaipatana (Cardenas) Ritter
Lepismium paranganiense (Cardenas) Barthlott . . . . BO
Acanthorhipsalis paranganiensis Cardenas
Pfeiffera paranganiensis (Cardenas) P.V. Heath
Rhipsalis paranganiensis (Cardenas) Kimnach
Lepismium warmingianum (Schumann) Barthlott . . . . BR, PY, AR
Rhipsalis gonocarpa Weber
Lepismium lineare (Schumann) Barthlott
Rhipsalis linearis Schumann
Rhipsalis warmingiana Schumann

Leptocereus arboreus Britton & Rose . . . . CU
Leptocereus assurgens (Wright ex Grisebach) Britton & Rose . . . . CU
Leptocereus carinatus Areces . . . . CU
Leptocereus ekmanii (Werdermann) F. Knuth . . . . CU
Leptocereus grantianus Britton . . . . PR
Neoabbottia grantiana (Britton) Buxbaum
Leptocereus leonii Britton & Rose . . . . CU, KY

Leptocereus maxonii Britton & Rose . . . . CU
Leptocereus paniculatus (Lamarck) Hunt . . . . HT
Neoabbottia paniculata (Lamarck) Britton & Rose
Leptocereus prostratus Britton & Rose . . . . CU
Leptocereus quadricostatus (Bello) Britton & Rose . . . . PR
Leptocereus santamarinae Areces . . . . CU
Leptocereus scopulophilus Areces . . . . CU
Leptocereus sylvestris Britton & Rose . . . . CU
Leptocereus weingartianus (Hartmann ex Dams) Britton & Rose . . . . HT, DO
Leptocereus wrightii Leon . . . . CU

Leuchtenbergia principis Hooker . . . . MX

Lophophora diffusa (Croizat) H. Bravo . . . . MX
Lophophora lutea (Rouhier) Backeberg
Lophophora williamsii (Lemaire ex Salm-Dyck) J. Coulter . . . . US, MX
Lophophora diffusa ssp. *fricii* (Habermann) Halda
Lophophora diffusa ssp. *viridescens* Halda
Lophophora echinata Croizat
Lophophora fricii Habermann
Lophophora jourdaniana Habermann

Maihuenia patagonica (Philippi) Spegazzini . . . . AR
Maihuenia albolanata Ritter
Maihuenia brachydelphys (Schumann) Schumann
Maihuenia cumulata Ritter
Maihuenia latispina Ritter
Opuntia patagonica Philippi
Maihuenia valentinii Spegazzini
Maihuenia poeppigii (Pfeiffer) Schumann . . . . AR, CL
Maihuenia philippii (Weber) Schumann

Mammillaria albicans (Britton & Rose) Berger . . . . MX
Mammillaria albicans ssp. **albicans** . . . . MX
Mammillaria slevinii (Britton & Rose) Boedeker
Mammillaria albicans ssp. **fraileana** (Britton & Rose) Hunt . . . . MX
Mammillaria fraileana (Britton & Rose) Boedeker
Mammillaria albicoma Boedeker . . . . MX
Mammillaria albiflora (Werdermann) Backeberg . . . . MX
Mammillaria albilanata Backeberg . . . . MX
Mammillaria albilanata ssp. **albilanata** . . . . MX
Mammillaria fuauxiana Backeberg
Mammillaria igualensis Reppenhagen
Mammillaria albilanata ssp. **oaxacana** Hunt . . . . MX
Mammillaria ignota Reppenhagen
Mammillaria lanigera Reppenhagen
Mammillaria monticola Reppenhagen
Mammillaria noureddineana Reppenhagen
Mammillaria albilanata ssp. **reppenhagenii** (Hunt) Hunt . . . . MX
Mammillaria reppenhagenii Hunt
Mammillaria albilanata ssp. **tegelbergiana** (Gates ex Lindsay) Hunt . . . . MX
Mammillaria tegelbergiana Gates ex Lindsay
Mammillaria amajacensis Brachet & Lacoste . . . . MX
Mammillaria anniana Glass & Foster . . . . MX
Mammillaria armillata K. Brandegee . . . . MX
Mammillaria armillata ssp. **armillata** . . . . MX
Mammillaria armillata ssp. **cerralboa** (Britton & Rose) Hunt . . . . MX
Mammillaria cerralboa (Britton & Rose) Orcutt
Mammillaria aureilanata Backeberg . . . . MX
Mammillaria backebergiana F.G. Buchenau . . . . MX

Mammillaria backebergiana ssp. **backebergiana** . . . . MX
Mammillaria backebergiana ssp. **ernestii** (Fittkau) Hunt . . . . MX
Mammillaria ernestii Fittkau
Mammillaria isotensis Reppenhagen
Mammillaria barbata Engelmann . . . . MX
Mammillaria garessii Cowper
Mammillaria morricalii Cowper
Mammillaria orestera L. Benson
Mammillaria santaclarensis Cowper
Mammillaria viridiflora (Britton & Rose) Boedeker
®*Mammillaria chavezei* Cowper
Mammillaria baumii Boedeker . . . . MX
Dolichothele baumii (Boedeker) Werdermann & Buxbaum
Mammillaria radiaissima Lindsay ex Craig
Mammillaria beneckei Ehrenberg . . . . MX
Dolichothele balsasoides (Craig) Backeberg
Mammillaria balsasoides Craig
Dolichothele beneckei (Ehrenberg) Backeberg
Oehmea beneckei (Ehrenberg) Buxbaum
Mammillaria colonensis Craig
Mammillaria guiengolensis Bravo & MacDougall
Dolichothele nelsonii (Britton & Rose) Backeberg
®*Mammillaria barkeri* Shurly ex Backeberg
Mammillaria berkiana Lau . . . . MX
Mammillaria blossfeldiana Boedeker . . . . MX
Mammillaria shurliana Gates
Mammillaria bocasana Poselger . . . . MX
Mammillaria bocasana ssp. **bocasana** . . . . MX
Mammillaria bocasana ssp. **eschauzieri** (Coulter) Fitz Maurice . . . . MX
Mammillaria eschauzieri (Coulter) Craig
Mammillaria hirsuta Boedeker
Mammillaria knebeliana Boedeker
Mammillaria kunzeana Boedeker & Quehl
Mammillaria longicoma (Britton & Rose) Berger
Mammillaria bocensis Craig . . . . MX
Mammillaria neoschwarzeana Schwarz ex Backeberg
Mammillaria rubida Schwarz ex Backeberg
Mammillaria boelderliana Wohlschlager . . . . MX
Mammillaria bombycina Quehl . . . . MX
Mammillaria bombycina ssp. **bombycina** . . . . MX
Mammillaria bombycina ssp. **perezdelarosae** (Bravo & Scheinvar) Hunt . . . . MX
Mammillaria perezdelarosae Bravo & Scheinvar
Mammillaria boolii Lindsay . . . . MX
Mammillaria brandegeei (Coulter) K. Brandegee . . . . MX
Mammillaria brandegeei ssp. **brandegeei** . . . . MX
Mammillaria brandegeei ssp. **gabbii** (J. Coulter) Hunt . . . . MX
Mammillaria gabbii (Coulter) K. Brandegee
Mammillaria brandegeei ssp. **glareosa** (Boedeker) Hunt . . . . MX
Mammillaria dawsonii (Houghton) Craig
Mammillaria glareosa Boedeker
Mammillaria brandegeei ssp. **lewisiana** (Gates) Hunt . . . . MX
Mammillaria lewisiana Gates
Mammillaria canelensis Craig . . . . MX
Mammillaria bellacantha Craig
Mammillaria capensis (Gates) Craig . . . . MX
Mammillaria carmenae Castaneda & Nunez . . . . MX
Mammillaria carnea Zuccarini ex Pfeiffer . . . . MX
¶*Mammillaria orcuttii* Boedeker
Mammillaria carretii Rebut ex Schumann . . . . MX
Mammillaria saffordii (Britton & Rose) Bravo

Mammillaria coahuilensis (Boedeker) Moran . . . . MX
Mammillaria coahuilensis ssp. **coahuilensis** . . . . MX
Mammillaria heyderi ssp. *coahuilensis* (Boedeker) Luethy
®*Mammillaria schwartzii* (Fric) Backeberg
Mammillaria coahuilensis ssp. **albiarmata** (Boedeker) Hunt . . . . MX
Mammillaria albiarmata Boedeker
Mammillaria columbiana Salm-Dyck . . . . MX, GT, HN, JM, CO, VE
Mammillaria columbiana ssp. **columbiana** . . . . CO, VE
Mammillaria hennisii Boedeker
Mammillaria soehlemannii Haage & Backeberg
Mammillaria columbiana ssp. **yucatanensis** (Britton & Rose) Hunt MX, GT, HN, JM
Mammillaria chiapensis Reppenhagen
Mammillaria ruestii Quehl
Mammillaria yucatanensis (Britton & Rose) Orcutt
Mammillaria compressa De Candolle . . . . MX
Mammillaria esseriana Boedeker
Mammillaria compressa ssp. **compressa** . . . . MX
Mammillaria bernalensis Reppenhagen
Mammillaria tolimensis Craig
Mammillaria compressa ssp. **centralifera** (Reppenhagen) Hunt . . . . MX
Mammillaria centralifera Reppenhagen
Mammillaria craigii Lindsay . . . . MX
Mammillaria crinita De Candolle . . . . MX
Mammillaria criniformis De Candolle
Mammillaria pubispina Boedeker
Mammillaria schelhasii Pfeiffer
Mammillaria crinita ssp. **crinita** . . . . MX
Mammillaria cadereytana Schmoll ex Backeberg
Mammillaria mollihamata Shurly
Mammillaria pygmaea (Britton & Rose) Berger
Mammillaria crinita ssp. **leucantha** (Boedeker) Hunt . . . . MX
Mammillaria aureoviridis Heinrich
Mammillaria aurihamata Boedeker
Mammillaria brevicrinita Reppenhagen
Mammillaria erectohamata Boedeker
Mammillaria leucantha Boedeker
Mammillaria moeller-valdeziana Appenzeller
Mammillaria puberula Reppenhagen
Mammillaria tezontle Fitz Maurice & Fitz Maurice
Mammillaria crinita ssp. **wildii** (A. Dietrich) Hunt . . . . MX
Mammillaria calleana Backeberg
Mammillaria wildii A. Dietrich
Mammillaria crucigera Martius . . . . MX
Mammillaria crucigera ssp. **crucigera** . . . . MX
Mammillaria buchenaui Backeberg
Mammillaria crucigera ssp. **tlalocii** (Reppenhagen) Hunt . . . . MX
Mammillaria tlalocii Reppenhagen
Mammillaria decipiens Scheidweiler . . . . MX
Mammillaria decipiens ssp. **decipiens** . . . . MX
Dolichothele decipiens (Scheidweiler) Tiegel
Mammillaria decipiens ssp. **albescens** (Tiegel) Hunt . . . . MX
Dolichothele albescens (Tiegel) Backeberg
Mammillaria albescens Tiegel
Mammillaria decipiens ssp. **camptotricha** (Dams) Hunt . . . . MX
Dolichothele camptotricha (Dams) Tiegel
Mammillaria camptotricha Dams
Mammillaria deherdtiana Farwig . . . . MX
Mammillaria deherdtiana ssp. **deherdtiana** . . . . MX
Mammillaria deherdtiana ssp. **dodsonii** (Bravo) Hunt . . . . MX
Mammillaria dodsonii Bravo

Mammillaria densispina (Coulter) Orcutt .... MX
Mammillaria buxbaumiana Reppenhagen
Mammillaria dioica K. Brandegee .... US, MX
Mammillaria rectispina (Dawson) Reppenhagen
Mammillaria dioica ssp. **dioica** .... US, MX
Mammillaria dioica ssp. **angelensis** (Craig) Hunt .... MX
Mammillaria angelensis Craig
Mammillaria dioica ssp. **estebanensis** (Lindsay) Hunt .... MX
Mammillaria estebanensis Lindsay
Mammillaria discolor Haworth .... MX
Mammillaria ochoterenae (Bravo) Werdermann ex Backeberg
Mammillaria pachyrhiza Backeberg
Mammillaria schmollii (Bravo) Werdermann ex Backeberg
Mammillaria discolor ssp. **discolor** .... MX
Mammillaria discolor ssp. **esperanzaensis** (Boedeker) Hunt .... MX
Mammillaria esperanzaensis Boedeker
Mammillaria dixanthocentron Backeberg ex Mottram .... MX
Mammillaria duoformis Craig & Dawson .... MX
Mammillaria claviformis Reppenhagen
Mammillaria erythrocalyx F.G. Buchenau
?*Mammillaria rossiana* Heinrich
Mammillaria duwei Rogozinski & Appenzeller .... MX
Mammillaria ekmanii Werdermann .... HT
Mammillaria elongata De Candolle .... MX
Mammillaria elongata ssp. **elongata** .... MX
Mammillaria elongata ssp. **echinaria** (De Candolle) Hunt .... MX
Mammillaria echinaria De Candolle
Mammillaria eriacantha Pfeiffer .... MX
Mammillaria erythrosperma Boedeker .... MX
Mammillaria multiformis (Britton & Rose) Backeberg
Mammillaria evermanniana (Britton & Rose) Orcutt .... MX
Mammillaria fittkaui Glass & Foster .... MX
Mammillaria variabilis Reppenhagen
Mammillaria fittkaui ssp. **fittkaui** .... MX
Mammillaria fittkaui ssp. **limonensis** (Reppenhagen) Luethy .... MX
Mammillaria limonensis Reppenhagen
Mammillaria flavicentra Backeberg ex Mottram .... MX
Mammillaria formosa Galeotti ex Scheidweiler .... MX
Mammillaria arroyensis Reppenhagen
Mammillaria formosa ssp. **formosa** .... MX
Mammillaria formosa ssp. **chionocephala** (J.A. Purpus) Hunt .... MX
Mammillaria caerulea Craig
Mammillaria chionocephala Purpus
Mammillaria ritteriana Boedeker
Mammillaria formosa ssp. **microthele** (Muehlenpfordt) Hunt .... MX
Mammillaria microthele Muehlenpfordt
Mammillaria formosa ssp. **pseudocrucigera** (Craig) Hunt .... MX
Mammillaria pseudocrucigera Craig
Mammillaria geminispina Haworth .... MX
¶*Mammillaria elegans* De Candolle
Mammillaria geminispina ssp. **geminispina** .... MX
Mammillaria geminispina ssp. **leucocentra** (Berg) Hunt .... MX
Mammillaria leucocentra Berg
Mammillaria albata Reppenhagen
Mammillaria gigantea Hildmann ex Schumann .... MX
Mammillaria armatissima Craig
Mammillaria hamiltonhoytea (Bravo) Werdermann ex Backeberg
Mammillaria hastifera Krainz & Keller
Mammillaria ocotillensis Craig
Mammillaria saint-pieana Backeberg ex Mottram

Mammillaria glassii R.A. Foster . . . . MX
Mammillaria glassii ssp. **glassii** . . . . MX
Mammillaria glassii ssp. **ascensionis** (Reppenhagen) Hunt . . . . MX
Mammillaria ascensionis Reppenhagen
Mammillaria glochidiata Martius . . . . MX
Mammillaria goodridgii Scheer ex Salm-Dyck . . . . MX
Mammillaria grahamii Engelmann . . . . US
Mammillaria milleri (Britton & Rose) Boedeker
Mammillaria oliviae Orcutt
®*Mammillaria microcarpa* Engelmann
®*Mammillaria microcarpa* ssp. *grahamii* (Engelmann) Mottram
Mammillaria grusonii Runge . . . . MX
Mammillaria durangicola Reppenhagen
Mammillaria pachycylindrica Backeberg
Mammillaria papasquiarensis (Bravo) Reppenhagen
Mammillaria zeyeriana Haage f. ex Schumann
Mammillaria guelzowiana Werdermann . . . . MX
Krainzia guelzowiana (Werdermann) Backeberg
Mammillaria guerreronis (Bravo) Boedeker ex Backeberg & Knuth . . . . MX
Mammillaria zopilotensis Craig
Mammillaria guillauminiana Backeberg . . . . MX
Mammillaria haageana Pfeiffer . . . . MX
Mammillaria dealbata Dietrich
Mammillaria donatii Berge ex Schumann
Mammillaria dyckiana Zuccarini ex Pfeiffer
Mammillaria kunthii Ehrenberg
Mammillaria vaupelii Tiegel
Mammillaria haageana ssp. **haageana** . . . . MX
Mammillaria haageana ssp. **acultzingensis** (Linzen et al.) Hunt . . . . MX
Mammillaria acultzingensis Linzen et al.
Mammillaria haageana ssp. **conspicua** (J. Purpus) Hunt . . . . MX
Mammillaria conspicua J. Purpus
®*Mammillaria albidula* Backeberg
Mammillaria haageana ssp. **elegans** Hunt . . . . MX
Mammillaria collina J. Purpus
®*Mammillaria elegans* auctt
Mammillaria haageana ssp. **san-angelensis** (Sanchez-Mejorada) Hunt . . . . MX
Mammillaria san-angelensis Sanchez-Mejorada
Mammillaria haageana ssp. **schmollii** (Craig) Hunt . . . . MX
Mammillaria meissneri Ehrenberg
Mammillaria hahniana Werdermann . . . . MX
Mammillaria hahniana ssp. **hahniana** . . . . MX
Mammillaria hahniana ssp. **bravoae** (Craig) Hunt . . . . MX
Mammillaria bravoae Craig
Mammillaria hahniana ssp. **mendeliana** (Bravo) Hunt . . . . MX
Mammillaria mendeliana (Bravo) Werdermann ex Backeberg
Mammillaria hahniana ssp. **woodsii** (Craig) Hunt . . . . MX
Mammillaria woodsii Craig
Mammillaria halbingeri Boedeker . . . . MX
Mammillaria halei T. Brandegee . . . . MX
Cochemiea halei (T. Brandegee) Walton
Mammillaria hamata Lehmann ex Pfeiffer . . . . MX
Mammillaria heidiae Krainz . . . . MX
Mammillaria hernandezii Glass & Foster . . . . MX
Mammillaria herrerae Werdermann . . . . MX
Mammillaria hertrichiana Craig . . . . MX
Mammillaria heyderi Muehlenpfordt . . . . US, MX
Mammillaria parrasensis Reppenhagen
Mammillaria heyderi ssp. **heyderi** . . . . MX
Mammillaria heyderi ssp. **gaumeri** (Britton & Rose) Hunt . . . . MX
Mammillaria gaumeri (Britton & Rose) Orcutt

Mammillaria heyderi ssp. **gummifera** (Engelmann) Hunt . . . . MX
Mammillaria gummifera Engelmann
Mammillaria heyderi ssp. **hemisphaerica** (Engelmann) Hunt . . . . US, MX
Mammillaria applanata Engelmann ex Salm-Dyck
Mammillaria hemisphaerica Engelmann
Mammillaria heyderi ssp. **macdougalii** (Rose ex Bailey) Hunt . . . . US
Mammillaria macdougalii Rose ex Bailey
Mammillaria heyderi ssp. **meiacantha** (Engelmann) Hunt . . . . US, MX
Mammillaria meiacantha Engelmann
Mammillaria huitzilopochtli Hunt . . . . MX
Mammillaria humboldtii Ehrenberg . . . . MX
Mammillaria hutchisoniana (Gates) Boedeker . . . . MX
Mammillaria hutchisoniana ssp. **hutchisoniana** . . . . MX
Mammillaria bullardiana (Gates) Boedeker
Mammillaria hutchisoniana ssp. **louisae** (Lindsay) Hunt . . . . MX
Mammillaria louisae Lindsay
Mammillaria insularis Gates . . . . MX
Mammillaria jaliscana (Britton & Rose) Boedeker . . . . MX
Mammillaria fuscohamata Backeberg
Mammillaria kleiniorum Appenzeller
Mammillaria jaliscana ssp. **jaliscana** . . . . MX
Mammillaria jaliscana ssp. **zacatecasensis** (Shurly) Hunt . . . . MX
Mammillaria zacatecasensis Shurly
Mammillaria johnstonii (Britton & Rose) Orcutt . . . . MX
Mammillaria karwinskiana Martius . . . . MX
Mammillaria jozef-bergeri Wojnowski & Prajer
Mammillaria multiseta Ehrenberg
Mammillaria neomystax Backeberg
Mammillaria praelii Muehlenpfordt
Mammillaria strobilina Tiegel
®*Mammillaria ebenacantha* Shurly ex Backeberg
Mammillaria karwinskiana ssp. **karwinskiana** . . . . MX
Mammillaria confusa (Britton & Rose) Orcutt
Mammillaria karwinskiana ssp. **beiselii** (Diers) Hunt . . . . MX
Mammillaria beiselii Diers & Esteves Pereira
Mammillaria karwinskiana ssp. **collinsii** (Britton & Rose) Hunt . . . . MX
Mammillaria collinsii (Britton & Rose) Orcutt
Mammillaria nagliana Reppenhagen
Mammillaria tropica Reppenhagen
Mammillaria karwinskiana ssp. **nejapensis** (Craig & Dawson) Hunt . . . . MX
Mammillaria nejapensis Craig & Dawson
Mammillaria klissingiana Boedeker . . . . MX
Mammillaria brauneana Boedeker
Mammillaria knippeliana Quehl . . . . MX
Mammillaria kraehenbuehlii (Krainz) Krainz . . . . MX
Pseudomammillaria kraehenbuehlii Krainz
Mammillaria lasiacantha Engelmann . . . . US, MX
Mammillaria lasiacantha ssp. **lasiacantha** . . . . US, MX
Mammillaria denudata (Engelmann) Berger
Mammillaria lasiacantha ssp. **egregia** (Backeberg ex Rogozinski & Appenzeller) Hunt
. . . . MX
Mammillaria egregia Backeberg ex Rogozinski & Appenzeller
Mammillaria lasiacantha ssp. **hyalina** Hunt . . . . MX
Mammillaria wohlschlageri Reppenhagen
Mammillaria lasiacantha ssp. **magallanii** (Schmoll ex Craig) Hunt . . . . MX
Mammillaria magallanii Schmoll ex Craig
Mammillaria neobertrandiana Backeberg
Mammillaria lengdobleriana Boedeker
Mammillaria laui Hunt . . . . MX
Mammillaria laui ssp. **laui** . . . . MX

Mammillaria laui ssp. **dasyacantha** (Hunt) Hunt . . . . MX
®*Mammillaria dasyacantha* (Hunt) Reppenhagen
Mammillaria laui ssp. **subducta** (Hunt) Hunt . . . . MX
Mammillaria subducta (Hunt) Reppenhagen
Mammillaria lenta K. Brandegee . . . . MX
Mammillaria lindsayi Craig . . . . MX
Mammillaria lloydii (Britton & Rose) Orcutt . . . . MX
Mammillaria longiflora (Britton & Rose) Berger . . . . MX
Mammillaria longiflora ssp. **longiflora** . . . . MX
Krainzia longiflora (Britton & Rose) Backeberg
Mammillaria longiflora ssp. **stampferi** (Reppenhagen) Hunt . . . . MX
Mammillaria stampferi Reppenhagen
Mammillaria longimamma De Candolle . . . . MX
Dolichothele longimamma (De Candolle) Britton & Rose
Dolichothele longimamma ssp. *uberiformis* (Zuccarini) Krainz
Dolichothele uberiformis (Zuccarini) Britton & Rose
Mammillaria uberiformis Zuccarini
Mammillaria luethyi G.S. Hinton . . . . MX
Mammillaria magnifica F.G. Buchenau . . . . MX
Mammillaria magnimamma Haworth . . . . MX
Mammillaria bucareliensis Craig
Mammillaria centricirrha Lemaire
Mammillaria macracantha De Candolle
Mammillaria priessnitzii Reppenhagen
Mammillaria saxicola Reppenhagen
Mammillaria vagaspina Craig
Mammillaria vallensis Reppenhagen
Mammillaria zuccariniana Martius
Mammillaria mainiae K. Brandegee . . . . US, MX
Mammillaria mammillaris (Linnaeus) Karsten . . . . WI, TT, AN, VE
Mammillaria simplex Haworth
®*Mammillaria pseudosimplex* W. Haage & Backeberg
Mammillaria marcosii Fitz Maurice & Glass . . . . MX
?*Mammillaria multihamata* Boedeker
Mammillaria marksiana Krainz . . . . MX
Mammillaria mathildae Kraehenbuehl & Krainz . . . . MX
Mammillaria fittkaui ssp. *mathildae* (Kraehenbuehl & Krainz) Luethy
Mammillaria matudae Bravo . . . . MX
Mammillaria compacticaulis Reppenhagen
Mammillaria mazatlanensis Schumann ex Guerke . . . . MX
Mammillaria mazatlanensis ssp. **mazatlanensis** . . . . MX
Mammillaria occidentalis (Britton & Rose) Boedeker
Mammillaria mazatlanensis ssp. **patonii** (Bravo) Hunt . . . . MX
Mammillaria patonii (Bravo) Werdermann ex Backeberg
Mammillaria melaleuca Karwinski ex Salm-Dyck . . . . MX
Dolichothele melaleuca (Karwinski ex Salm-Dyck) Boedeker
Mammillaria melanocentra Poselger . . . . MX
Mammillaria melanocentra ssp. **melanocentra** . . . . MX
Mammillaria melanocentra ssp. **linaresensis** (R. & F. Wolf) Hunt . . . . MX
Mammillaria linaresensis R. & F. Wolf
Mammillaria melanocentra ssp. **rubrograndis** (Reppenhagen & Lau) Hunt . . . . . MX
Mammillaria rubrograndis Reppenhagen & Lau
Mammillaria mercadensis Patoni . . . . MX
Mammillaria brachytrichion Luethy
Mammillaria meyranii Bravo . . . . MX
Mammillaria microhelia Werdermann . . . . MX
Leptocladodia microhelia (Werdermann) Buxbaum
Leptocladodia microheliopsis (Werdermann) Buxbaum
Mammillaria microheliopsis Werdermann
®*Mammillaria droegeana* Craig

Mammillaria miegiana Earle . . . . MX
Mammillaria mieheana Tiegel . . . . MX
Mammillaria moelleriana Boedeker . . . . MX
Mammillaria cowperae Shurly
Mammillaria morganiana Tiegel . . . . MX
Mammillaria muehlenpfordtii Foerster . . . . MX
Mammillaria neopotosina Craig
®*Mammillaria celsiana* auctt
Mammillaria multidigitata Radley ex Lindsay . . . . MX
Mammillaria mystax Martius . . . . MX
Mammillaria casoi Bravo
Mammillaria crispiseta Craig
Mammillaria erythra Reppenhagen
Mammillaria huajuapensis Bravo
Mammillaria mixtecensis Bravo
®*Mammillaria atroflorens* Backeberg
Mammillaria nana Backeberg ex Mottram . . . . MX
Mammillaria felipensis Reppenhagen
®*Mammillaria monancistracantha* Backeberg
Mammillaria napina J. Purpus . . . . MX
Mammillaria neopalmeri Craig . . . . MX
Mammillaria nivosa Link ex Pfeiffer . . . . BS, PR, VI, WI
?*Mammillaria flavescens* Haworth
Mammillaria nunezii (Britton & Rose) Orcutt . . . . MX
Mammillaria silvatica Reppenhagen
Mammillaria supraflumen Reppenhagen
Mammillaria nunezii ssp. **nunezii** . . . . MX
Mammillaria hubertmulleri Reppenhagen
Mammillaria solisii (Britton & Rose) Boedeker
Mammillaria wuthenauiana Backeberg
Mammillaria nunezii ssp. **bella** (Backeberg) Hunt . . . . MX
Mammillaria bella Backeberg
Mammillaria oteroi Glass & Foster . . . . MX
Mammillaria painteri Rose ex Quehl . . . . MX
Mammillaria parkinsonii Ehrenberg . . . . MX
Mammillaria auriareolis Tiegel
Mammillaria neocrucigera Backeberg
Mammillaria rosensis Craig
Mammillaria pectinifera Weber . . . . ♦MX
Solisia pectinata (B. Stein) Britton & Rose♦
Mammillaria peninsularis (Britton & Rose) Orcutt . . . . MX
Mammillaria pennispinosa Krainz . . . . MX
Mammillaria pennispinosa ssp. **pennispinosa** . . . . MX
Mammillaria pennispinosa ssp. **nazasensis** (Glass & Foster) Hunt . . . . MX
Mammillaria nazasensis (Glass & Foster) Reppenhagen
Mammillaria perbella Hildmann ex Schumann . . . . MX
Mammillaria cadereytensis Craig
Mammillaria infernillensis Craig
Mammillaria queretarica Craig
®*Mammillaria avila-camachoi* Shurly
Mammillaria petrophila K. Brandegee . . . . MX
Mammillaria petrophila ssp. **petrophila** . . . . MX
Mammillaria gatesii M.E. Jones
Mammillaria petrophila ssp. **arida** (Rose ex Quehl) Hunt . . . . MX
Mammillaria arida Rose ex Quehl
Mammillaria petrophila ssp. **baxteriana** (Gates) Hunt . . . . MX
Mammillaria baxteriana (Gates) Boedeker ex Backeberg & F. Knuth
Mammillaria marshalliana (Gates) Boedeker
Mammillaria pacifica (Gates) Boedeker

Mammillaria petterssonii Hildmann . . . . MX
Mammillaria apozolensis Reppenhagen
Mammillaria huiguerensis Reppenhagen
Mammillaria obscura Hildmann
Mammillaria pilensis Shurly ex Eggli
Mammillaria phitauiana (Baxter) Werdermann ex Backeberg . . . . MX
?*Mammillaria verhaertiana* Boedeker
Mammillaria picta Meinshausen . . . . MX
Mammillaria picta ssp. **picta** . . . . MX
Mammillaria schieliana Schick
®*Mammillaria aurisaeta* Backeberg
Mammillaria picta ssp. **viereckii** (Boedeker) Hunt . . . . MX
Mammillaria viereckii Boedeker
Mammillaria pilispina J. Purpus . . . . MX
Mammillaria rayonensis Reppenhagen
Mammillaria sanluisensis Shurly
Mammillaria subtilis Backeberg
¶*Neolloydia pilispina* (J. Purpus) Britton & Rose
Mammillaria plumosa Weber . . . . MX
Mammillaria polyedra Martius . . . . MX
Mammillaria polythele Martius . . . . MX
Mammillaria polythele ssp. **polythele** . . . . MX
Mammillaria hidalgensis J. Purpus
Mammillaria hoffmanniana (Tiegel) Bravo
Mammillaria kewensis Salm-Dyck
Mammillaria neophaeacantha Schwarz ex Backeberg
Mammillaria tetracantha Salm-Dyck ex Pfeiffer
Mammillaria xochipilli Reppenhagen
Mammillaria polythele ssp. **durispina** (Boedeker) Hunt . . . . MX
Mammillaria durispina Boedeker
Mammillaria kelleriana Schmoll ex Craig
Mammillaria subdurispina Schwarz ex Backeberg
Mammillaria polythele ssp. **obconella** (Scheidweiler) Hunt . . . . MX
Mammillaria ingens Backeberg
Mammillaria obconella Scheidweiler
Mammillaria pondii Greene . . . . MX
Mammillaria pondii ssp. **pondii** . . . . MX
Cochemiea pondii (Greene) Walton
Mammillaria pondii ssp. **maritima** (Lindsay) Hunt . . . . MX
Cochemiea maritima Lindsay
Mammillaria maritima (Lindsay) Hunt
Mammillaria pondii ssp. **setispina** (J. Coulter) Hunt . . . . MX
Cochemiea setispina (Coulter) Walton
Mammillaria setispina (Coulter) K. Brandegee
Mammillaria poselgeri Hildmann . . . . MX
Cochemiea poselgeri (Hildmann) Britton & Rose
Mammillaria pottsii Scheer ex Salm-Dyck . . . . US, MX
Mammillaria leona Poselger
Mammillaria prolifera (Miller) Haworth . . . . US, MX, CU, HT, DO
Mammillaria glomerata (Lamarck) De Candolle
Mammillaria prolifera ssp. **prolifera** . . . . CU
Mammillaria prolifera ssp. **arachnoidea** (Hunt) Hunt . . . . MX
Mammillaria prolifera ssp. **haitiensis** (Schumann) Hunt . . . . HT, DO
Mammillaria prolifera ssp. **texana** (Engelmann) Hunt . . . . US, MX
Mammillaria multiceps Salm-Dyck
Mammillaria prolifera ssp. **zublerae** (Reppenhagen) Hunt . . . . MX
Mammillaria zublerae Reppenhagen
Mammillaria rekoi (Britton & Rose) Vaupel . . . . MX
Mammillaria albrechtiana Wohlschlager
Mammillaria krasuckae Reppenhagen

Mammillaria mitlensis Bravo
Mammillaria pseudorekoi Boedeker
Mammillaria pullihamata Backeberg ex Reppenhagen
Mammillaria sanjuanensis Reppenhagen
Mammillaria rekoi ssp. **rekoi** . . . . MX
Mammillaria rekoiana Craig
Mammillaria rekoi ssp. **aureispina** (Lau) Hunt . . . . MX
Mammillaria aureispina (Lau) Reppenhagen
Mammillaria rekoi ssp. **leptacantha** (Lau) Hunt . . . . MX
Mammillaria leptacantha (Lau) Reppenhagen
Mammillaria rettigiana Boedeker . . . . MX
Mammillaria gilensis Boedeker
Mammillaria posseltiana Boedeker
®*Mammillaria flavihamata* Backeberg
Mammillaria rhodantha Link & Otto . . . . MX
Mammillaria bonavitii Reppenhagen
Mammillaria calacantha Tiegel
Mammillaria rhodantha ssp. **rhodantha** . . . . MX
Mammillaria rhodantha ssp. **aureiceps** (Lemaire) Hunt . . . . MX
Mammillaria aureiceps Lemaire
Mammillaria rhodantha ssp. **fera-rubra** (Schmoll ex Craig) Hunt . . . . MX
Mammillaria fera-rubra Schmoll ex Craig
Mammillaria rhodantha ssp. **mccartenii** Hunt . . . . MX
Mammillaria verticealba Reppenhagen
Mammillaria rhodantha ssp. **mollendorffiana** (Shurly) Hunt . . . . MX
Mammillaria mollendorffiana Shurly
Mammillaria rhodantha ssp. **pringlei** (J. Coulter) Hunt . . . . MX
Mammillaria parensis Craig
Mammillaria pringlei (Coulter) K. Brandegee
Mammillaria roseoalba Boedeker . . . . MX
Mammillaria saboae Glass . . . . MX
Mammillaria saboae ssp. **saboae** . . . . MX
Mammillaria saboae ssp. **goldii** (Glass & Foster) Hunt . . . . MX
Mammillaria goldii Glass & Foster
Mammillaria saboae ssp. **haudeana** (Lau & Wagner) Hunt . . . . MX
Mammillaria haudeana Lau & Wagner
Mammillaria sanchez-mejoradae R. Gonzalez G. . . . . MX
Mammillaria sartorii J. Purpus . . . . MX
Mammillaria tenampensis (Britton & Rose) Berger
Mammillaria scheinvariana Ortega-Varela & Glass . . . . MX
Mammillaria crinita ssp. *scheinvariana* (Ortega-Varela & Glass) Fitz Maurice
Mammillaria schiedeana Ehrenberg . . . . MX
Mammillaria schiedeana ssp. **schiedeana** . . . . MX
Mammillaria schiedeana ssp. **dumetorum** (J. Purpus) Hunt . . . . MX
Mammillaria dumetorum J. Purpus
Mammillaria schiedeana ssp. **giselae** (Martinez-Avalos & Glass) Luethy . . . . MX
Mammillaria giselae Martinez-Avalos & Glass . . . . MX
Mammillaria schumannii Hildmann . . . . MX
Bartschella schumannii (Hildmann) Britton & Rose
Mammillaria schwarzii Shurly . . . . MX
Mammillaria scrippsiana (Britton & Rose) Orcutt . . . . MX
Mammillaria pseudoscrippsiana Backeberg
Mammillaria sempervivi De Candolle . . . . MX
Mammillaria senilis Loddiges ex Salm-Dyck . . . . MX
Mamillopsis diguetii (Weber) Britton & Rose
Mammillaria diguetii (Weber) Hunt
Mamillopsis senilis (Loddiges ex Salm-Dyck) Britton & Rose
Mammillaria sheldonii (Britton & Rose) Boedeker . . . . MX
Mammillaria alamensis Craig
Mammillaria gueldemanniana Backeberg

Mammillaria guirocobensis Craig
Mammillaria inae ['inaiae'] Craig
Mammillaria marnieriana Backeberg
Mammillaria swinglei (Britton & Rose) Boedeker
Mammillaria sinistrohamata Boedeker . . . . MX
Mammillaria solisioides Backeberg . . . . ♦MX
Mammillaria sonorensis Craig . . . . MX
Mammillaria bellisiana Craig
Mammillaria movensis Craig
Mammillaria tesopacensis Craig
Mammillaria sphacelata Martius . . . . MX
Mammillaria sphacelata ssp. **sphacelata** . . . . MX
Mammillaria sphacelata ssp. **viperina** (J. Purpus) Hunt . . . . MX
Mammillaria viperina J. Purpus
Mammillaria sphaerica Dietrich ex Engelmann . . . . US, MX
Dolichothele sphaerica (Dietrich) Britton & Rose
Mammillaria spinosissima Lemaire . . . . MX
Mammillaria gasterantha Reppenhagen
Mammillaria spinosissima ssp. **spinosissima** . . . . MX
Mammillaria centraliplumosa Fittkau
Mammillaria haasii J. Meyran
Mammillaria virginis Fittkau & Kladiwa
Mammillaria spinosissima ssp. **pilcayensis** (Bravo) Hunt . . . . MX
Mammillaria pilcayensis Bravo
®*Mammillaria pitcayensis* Bravo
Mammillaria spinosissima ssp. **tepoxtlana** Hunt . . . . MX
Mammillaria auricoma Ehrenberg
Mammillaria crassior Reppenhagen
Mammillaria standleyi (Britton & Rose) Orcutt . . . . MX
Mammillaria auricantha Craig
Mammillaria auritricha Craig
Mammillaria floresii Schwarz ex Backeberg
Mammillaria lanisumma Craig
Mammillaria mayensis Craig
Mammillaria montensis Craig
Mammillaria xanthina (Britton & Rose) Boedeker
Mammillaria stella-de-tacubaya Heese . . . . MX
Mammillaria chica Reppenhagen
Mammillaria gasseriana Boedeker
Mammillaria viescensis Rogozinski & Appenzeller
Mammillaria supertexta Martius ex Pfeiffer . . . . MX
Mammillaria lanata (Britton & Rose) Orcutt
Mammillaria martinezii Backeberg
Mammillaria surculosa Boedeker . . . . MX
Dolichothele surculosa (Boedeker) Backeberg
Mammillaria tayloriorum Glass & Foster . . . . MX
Mammillaria tepexicensis Meyran . . . . MX
Mammillaria tetrancistra Engelmann . . . . US
Phellosperma tetrancistra (Engelmann) Britton & Rose
Mammillaria theresae Cutak . . . . MX
Mammillaria thornberi Orcutt . . . . US, MX
Mammillaria thornberi ssp. **thornberi** . . . . US, MX
®*Mammillaria fasciculata* Britton & Rose
Mammillaria thornberi ssp. **yaquensis** (Craig) Hunt . . . . MX
Mammillaria yaquensis Craig
Mammillaria tonalensis Hunt . . . . MX
Mammillaria uncinata Zuccarini ex Pfeiffer . . . . MX
Mammillaria vari(e)aculeata F.G. Buchenau . . . . MX
Mammillaria vetula Martius . . . . MX

Mammillaria vetula ssp. **vetula** . . . . MX
Mammillaria magneticola Meyran
®*Mammillaria kuentziana* Fearn & Fearn
Mammillaria vetula ssp. **gracilis** (Pfeiffer) Hunt . . . . MX
Mammillaria gracilis Pfeiffer
Mammillaria voburnensis Scheer . . . . MX, GT, HN, NI
®*Mammillaria woburnensis* Britton & Rose
Mammillaria voburnensis ssp. **voburnensis** . . . . MX, GT
Mammillaria felicis Schreier ex W. Haage
Mammillaria voburnensis ssp. **eichlamii** (Quehl) Hunt . . . . GT, HN, NI
Mammillaria eichlamii Quehl
Mammillaria wagneriana Boedeker . . . . MX
Mammillaria antesbergeriana Lau
Mammillaria crassa Reppenhagen
Mammillaria weingartiana Boedeker . . . . MX
Mammillaria unihamata Boedeker
Mammillaria wiesingeri Boedeker . . . . MX
Mammillaria wiesingeri ssp. **wiesingeri** . . . . MX
Mammillaria wiesingeri ssp. **apamensis** (Reppenhagen) Hunt . . . . MX
Mammillaria apamensis Reppenhagen
Mammillaria mundtii Schumann
®*Mammillaria erectacantha* Foerster
Mammillaria winterae Boedeker . . . . MX
Mammillaria winterae ssp. **winterae** . . . . MX
Mammillaria freudenbergeri Reppenhagen
Mammillaria zahniana Boedeker & Ritter
Mammillaria winterae ssp. **aramberri** Hunt . . . . MX
Mammillaria crassimammillis Reppenhagen
Mammillaria wrightii Engelmann . . . . US, MX
Mammillaria wrightii ssp. **wrightii** . . . . US, MX
Mammillaria wrightii ssp. **wilcoxii** (Toumey ex Schumann) Hunt . . . . US, MX
Mammillaria meridiorosei Castetter et al.
Mammillaria wilcoxii Toumey ex Schumann
Mammillaria xaltianguensis Sanchez-Mejorada . . . . MX
Mammillaria xaltianguensis ssp. **xaltianguensis** . . . . MX
Mammillaria xaltianguensis ssp. **bambusiphila** (Reppenhagen) Hunt . . . . MX
Mammillaria bambusiphila Reppenhagen
Mammillaria zeilmanniana Boedeker . . . . MX
Mammillaria zephyranthoides Scheidweiler . . . . MX
Dolichothele zephyranthoides (Scheidweiler) Backeberg

Mammilloydia candida (Scheidweiler) Buxbaum . . . . MX
Mammillaria candida Scheidweiler
Mammilloydia candida ssp. *ortizrubiana* (H. Bravo) Krainz
Mammillaria estanzuelensis Moeller ex Berger
Mammillaria ortizrubiana (Bravo) Werdermann ex Backeberg

Matucana aurantiaca (Vaupel) Buxbaum . . . . PE
Matucana aurantiaca ssp. **aurantiaca** . . . . PE
Submatucana aurantiaca (Vaupel) Backeberg
Borzicactus aurantiacus (Vaupel) Kimnach & Hutchison
Borzicactus calvescens Kimnach & Hutchison
Matucana calvescens (Kimnach & Hutchison) Buxbaum
Submatucana calvescens (Kimnach & Hutchison) Backeberg
Matucana pallarensis Ritter
Matucana aurantiaca ssp. **currundayensis** (Ritter) Mottram . . . . PE
Matucana currundayensis Ritter
Submatucana currundayensis (Ritter) Backeberg
Matucana hastifera Ritter

Matucana aureiflora Ritter . . . . PE
Submatucana aureiflora (Ritter) Backeberg
Matucana celendinensis Ritter
Matucana comacephala Ritter . . . . PE
Matucana formosa Ritter . . . . PE
Submatucana formosa (Ritter) Backeberg
Borzicactus formosus (Ritter) Donald
Loxanthocereus formosus (Ritter) Buxbaum
Matucana fruticosa Ritter . . . . PE
Borzicactus fruticosus (Ritter) Donald
Matucana haynei (Otto ex Salm-Dyck) Britton & Rose . . . . PE
Matucana haynei ssp. **haynei** . . . . PE
Matucana cereoides Rauh & Backeberg
Matucana elongata Rauh & Backeberg
Borzicactus haynei (Otto ex Salm-Dyck) Kimnach
Matucana supertexta Ritter
Borzicactus variabilis (Rauh & Backeberg) Donald
Matucana variabilis Rauh & Backeberg
Matucana haynei ssp. **herzogiana** (Backeberg) Mottram . . . . PE
Matucana blancii Backeberg
Matucana crinifera Ritter
Matucana herzogiana Backeberg
Matucana megalantha Ritter
Matucana yanganucensis Rauh & Backeberg
Matucana haynei ssp. **hystrix** (Rauh & Backeberg) Mottram . . . . PE
Matucana breviflora Rauh & Backeberg
Matucana hystrix Rauh & Backeberg
Matucana multicolor Rauh & Backeberg
Matucana haynei ssp. **myriacantha** (Vaupel) Mottram . . . . PE
Matucana calocephala Skarupke
Borzicactus calocephalus (Skarupke) Donald
Arequipa myriacantha (Vaupel) Britton & Rose
Matucana myriacantha (Vaupel) Buxbaum
Submatucana myriacantha (Vaupel) Backeberg
Borzicactus myriacanthus (Vaupel) Donald
Matucana purpureoalba Ritter
Matucana winteri Ritter
Matucana huagalensis (Donald & Lau) Bregman et al. . . . . PE
Borzicactus huagalensis Donald & Lau
Matucana intertexta Ritter . . . . PE
Submatucana intertexta (Ritter) Backeberg
Borzicactus intertextus (Ritter) Donald
Matucana krahnii (Donald) Bregman . . . . PE
Matucana calliantha Ritter
Borzicactus krahnii Donald
Matucana madisoniorum (Hutchison) Rowley . . . . PE
Borzicactus madisoniorum Hutchison
Eomatucana madisoniorum (Hutchison) Ritter
Loxanthocereus madisoniorum (Hutchison) Buxbaum
Submatucana madisoniorum (Hutchison) Backeberg
Matucana oreodoxa (Ritter) Slaba . . . . PE
Eomatucana oreodoxa Ritter
Borzicactus oreodoxus (Ritter) Donald
Matucana paucicostata Ritter . . . . PE
Submatucana paucicostata (Ritter) Backeberg
Borzicactus paucicostatus (Ritter) Donald
Matucana polzii Diers et al. . . . . PE
Matucana pujupatii (Donald & Lau) Bregman . . . . PE
Matucana ritteri Buining . . . . PE
Borzicactus ritteri (Buining) Donald
Submatucana ritteri (Buining) Backeberg

Matucana tuberculata (Donald) Bregman et al. . . . . PE
Borzicactus tuberculatus Donald
Matucana tuberculosa Ritter
Matucana weberbaueri (Vaupel) Backeberg . . . . PE
Borzicactus weberbaueri (Vaupel) Donald

Melocactus ×albicephalus Buining & Brederoo . . . . BR
Melocactus andinus Gruber ex Taylor . . . . CO, VE
Melocactus azureus Buining & Brederoo . . . . BR
Melocactus azureus ssp. **azureus** . . . . BR
Melocactus krainzianus Buining & Brederoo
Melocactus azureus ssp. **ferreophilus** (Buining & Brederoo) Taylor . . . . BR
Melocactus ferreophilus Buining & Brederoo
Melocactus bahiensis (Britton & Rose) Luetzelburg . . . . BR
Melocactus bahiensis ssp. **bahiensis** . . . . BR
Melocactus acispinosus Buining & Brederoo
Melocactus brederooianus Buining
¶*Melocactus inconcinnus* Buining & Brederoo
Melocactus bahiensis ssp. **amethystinus** (Buining & Brederoo) Taylor . . . . BR
Melocactus amethystinus Buining & Brederoo
Melocactus ammotrophus Buining et al.
Melocactus glauxianus Brederoo
Melocactus griseoloviridis Buining & Brederoo
Melocactus lensselinkianus Buining & Brederoo
Melocactus bellavistensis Rauh & Backeberg . . . . EC, PE
Melocactus bellavistensis ssp. **bellavistensis** . . . . EC, PE
Melocactus bellavistensis ssp. **onychacanthus** (Ritter) Taylor . . . . PE
Melocactus onychacanthus Ritter
Melocactus broadwayi (Britton & Rose) Berger . . . . WI, TT
Melocactus caroli-linnaei Taylor . . . . JM
⊗*Melocactus coronatus* (Lamarck) Backeberg
Melocactus concinnus Buining & Brederoo . . . . BR
Melocactus axiniphorus Buining & Brederoo
Melocactus robustispinus Buining & Brederoo
Melocactus zehntneri ssp. *robustispinus* (Buining et al.) Braun
?*Melocactus pruinosus* Werdermann
Melocactus conoideus Buining & Brederoo . . . . ♦BR
Melocactus curvispinus Pfeiffer . . . . MX, GT, HN, CR, PA, CU, CO, VE
Melocactus guitartii Leon
Melocactus holguinensis Areces
Melocactus jakusii Meszaros
Melocactus curvispinus ssp. **curvispinus** . . . . . . MX, GT, HN, CR, PA, CU, CO, VE
Melocactus delessertianus Lemaire
Melocactus loboguerreroi Cardenas
Melocactus maxonii (Rose) Guerke
Melocactus oaxacensis (Britton & Rose) Backeberg
Melocactus ruestii Schumann
Melocactus ruestii ssp. *cintalapensis* Elizondo
Melocactus ruestii ssp. *maxonii* (Rose) Elizondo
Melocactus ruestii ssp. *oaxacensis* (Britton & Rose) Elizondo
Melocactus ruestii ssp. *sanctae-rosae* (L.D. Gomez) Elizondo
Melocactus salvador Murillo
?*Melocactus obtusipetalus* Lemaire
Melocactus curvispinus ssp. **caesius** (Wendland) Taylor . . . . CO, VE
Melocactus caesius H.L. Wendland
Melocactus lobelii Suringar
⊗*Melocactus amoenus* (Hoffmannsegg) Pfeiffer
Melocactus curvispinus ssp. **dawsonii** (Bravo) Taylor . . . . MX
Melocactus dawsonii Bravo
Melocactus deinacanthus Buining & Brederoo . . . . ♦BR

Melocactus ernestii Vaupel . . . . BR
Melocactus ernestii ssp. **ernestii** . . . . BR
Melocactus azulensis Buining & Brederoo
Melocactus erythracanthus Bruining & Brederoo
Melocactus interpositus Ritter
Melocactus longispinus Buining & Brederoo
Melocactus nitidus Ritter
Melocactus oreas ssp. *ernestii* (Vaupel) Braun
®*Melocactus erythranthus* W. Haage
Melocactus ernestii ssp. **longicarpus** (Buining & Brederoo) Taylor . . . . BR
Melocactus deinacanthus ssp. *florschuetzianus* (Buining & Brederoo) Braun
Melocactus deinacanthus ssp. *longicarpus* (Buining & Brederoo) Braun
Melocactus florschuetzianus Buining & Brederoo
Melocactus longicarpus Buining & Brederoo
Melocactus montanus Ritter
Melocactus mulequensis Buining & Brederoo
Melocactus neomontanus Heek & Hovens
Melocactus estevesii Braun . . . . BR
Melocactus glaucescens Buining & Brederoo . . . . ♦BR
Melocactus harlowii (Britton & Rose) Vaupel . . . . CU
Melocactus acunae Leon
Melocactus acunae ssp. *lagunaensis* Meszaros
Melocactus borhidii Meszaros
Melocactus evae Meszaros
Melocactus nagyi Meszaros
Melocactus radoczii Meszaros
®*Melocactus ocujalii* Riha ex W. Haage
Melocactus ×horridus Werdermann . . . . BR
Melocactus intortus (Miller) Urban . . . . BS, DO, PR, VI, WI
Melocactus communis Link & Otto
¶*Melocactus coronatus* (Lamarck) Backeberg
Melocactus intortus ssp. **intortus** . . . . BS, PR, VI, WI
Melocactus intortus ssp. **domingensis** Areces . . . . DO
Melocactus pedernalensis M. Mejia & Ricardo Garcia
Melocactus lanssensianus Braun . . . . BR
Melocactus lemairei (Monville ex Lemaire) Miquel ex Lemaire . . . . HT, DO
?*Melocactus hispaniolicus* Vaupel
Melocactus levitestatus Buining & Brederoo . . . . BR
Melocactus diersianus Buining & Brederoo
Melocactus rubrispinus Ritter
Melocactus securituberculatus Buining & Brederoo
Melocactus uebelmannii Braun
Melocactus warasii Pereira & Buenecker
Melocactus macracanthos (Salm-Dyck) Link & Otto . . . . AN
Melocactus barbarae Antesberger
Melocactus bozsingianus Antesberger
Melocactus inclinatus Antesberger
Melocactus laui Antesberger
Melocactus matanzanus Leon . . . . CU
Melocactus actinacanthus Areces
Melocactus mazelianus Riha . . . . CO, VE
Melocactus neryi Schumann . . . . GY, SR, BR, VE
Melocactus guaricensis Croizat
Melocactus schulzianus Buining & Brederoo
Melocactus oreas Miquel . . . . BR
Melocactus oreas ssp. **oreas** . . . . BR
Melocactus oreas ssp. *rubrisaetosus* (Buining) Braun
Melocactus rubrisaetosus Buining & Brederoo
®*Melocactus oreas* ssp. *bahiensis* (Britton & Rose) Rizzini
Melocactus oreas ssp. **cremnophilus** (Buining & Brederoo) Braun . . . . BR
Melocactus cremnophilus Buining & Brederoo

Melocactus pachyacanthus Buining & Brederoo . . . . BR
Melocactus pachyacanthus ssp. **pachyacanthus** . . . . BR
Melocactus pachyacanthus ssp. **viridis** Taylor . . . . BR
Melocactus paucispinus Heimen & Paul . . . . ♦BR
Melocactus perezassoi Areces . . . . CU
Melocactus peruvianus Vaupel . . . . EC, PE
Melocactus amstutziae Rauh & Backeberg
Melocactus fortalezensis Rauh & Backeberg
Melocactus huallanc(a)ensis Rauh & Backeberg
Melocactus jansenianus Backeberg
Melocactus trujilloensis Rauh & Backeberg
Melocactus unguispinus Backeberg
Melocactus salvadorensis Werdermann . . . . BR
⊗*Melocactus inconcinnus* Ritter
Melocactus schatzlii Till & Gruber . . . . CO, VE
Melocactus smithii (Alexander) Buining ex Rowley . . . . GY, BR
Melocactus roraimensis Braun & Esteves Pereira
Melocactus violaceus Pfeiffer . . . . BR
?*Melocactus melocactoides* (Hoffmannsegg) De Candolle
Melocactus violaceus ssp. **violaceus** . . . . BR
Melocactus depressus Hooker
Melocactus violaceus ssp. *natalensis* Braun & Esteves Pereira
Melocactus violaceus ssp. **margaritaceus** Taylor . . . . BR
Melocactus ellemeetii Miquel
⊗*Melocactus margaritaceus* Rizzini
Melocactus violaceus ssp. **ritteri** Taylor . . . . BR
Melocactus zehntneri (Britton & Rose) Luetzelburg . . . . BR
Melocactus arcuatispinus Brederoo & Eerkens
Melocactus canescens Ritter
Melocactus curvicornis Buining & Brederoo
Melocactus douradaensis Hovens & Strecker
Melocactus giganteus Buining & Brederoo
Melocactus helvolilanatus Buining & Brederoo
Melocactus macrodiscus Werdermann
Melocactus saxicola Diers & Esteves Pereira
Melocactus zehntneri ssp. *canescens* (Ritter) Braun

Micranthocereus albicephalus (Buining & Brederoo) Ritter . . . . BR
Austrocephalocereus albicephalus Buining & Brederoo
Coleocephalocereus albicephalus (Buining & Brederoo) Brandt
Micranthocereus aureispinus Ritter
Micranthocereus monteazulensis Ritter
Micranthocereus auriazureus Buining & Brederoo . . . . BR
Micranthocereus dolichospermaticus (Buining & Brederoo) Ritter . . . . BR
Austrocephalocereus dolichospermaticus Buining & Brederoo
Siccobaccatus dolichospermaticus (Buining & Brederoo) Braun & Esteves Pereira
Micranthocereus estevesii (Buining & Brederoo) Ritter . . . . BR
Austrocephalocereus estevesii Buining & Brederoo
Siccobaccatus estevesii (Buining & Brederoo) Braun & Esteves Pereira
Austrocephalocereus estevesii ssp. *grandiflorus* Diers & Esteves Pereira
Siccobaccatus estevesii ssp. *grandiflorus*
(Diers & Esteves Pereira) Braun & Esteves Pereira
Austrocephalocereus estevesii ssp. *insigniflorus* Diers & Esteves Pereira
Siccobaccatus estevesii ssp. *insigniflorus*
(Diers & Esteves Pereira) Braun & Esteves Pereira
Micranthocereus flaviflorus Buining & Brederoo . . . . BR
Micranthocereus densiflorus Buining & Brederoo
Micranthocereus flaviflorus ssp. *densiflorus*
(Buining & Brederoo) Braun & Esteves Pereira
Micranthocereus uilianus Brederoo & Bercht

Micranthocereus polyanthus (Werdermann) Backeberg . . . . BR
Arrojadoa polyantha (Werdermann) Hunt
Micranthocereus purpureus (Guerke) Ritter . . . . BR
Micranthocereus haematocarpus Ritter
Austrocephalocereus lehmannianus (Werdermann) Backeberg
Coleocephalocereus lehmannianus (Werdermann) Brandt
Micranthocereus lehmannianus (Werdermann) Ritter
Austrocephalocereus purpureus (Guerke) Backeberg
Micranthocereus ruficeps Ritter
Micranthocereus streckeri Heek & Criekinge . . . . BR
Micranthocereus violaciflorus Buining . . . . BR

Mila caespitosa Britton & Rose . . . . PE
Mila pugionifera Rauh & Backeberg . . . . PE
Mila albisaetacens Rauh & Backeberg
Mila cereoides Rauh & Backeberg
Mila fortalezensis Rauh & Backeberg
Mila sublanata Rauh & Backeberg
Mila colorea Ritter . . . . PE
Mila nealeana Backeberg . . . . PE
Mila alboareolata Akers
Mila breviseta Rauh & Backeberg
Mila caespitosa ssp. *nealeana* (Backeberg) Donald
Mila densiseta Rauh & Backeberg
Mila kubeana Werdermann & Backeberg
Mila lurinensis Rauh & Backeberg

×***Myrtgerocactus lindsayi*** Moran . . . . MX

Myrtillocactus cochal (Orcutt) Britton & Rose . . . . MX
Myrtillocactus eichlamii Britton & Rose . . . . GT
Myrtillocactus geometrizans (Martius) Console . . . . MX
Myrtillocactus grandiareolatus Bravo
Myrtillocactus schenckii (J. Purpus) Britton & Rose . . . . MX

Neobuxbaumia euphorbioides (Haworth) Buxbaum ex Bravo . . . . MX
Carnegiea euphorbioides (Haworth) Backeberg
Rooksbya euphorbioides (Haworth) Backeberg
Neobuxbaumia laui (Heath) Hunt . . . . MX
Carnegiea laui P.V. Heath
Neobuxbaumia sanchezmejoradae Lau
Neobuxbaumia macrocephala (Weber ex Schumann) Dawson . . . . MX
Carnegiea macrocephala (Schumann) P.V. Heath
Cephalocereus macrocephalus (Weber) Schumann
Mitrocereus ruficeps (Weber) Backeberg
Pachycereus ruficeps (Weber) Britton & Rose
Neobuxbaumia mezcalaensis (Bravo) Backeberg . . . . MX
Carnegiea mezcalaensis (Bravo) P.V. Heath
Carnegiea nova P.V. Heath
Neobuxbaumia multiareolata (Dawson) Bravo et al. . . . . MX
Neobuxbaumia polylopha (De Candolle) Backeberg . . . . MX
Carnegiea polylopha (De Candolle) Hunt
Cephalocereus polylophus (De Candolle) Britton & Rose
Neobuxbaumia scoparia (Poselger) Backeberg . . . . MX
Carnegiea scoparia (Poselger) P.V. Heath
Cephalocereus scoparius (Poselger) Britton & Rose
Neobuxbaumia squamulosa Scheinvar & Sanchez-Mejorada . . . . MX
Carnegiea squamulosa (Scheinvar & Sanchez-Mejorada) P.V. Heath
Neobuxbaumia tetetzo (Weber ex Coulter) Backeberg . . . . MX
Carnegiea tetetzo (Coulter) P.V. Heath
®*Neobuxbaumia tetazo* (Weber ex Coulter) Scheinvar

Neolloydia conoidea (De Candolle) Britton & Rose . . . . US, MX
Neolloydia ceratites (Quehl) Britton & Rose
Pediocactus conoideus (De Candolle) Halda
Neolloydia grandiflora (Otto ex Pfeiffer) F. Knuth
Neolloydia texensis Britton & Rose
Neolloydia matehualensis Backeberg . . . . MX

Neoraimondia arequipensis (Meyen) Backeberg . . . . PE
Neoraimondia aticensis Rauh & Backeberg
Neoraimondia macrostibas (Schumann) Britton & Rose
Neoraimondia roseiflora (Werdermann & Backeberg) Backeberg
Ⓤ*Neoraimondia gigantea* Backeberg
Ⓜ*Neoraimondia peruviana* (Linnaeus) Ritter
Neoraimondia herzogiana (Backeberg) Buxbaum & Krainz . . . . BO
Neocardenasia herzogiana Backeberg

Neowerdermannia chilensis Backeberg . . . . PE, CL
Neowerdermannia chilensis ssp. **chilensis** CL
Sulcorebutia chilensis (Backeberg) Brandt
Weingartia chilensis (Backeberg) Backeberg
Neowerdermannia chilensis ssp. **peruviana** (Ritter) Ostolaza . . . . PE
Neowerdermannia peruviana Ritter
Neowerdermannia vorwerkii (Fric) Backeberg . . . . BO, AR
Sulcorebutia vorwerkii (Fric) Brandt
Weingartia vorwerkii (Fric) Backeberg

Obregonia denegrii Fric . . . . ♦MX
Strombocactus denegrii (Fric) Rowley♦

Opuntia abyssi Hester . . . . US
Cylindropuntia abyssi (Hester) Backeberg
Opuntia acanthocarpa Engelmann & Bigelow . . . . US, MX
Cylindropuntia acanthocarpa (Engelmann & Bigelow) F. Knuth
Opuntia thornberi Thornber & Bonker
Opuntia acanthocarpa ssp. ***ganderi*** Wolf . . . . US
Opuntia ganderi (C.B. Wolf) Pinkava
Opuntia acaulis Ekman & Werdermann . . . . HT
Opuntia aciculata Griffiths . . . . US
Opuntia ×aequatorialis Britton & Rose . . . . EC
Opuntia aggeria B.E. Ralston & R.A. Hilsenbeck . . . . US
Opuntia agglomerata Berger . . . . MX
Corynopuntia agglomerata (Berger) F. Knuth
Opuntia albisaetacens Backeberg . . . . BO
Platyopuntia albisaetacens (Backeberg) Ritter
Opuntia alcahes Weber . . . . MX, US
Cylindropuntia brevispina (Gates) Backeberg
Opuntia brevispina Gates
Opuntia alcerrecensis Iliff CL
Platyopuntia chilensis Ritter
Opuntia alexanderi Britton & Rose . . . . AR
Tephrocactus alexanderi (Britton & Rose) Backeb.
Opuntia bruchii Spegazzini
Opuntia geometrica Castellanos
Tephrocactus geometricus (Castellanos) Backeberg
Tephrocactus microsphaericus Backeberg
Opuntia alko-tuna Cardenas . . . . BO
Opuntia ammophila Small . . . . US
Opuntia lata Small
Opuntia turgida Small ex Britton & Rose
Opuntia amyclaea Tenore . . . . MX

Opuntia anacantha Spegazzini . . . . PY, BO, AR
Opuntia canina Spegazzini
Opuntia kiska-loro Spegazzini
Platyopuntia kiska-loro (Spegazzini) Ritter
Opuntia retrorsa Spegazzini
Platyopuntia retrorsa (Spegazzini) Ritter
Opuntia utkilio Spegazzini
®*Opuntia bispinosa* Backeberg
Opuntia anteojoensis Pinkava . . . . MX
Opuntia antillana Britton & Rose . . . . DO, PR, VI, WI
Opuntia domingensis Britton & Rose ex Urban
Opuntia aoracantha Lemaire . . . . AR
Tephrocactus ovatus (Loudon) Ritter
Opuntia p(a)ediophila Castellanos
Tephrocactus p(a)ediophilus (Castellanos) Ritter
Opuntia apurimacensis (Ritter) Crook & Mottram . . . . PE
Platyopuntia apurimacensis Ritter
Opuntia arbuscula Engelmann . . . . US, MX
Cylindropuntia arbuscula (Engelmann) F. Knuth
Cylindropuntia vivipara (Rose) F. Knuth
Opuntia arcei Cardenas . . . . BO
Opuntia archiconoidea (Ritter) Hunt CL
Maihueniopsis archiconoidea Ritter
Opuntia armata Backeberg . . . . AR
Opuntia articulata (Pfeiffer) Hunt . . . . AR
Tephrocactus articulatus (Pfeiffer) Backeberg
Opuntia papyracantha Philippi
⊗*Opuntia glomerata* Schumann
Opuntia assumptionis Schumann . . . . PY
Opuntia atacamensis Philippi CL
Tephrocactus atacamensis (Philippi) Backeberg
Opuntia atrispina Griffiths . . . . US, MX
Opuntia atropes Rose . . . . MX
Opuntia atrovirens Spegazzini . . . . UY, AR
Opuntia auberi Pfeiffer . . . . MX, CU
Nopalea auberi (Pfeiffer) Salm-Dyck
Opuntia aurantiaca Lindley . . . . PY, UY, AR
Opuntia aurea Baxter . . . . US
Opuntia aureispina (Brack & Heil) Pinkava & Parfitt . . . . US
Opuntia austrina Small . . . . US
Opuntia pollardii Britton & Rose
Opuntia polycarpa Small
Opuntia azurea Rose . . . . MX
Opuntia backebergii Rowley . . . . BO
Tephrocactus minor Backeberg
Opuntia ×bakeri Madsen . . . . EC
Opuntia basilaris Engelmann & Bigelow . . . . US, MX
Opuntia whitneyana Baxter
Opuntia bella Britton & Rose . . . . CO
Opuntia bensonii Sanchez-Mejorada . . . . MX
Opuntia bigelovii Engelmann . . . . US, MX
Cylindropuntia bigelowii (Engelmann) F. Knuth
Cylindropuntia ciribe (Engelmann ex Coulter) F. Knuth
Opuntia ciribe Engelmann ex Coulter
Opuntia bisetosa Pittier . . . . VE
Opuntia blancii (Backeberg) Rowley . . . . PE
Tephrocactus blancii Backeberg
Opuntia boldinghii Britton & Rose . . . . TT, AN, VE
Opuntia boliviana Salm-Dyck . . . . PE, BO, AR, CL
Opuntia asplundii (Backeberg) Rowley

Tephrocactus asplundii Backeberg
Maihueniopsis boliviana (Salm-Dyck) Kiesling
Tephrocactus bolivianus (Salm-Dyck) Backeberg
Cumulopuntia echinacea (Ritter) Ritter
Opuntia echinacea (Ritter) A.E. Hoffmann
Tephrocactus echinaceus Ritter
Cumulopuntia famatinensis Ritter
Tephrocactus melanacanthus Backeberg
Cumulopuntia pampana Ritter
®*Tephrocactus albiscoparius* Backeberg
¶*Cumulopuntia boliviana* (Salm-Dyck) Ritter
Opuntia bonplandii (Kunth) Weber . . . . EC
Opuntia borinquensis Britton & Rose . . . . PR
Opuntia brachyarthra Engelmann & Bigelow . . . . US
Opuntia fragilis ssp. *brachyarthra* (Engelmann & Bigelow) W.A. Weber
Opuntia brachyclada Griffiths . . . . US
Opuntia bradleyi Rowley . . . . PE
Opuntia bradtiana (Coulter) K. Brandegee . . . . MX
Grusonia bradtiana (Coulter) Britton & Rose
Opuntia brasiliensis (Willdenow) Haworth . . . . BR, PY, PE, BO, AR
Opuntia argentina Grisebach
Brasiliopuntia brasiliensis (Willdenow) Berger
Brasiliopuntia neoargentina Backeberg
Opuntia neoargentina (Backeberg) Rowley
Brasiliopuntia schulzii (Castellanos & Lelong) Backeberg
Opuntia schulzii Castellanos & Lelong
Brasiliopuntia bahiensis (Britton & Rose) Berger
Opuntia bahiensis Britton & Rose
Opuntia brasiliensis ssp. *bahiensis* (Britton & Rose) Braun & Esteves Pereira
Opuntia brasiliensis ssp. *subacarpa* (Rizzini & Mattos) Braun & Esteves Pereira
Brasiliopuntia subacarpa Rizzini & Mattos
Opuntia bravoana Baxter . . . . MX
Opuntia bulbispina Engelmann . . . . MX
Corynopuntia bulbispina (Engelmann) F. Knuth
Grusonia bulbispina (Engelmann) H. Robinson
Opuntia burrageana Britton & Rose . . . . MX
Cylindropuntia burrageana (Britton & Rose) Backeberg
Opuntia californica Torrey & Gray . . . . US, MX
Cylindropuntia californica (Torrey & Gray) F. Knuth
Opuntia serpentina Engelmann
Opuntia camachoi Espinosa CL
Maihueniopsis camachoi (Espinosa) Ritter
Tephrocactus camachoi (Espinosa) Backeberg
Opuntia camanchica Engelmann & Bigelow . . . . US
Opuntia canterae Arechavaleta . . . . UY
Opuntia caracassana Salm-Dyck . . . . WI, TT, AN, VE
Opuntia wentiana Britton & Rose
Opuntia cardenche Griffiths . . . . MX
Opuntia cardiosperma Schumann . . . . PY
Platyopuntia cardiosperma (Schumann) Ritter
Opuntia caribaea Britton & Rose . . . . HT, DO, TT, AN, VE
Cylindropuntia caribaea (Britton & Rose) F. Knuth
Cylindropuntia metuenda (Pittier) Backeberg
Opuntia chaffeyi Britton & Rose . . . . MX
Opuntia chakensis Spegazzini . . . . PY, AR
Opuntia chavena Griffiths . . . . MX
Opuntia chichensis (Cardenas) Rowley . . . . BO, AR
Tephrocactus chichensis Cardenas
Opuntia ferocior (Backeberg) Rowley
Tephrocactus ferocior Backeberg
®*Cumulopuntia boliviana* Ritter

Opuntia chihuahuensis Rose ... MX
Opuntia chisosensis (Anthony) D.J. Ferguson ... US
Opuntia chlorotica Engelmann & Bigelow ... US, MX
   Opuntia palmeri Engelmann ex Coulter
Opuntia cholla Weber ... MX
   Cylindropuntia cholla (Weber) F. Knuth
Opuntia cineracea Wiggins ... MX
Opuntia clavarioides Pfeiffer ... AR
   Austrocylindropuntia clavarioides (Pfeiffer) Backeberg
   Puna clavarioides (Pfeiffer) Kiesling
Opuntia clavata Engelmann ... US
   Corynopuntia clavata (Engelmann) F. Knuth
Opuntia cochabambensis Cardenas ... BO
Opuntia cochenillifera (Linnaeus) Miller ... XC (MX, GY etc)
   Nopalea cochenillifera (Linnaeus) Salm-Dyck
Opuntia cognata (Ritter) Braun & Esteves Pereira ... PY
   Platyopuntia cognata Ritter
Opuntia colorea (Ritter) Hunt ... CL
   Maihueniopsis colorea (Ritter) Ritter
   Tephrocactus coloreus Ritter
Opuntia colubrina Castellanos ... AR
   Austrocylindropuntia colubrina (Castellanos) Backeberg
Opuntia ×columbiana Griffiths ... CA, US
Opuntia conjungens (Ritter) Braun & Esteves Pereira ... BO
   Platyopuntia conjungens Ritter
®***Opuntia conoidea*** (Backeberg) Rowley ... CL
   ®*Maihueniopsis conoidea* (Backeberg) Ritter
   ®*Tephrocactus conoideus* Backeberg
Opuntia corotilla Schumann ex Vaupel ... PE
   Tephrocactus corotilla (Schumann ex Vaupel) Backeberg
   Opuntia ignota Britton & Rose
   ®*Cumulopuntia ignota* (Britton & Rose) Ritter
Opuntia corrugata (Gillies ex Loudon) Salm-Dyck ... AR
   Platyopuntia corrugata (Salm-Dyck) Ritter
Opuntia crassa Haworth ... MX
Opuntia crassicylindrica (Rauh & Backeberg) Rowley ... PE
   Tephrocactus crassicylindricus Rauh & Backeberg
   ®*Cumulopuntia crassicylindrica* (Rauh & Backeberg) Ritter
Opuntia crassipina (Ritter) Hunt ... CL
   Maihueniopsis crassispina Ritter
Opuntia crystalenia Griffiths ... MX
   Opuntia ochrocentra Small ex Britton & Rose
Opuntia ×cubensis Britton & Rose ... CU
Opuntia curassavica (Linnaeus) Miller ... WI, AN, CO, VE
Opuntia ×curvispina Griffiths ... US
Opuntia cylindrica (Lamarck) De Candolle ... EC, PE
   Austrocylindropuntia cylindrica (Lamarck) Backeberg
   Austrocylindropuntia intermedia Rauh & Backeberg
   Cylindropuntia intermedia (Rauh & Backeberg) Rauh & Backeberg
Opuntia cymochila Engelmann & Bigelow ... US
   Opuntia mackensenii Rose
Opuntia dactylifera Vaupel ... PE, BO
   Opuntia cylindrarticulata (Cardenas) Rowley
   Tephrocactus cylindrarticulatus Cardenas
   Tephrocactus dactylifer (Vaupel) Backeberg
   Opuntia noodtiae (Backeberg & Jacobsen) Rowley
   Tephrocactus noodtiae Backeberg & Jacobsen
Opuntia darrahiana Weber ex Roland-Gosselin ... BS
Opuntia darwinii Henslow ... AR
   Maihueniopsis albomarginata Ritter

Maihueniopsis darwinii (Henslow) Ritter
Tephrocactus darwinii (Henslow) [Backeberg ex] Fric
Opuntia hickenii Britton & Rose
Tephrocactus hickenii (Britton & Rose) Spegazzini
Maihueniopsis neuquensis (Borg) Ritter
Opuntia neuquensis Borg
Tephrocactus neuquensis (Borg) Backeberg
Opuntia platyacantha Pfeiffer
Tephrocactus platyacanthus (Pfeiffer) Lemaire
Opuntia davisii Engelmann & Bigelow . . . . US
Cylindropuntia davisii (Engelmann & Bigelow) F. Knuth
Opuntia deamii Rose . . . . MX, GT, HN
Opuntia decumbens Salm-Dyck . . . . MX, GT, HN, NI, PA
Opuntia dejecta Salm-Dyck . . . . CU
Nopalea dejecta (Salm-Dyck) Salm-Dyck
Opuntia delaetiana Weber . . . . PY, AR
Opuntia densispina B.E. Ralston & R.A. Hilsenbeck . . . . US
Opuntia depauperata Britton & Rose . . . . VE
Opuntia depressa Rose . . . . MX
Opuntia discolor Britton & Rose . . . . AR
Platyopuntia discolor (Britton & Rose) Ritter
Opuntia dillenii (Ker-Gawler) Haworth
. . . . US, MX, BS, CU, KY, JM, DO, PR, VI, WI, AN, EC
Opuntia anahuacensis Griffiths
Opuntia atrocapensis Small
Opuntia melanosperma Svenson
Opuntia nitens Small
Opuntia zebrina Small
Opuntia domeykoensis (Ritter) Hunt CL
Maihueniopsis domeykoensis Ritter
Opuntia dumetorum Berger . . . . MX
Corynopuntia dumetorum (Berger) F. Knuth
Platyopuntia dumetorum (Berger) Ritter
Opuntia durangensis Britton & Rose . . . . MX
Opuntia echinocarpa Engelmann & Bigelow . . . . US, MX
Cylindropuntia echinocarpa (Engelmann & Bigelow) F. Knuth
Opuntia echios Howell . . . . EC-GAL
Opuntia zacana Howell
Opuntia edwardsii Grant & Grant . . . . US
Opuntia eichlamii Rose . . . . MX, GT
Opuntia ekmanii Werdermann . . . . HT
Opuntia elata Link & Otto ex Salm-Dyck . . . . CU, PY
Opuntia elatior Miller . . . . CR, PA, WI, AN, CO, VE [widely introduced]
Opuntia bergeriana Weber ex Berger
Opuntia elizondoana Sanchez & Villasenor . . . . MX
Opuntia ×ellisiana Griffiths . . . . US
Opuntia emoryi Engelmann . . . . US, MX
Corynopuntia stanlyi (Engelmann) F. Knuth
Opuntia stanlyi Engelmann
Opuntia engelmannii Salm-Dyck ex Engelmann . . . . US, MX
Opuntia cantabrigiensis Lynch
Opuntia cuija (Griffiths & Hare) Britton & Rose
Opuntia discata Griffiths
Opuntia flexospina Griffiths
Opuntia lindheimeri Engelmann
Opuntia microcarpa Engelmann
Opuntia procumbens Engelmann & Bigelow
Opuntia subarmata Griffiths
Opuntia tardospina Griffiths
Opuntia tricolor Griffiths

Opuntia erectoclada Backeberg . . . . AR
Opuntia erinacea Engelmann & Bigelow . . . . US
Opuntia estevesii Braun . . . . BR
Opuntia excelsa Sanchez-Mejorada . . . . MX
Opuntia falcata Ekman & Werdermann . . . . HT, DO
Consolea falcata (Ekman & Werdermann) F. Knuth
Opuntia feracantha Britton & Rose . . . . MX
Opuntia ficus-indica (Linnaeus) Miller . . . . US, MX [widely introduced]
Opuntia cordobensis Spegazzini
Platyopuntia cordobensis (Spegazzini) Ritter
Opuntia tuna-blanca Spegazzini
¶*Opuntia vulgaris* Miller
Opuntia flexuosa (Backeberg) Rowley . . . . BO
Tephrocactus flexuosus Backeberg
Opuntia floccosa Salm-Dyck ex Winterfeld . . . . PE, BO
Opuntia atroviridis Werdermann & Backeberg
Tephrocactus atroviridis (Werdermann & Backeberg) Backeberg
Opuntia crispicrinita (Rauh & Backeberg) Rowley
Tephrocactus crispicrinitus Rauh & Backeberg
Opuntia cylindrolanata (Rauh & Backeberg) Rowley
Tephrocactus cylindrolanatus Rauh & Backeberg
Austrocylindropuntia floccosa (Salm-Dyck) Ritter
Tephrocactus floccosus (Salm-Dyck) Backeberg
Austrocylindropuntia lauliacoana Ritter
Austrocylindropuntia machacana Ritter
Opuntia pseudo-udonis (Rauh & Backeberg) Rowley
Tephrocactus pseudo-udonis Rauh & Backeberg
Opuntia rauhii (Backeberg) Rowley
Tephrocactus rauhii Backeberg
Austrocylindropuntia tephrocactoides Rauh & Backeberg
Opuntia tephrocactoides (Rauh & Backeberg) Rowley
Opuntia udonis Weingart
Tephrocactus udonis (Weingart) Backeberg
Opuntia verticosa Weingart
Tephrocactus verticosus (Weingart) Backeberg
Opuntia ×fosbergii C.B. Wolf . . . . US
Opuntia fragilis (Nuttall) Haworth . . . . CA, US
Opuntia frigida (Ritter) G. Navarro . . . . BO
Cumulopuntia frigida Ritter
Opuntia fulgida Engelmann . . . . US, MX
Cylindropuntia fulgida (Engelmann) F. Knuth
Opuntia fuliginosa Griffiths . . . . MX
Opuntia fulvicoma (Rauh & Backeberg) Rowley . . . . PE
Tephrocactus fulvicomus Rauh & Backeberg
Opuntia galapageia Henslow . . . . EC
Opuntia galerasensis (Ritter) Hunt . . . . PE
Cumulopuntia galerasensis Ritter
Opuntia glomerata Haworth . . . . BO, AR, CL
Maihueniopsis glomerata (Haworth) Kiesling
Tephrocactus glomeratus (Haworth) Backeberg
Maihueniopsis hypogaea (Werdermann) Ritter
Maihueniopsis leoncito (Werdermann) Ritter
Opuntia leoncito Werdermann
Maihueniopsis leptoclada Ritter
Maihueniopsis molfinoi Spegazzini
Maihueniopsis ovallei (Remy ex C. Gay) Ritter
Tephrocactus ovallei (Remy ex Gay) Backeberg
Opuntia reicheana Espinosa
Tephrocactus reicheanus (Espinosa) Backeberg
Opuntia gosseliniana Weber . . . . US

Opuntia grahamii Engelmann . . . . . US, MX
Corynopuntia grahamii (Engelmann) F. Knuth
Opuntia grosseana Weber ex Roland-Gosselin . . . . . PY
Opuntia guatemalensis Britton & Rose . . . . . GT, HN, NI, CR
Opuntia guatinensis Hunt . . . . . PE
Cumulopuntia tortispina Ritter
Opuntia guilanchi Griffiths . . . . . MX, GT
Opuntia halophila Spegazzini . . . . . AR
Opuntia heacockiae Arp . . . . . US
Opuntia helleri Schuman ex Robinson . . . . . EC
Opuntia heteromorpha Philippi . . . . . BO
Tephrocactus heteromorphus (Philippi) Backeberg
Opuntia hirschii (Backeberg) Rowley . . . . . PE
Tephrocactus hirschii Backeberg
Opuntia hitchcockii J.G. Ortega . . . . . MX
Opuntia hondurensis Standley . . . . . HN
Opuntia howeyi J. Purpus . . . . . MX
Opuntia huajuapensis Bravo . . . . . MX
Opuntia humifusa (Rafinesque) Rafinesque . . . . . US, MX
Opuntia allairei Griffiths
Opuntia calcicola Wherry
Ⓤ*Opuntia compressa* (Salisbury) Macbride
Opuntia cumulicola Small
Opuntia fuscoatra Engelmann
Opuntia impedata Small ex Britton & Rose
Opuntia italica Tenore ex Pfeiffer
Opuntia nemoralis Griffiths
Opuntia rafinesquei Engelmann
Opuntia rubiflora Griffiths
Ⓜ*Opuntia vulgaris* Miller
Opuntia hyptiacantha Weber . . . . . MX
Opuntia cretochaeta Griffiths
Opuntia matudae Scheinvar
Opuntia ianthinantha (Ritter) Iliff . . . . . AR
Platyopuntia ianthinantha Ritter
Opuntia ignescens Vaupel . . . . . PE, CL
Cumulopuntia ignescens (Vaupel) Ritter
Tephrocactus ignescens (Vaupel) Backeberg
Opuntia imbricata (Haworth) De Candolle . . . . . US, MX
Cylindropuntia imbricata (Haworth) F. Knuth
Opuntia inaequilateralis Berger . . . . . PE
Platyopuntia inaequilateralis (Berger) Ritter
Opuntia inamoena Schumann . . . . . BR
Platyopuntia inamoena (Schumann) Ritter
Opuntia inaperta (Schott ex Griffiths) Hunt . . . . . MX
Nopalea escuintlensis Matuda
Nopalea gaumeri Britton & Rose
Nopalea inaperta Schott ex Griffiths
Opuntia infesta (Ritter) Iliff . . . . . PE
Platyopuntia infesta Ritter
Opuntia insularis Stewart . . . . . EC
Opuntia invicta T. Brandegee . . . . . MX
Corynopuntia invicta (T. Brandegee) F. Knuth
Opuntia jaliscana Bravo . . . . . MX
Opuntia jamaicensis Britton & Harris . . . . . JM
Opuntia joconostle Weber ex Diguet . . . . . MX
Opuntia karwinskiana Salm-Dyck . . . . . MX
Nopalea karwinskiana (Salm-Dyck) Schumann
Opuntia ×kelvinensis Grant & Grant . . . . . US

Opuntia kleiniae De Candolle . . . . US, MX
Cylindropuntia kleiniae (De Candolle) F. Knuth
Cylindropuntia recondita (Griffiths) F. Knuth
Opuntia kunzei Rose . . . . US, MX
Grusonia wrightiana Baxter
Opuntia wrightiana (Baxter) Peebles
Opuntia laevis Coulter . . . . US
Opuntia lagopus Schumann . . . . PE, BO
Tephrocactus lagopus (Schumann) Backeberg
Austrocylindropuntia malyana (Rausch) Ritter
Tephrocactus malyanus Rausch
®*Austrocylindropuntia lagopus* (Schumann) Ritter
Opuntia lagunae Baxter ex Bravo . . . . MX
Opuntia larreyi Weber ex Coulter . . . . MX
Opuntia lasiacantha Hort. Vindob. ex Pfeiffer . . . . MX
Opuntia rzedowskii Scheinvar
Opuntia leptocaulis De Candolle . . . . US, MX
Cylindropuntia brittonii (J.G. Ortega) Backeberg
Cylindropuntia leptocaulis (De Candolle) F. Knuth
Opuntia leucophaea Philippi CL
Opuntia leucotricha De Candolle . . . . MX
Opuntia lilae Trujillo & Ponce . . . . VE
Opuntia limitata (Ritter) Braun & Esteves Pereira . . . . PY
Platyopuntia limitata Ritter
Opuntia lindsayi J.P. Rebman . . . . MX
Opuntia linguiformis Griffiths . . . . US
Opuntia littoralis (Engelmann) Cockerell . . . . US, MX
Opuntia llanos-de-huanta Hunt CL
Maihueniopsis grandiflora Ritter
Opuntia lloydii Rose . . . . MX
Cylindropuntia lloydii (Rose) F. Knuth
Opuntia longiareolata Clover & Jotter . . . . US
Opuntia longispina Haworth . . . . AR
Opuntia ×lucayana Britton . . . . BS
Opuntia lutea (Rose) Hunt . . . . MX, GT, HN, NI
Nopalea guatemalensis Rose
Nopalea lutea Rose
Opuntia macracantha Grisebach . . . . CU
Consolea macracantha (Grisebach) Berger
Opuntia macrocentra Engelmann . . . . US, MX
®*Opuntia violacea* Engelmann
Opuntia macrorhiza Engelmann . . . . US, MX
Opuntia ballii Rose
Opuntia delicata Rose
Opuntia pottsii Salm-Dyck
Opuntia setispina Engelmann ex Salm-Dyck
Opuntia tenuispina Engelmann & Bigelow
Opuntia mamillata Schott ex Engelmann . . . . US, MX
Opuntia marenae Parsons . . . . MX
Marenopuntia marenae (Parsons) Backeberg
Opuntia martiniana (Benson) Parfitt . . . . US
Opuntia megacantha Salm-Dyck . . . . MX
Opuntia megapotamica Arechavaleta . . . . UY
Opuntia megarhiza Rose . . . . MX
Opuntia megasperma Howell . . . . EC
Opuntia microdasys (Lehmann) Pfeiffer . . . . MX
Opuntia macrocalyx Griffiths
Opuntia microdisca Weber . . . . BO, AR
Platyopuntia microdisca (Weber) Ritter
®*Opuntia poecilacantha* Backeberg

Opuntia mieckleyi Schumann ... PY
Opuntia millspaughii Britton ... CU, BS
Consolea millspaughii (Britton) Berger
Opuntia minuscula (Backeberg) Rowley ... BO
Tephrocactus minusculus Backeberg
Opuntia minuta (Backeberg) Castellanos ... AR
Maihueniopsis mandragora (Backeberg) Ritter
Opuntia mandragora (Backeberg) Rowley
Tephrocactus mandragora Backeberg
Maihueniopsis minuta (Backeberg) Kiesling
Tephrocactus minutus Backeberg
Opuntia miquelii Monville ... CL
Austrocylindropuntia miquelii (Monville) Backeberg
Miqueliopuntia miquelii (Monville) Ritter
Opuntia mistiensis (Backeberg) Rowley ... PE
Tephrocactus mistiensis Backeberg
Opuntia moelleri Berger ... MX
Corynopuntia moelleriana (Berger ex Y. Wright) F. Knuth
Opuntia mojavensis Engelmann & Bigelow ... US
Opuntia molesta T. Brandegee ... MX
Cylindropuntia calmalliana (Coulter) F. Knuth
Cylindropuntia clavellina (Engelmann ex Coulter) F. Knuth
Cylindropuntia molesta (T. Brandegee) F. Knuth
Opuntia molinensis Spegazzini ... AR
Maihueniopsis molinensis (Spegazzini) Ritter
Tephrocactus molinensis (Spegazzini) Backeberg
Opuntia monacantha Haworth ... BR, PY, UY, AR
Opuntia arechavaletae Spegazzini
Opuntia brunneogemmia (Ritter) Schlindwein
Platyopuntia brunneogemmia Ritter
Opuntia monacantha ssp. *brunneogemmia* (Ritter) Braun & Esteves Pereira
⊗*Opuntia vulgaris* auctt
⊗*Platyopuntia vulgaris* (Miller) Ritter
Opuntia moniliformis (Linnaeus) Haworth ex Steudel ... CU, HT, DO, PR
Opuntia haitiensis Britton
Consolea moniliformis (Linnaeus) Berger
Consolea moniliformis ssp. *guantanamana* Areces
Opuntia picardae Urban
Opuntia testudinis-crus Weber ex Roland Gosselin
Opuntia montevideensis Spegazzini ... UY
Opuntia multigeniculata Clokey ... US
Cylindropuntia multigeniculata (Clokey) Backeberg
Opuntia munzii C.B. Wolf ... US
Cylindropuntia munzii (C.B. Wolf) Backeberg
Opuntia nashii Britton ... BS, CU, DO
Consolea nashii (Britton) Berger
Consolea nashii ssp. *gibarensis* A.E. Areces-Mallea
Opuntia nejapensis Bravo ... MX
Opuntia neochrysacantha Bravo ... MX
Opuntia nigrispina Schumann ... BO, AR
Maihueniopsis nigrispina (Schumann) Kiesling
Platyopuntia nigrispina (Schumann) Ritter
Tephrocactus nigrispinus (Schumann) Backeberg
Opuntia purpurea R. Fries
Opuntia atroglobosa (Backeberg) Crook & Mottram
◎*Platyopuntia atroglobosa* (Backeberg) Ritter
◎*Tephrocactus atroglobosus* Backeberg
Opuntia nuda (Backeberg) Rowley ... MX
Nopalea nuda Backeberg

Opuntia orbiculata Salm-Dyck ex Pfeiffer . . . . MX
Opuntia oricola Philbrick . . . . US, MX
Opuntia orurensis Cardenas . . . . BO
Platyopuntia orurensis (Cardenas) Ritter
Opuntia ovata Pfeiffer . . . . AR, CL
Maihueniopsis ovata (Pfeiffer) Ritter
Tephrocactus ovatus (Pfeiffer) Backeberg
Opuntia russellii Britton & Rose
Tephrocactus russellii (Britton & Rose) Backeberg
Opuntia pachona Griffiths . . . . MX
Opuntia pachypus Schumann . . . . PE
Austrocylindropuntia pachypus (Schumann) Backeberg
Opuntia pailana Weingart . . . . MX
Opuntia palmadora Britton & Rose . . . . BR
Opuntia paraguayensis Schumann . . . . PY, UY, AR
Opuntia bonaerensis Spegazzini
Opuntia parishii Orcutt . . . . US
Opuntia parryi Engelmann . . . . US, MX
Cylindropuntia parryi (Engelmann) F. Knuth
Opuntia parviclada S. Arias & S. Gama . . . . MX
Opuntia penicilligera Spegazzini . . . . AR
Opuntia pennellii Britton & Rose . . . . CO
Opuntia pentlandii Salm-Dyck . . . . PE, BO, AR
Cumulopuntia pentlandii (Salm-Dyck) Ritter
Maihueniopsis pentlandii (Salm-Dyck) Kiesling
Tephrocactus pentlandii (Salm-Dyck) Backeberg
Tephrocactus rarissimus Backeberg
Tephrocactus subinermis (Backeberg) Backeberg
Tephrocactus wilkeanus Backeberg
Opuntia phaeacantha Engelmann . . . . US, MX
Opuntia angustata Engelmann & Bigelow
Opuntia woodsii Backeberg
Opuntia picardoi Marnier-Lapostolle . . . . AR
Opuntia pilifera Weber . . . . MX
Opuntia pittieri Britton & Rose . . . . CO
Opuntia pituitosa (Ritter) Iliff . . . . AR
Platyopuntia pituitosa Ritter
Opuntia plumbea Rose . . . . MX
Opuntia polyacantha Haworth . . . . CA, US
Opuntia arenaria Engelmann
Opuntia hystricina Engelmann & Bigelow
Opuntia nicholii L. Benson
Opuntia rhodantha Schumann
Opuntia prasina Spegazzini . . . . AR
Opuntia prolifera Engelmann . . . . US, MX
Cylindropuntia alcahes (Weber) F. Knuth
Cylindropuntia prolifera (Engelmann) F. Knuth
Opuntia puberula Pfeiffer. . . . . MX, GT
Opuntia heliae Matuda
Opuntia maxonii J.G. Ortega
®*Opuntia scheinvariana* Paniagua
Opuntia pubescens Wendland ex Pfeiffer . . . . MX, GT, EC, BO, PY
Opuntia hoffmannii Bravo
Platyopuntia nana (Kunth) Ritter
Opuntia pascoensis Britton & Rose
Opuntia pestifer Britton & Rose
Opuntia tayapayensis Cardenas
Opuntia pulchella Engelmann . . . . US
Micropuntia barkleyana Daston

Micropuntia brachyrhopalica Daston
Micropuntia gracilicylindrica Wiegand & Backeberg
Corynopuntia pulchella (Engelmann) F. Knuth
Micropuntia pulchella (Engelmann) C. Holland
Micropuntia pygmaea Wiegand & Backeberg
Micropuntia tuberculosirhopalica Wiegand & Backeberg
Micropuntia wiegandii Backeberg
Opuntia punta-caillan (Rauh & Backeberg) Rowley . . . . PE
Tephrocactus punta-caillan Rauh & Backeberg
Opuntia pumila Rose . . . . MX
Opuntia pusilla (Haworth) Haworth . . . . US
Opuntia drummondii Graham
Opuntia macateei Britton & Rose
Opuntia pes-corvi Leconte ex Engelmann
Opuntia tracyi Britton
Opuntia pycnantha Engelmann ex Coulter . . . . MX
Opuntia pyriformis Rose . . . . MX
Opuntia pyrrhacantha Schumann . . . . PE
Cumulopuntia pyrrhacantha (Schumann) Ritter
Tephrocactus pyrrhacanthus (Schumann) Backeberg
Opuntia pyrrhantha (Ritter) Braun & Esteves Pereira . . . . BO
Platyopuntia pyrrhantha Ritter
Opuntia quimilo Schumann . . . . PY, BO, AR
Opuntia distans Britton & Rose
Platyopuntia quimilo (Schumann) Ritter
Opuntia ×quipa Weber . . . . BR
Opuntia catingicola Werdermann
Opuntia palmadora ssp. *catingicola* (Werdermann) Braun & Esteves Pereira
Opuntia quitensis Weber . . . . EC, PE
Opuntia macbridei Britton & Rose
Platyopuntia quitensis (Weber) Ritter
Opuntia rahmeri Philippi CL
Maihueniopsis rahmeri (Philippi) Ritter
Opuntia ramosissima Engelmann . . . . US, MX
Cylindropuntia ramosissima (Engelmann) F. Knuth
Opuntia rastrera Weber . . . . MX
Opuntia reflexispina Wiggins & Rollins . . . . MX
Corynopuntia reflexispina (Wiggins & Rollins) Backeberg
Opuntia repens Bello . . . . PR
Opuntia rileyi J.G. Ortega . . . . MX
Opuntia ritteri Berger . . . . MX
Opuntia robinsonii J.G. Ortega . . . . MX
Opuntia roborensis Cardenas . . . . BO
Opuntia robusta Wendland ex Pfeiffer . . . . MX
Opuntia guerrana Griffiths
Opuntia rosarica Lindsay . . . . MX
Grusonia hamiltonii Gates ex Marshall & Bock
Opuntia hamiltonii (Gates ex Backeberg) Rowley
Cylindropuntia rosarica (Lindsay) Backeberg
Opuntia rosea De Candolle . . . . MX
Opuntia pallida Rose
Cylindropuntia rosea (De Candolle) Backeberg
Opuntia rossiana (Heinrich & Backeberg) Hunt . . . . PE, BO, AR, CL
Cumulopuntia rossiana (Heinrich & Backeberg) Ritter
®*Tephrocactus microclados* Backeberg
Opuntia rubescens Salm-Dyck ex De Candolle . . . . PR, VI, WI
Consolea rubescens (Salm-Dyck) Lemaire
Opuntia rufida Engelmann . . . . US, MX
Opuntia herrfeldtii Kupper
Opuntia lubrica Griffiths
Opuntia rutila Nuttall ex Torrey & Gray . . . . US

Opuntia salagria Castellanos ... PY, AR
Opuntia salmiana Parmentier ex Pfeiffer ... BR, PY, BO, AR
Austrocylindropuntia ipatiana (Cardenas) Backeberg
Austrocylindropuntia salmiana (Parmentier ex Pfeiffer) Backeberg
Platyopuntia salmiana (Parmentier ex Pfeiffer) Ritter
Opuntia salvadorensis Britton & Rose ex Standley & Calderon ... SV
Opuntia sanctae-barbarae Hunt ... CL
Cumulopuntia hystrix Ritter
Opuntia sanguinea Proctor ... JM
Opuntia santa-rita (Griffiths & Hare) Rose ... US
Opuntia shreveana C.Z. Nelson
Opuntia santamaria (Baxter) Wiggins ... MX
Grusonia santamaria Baxter
Opuntia saxatilis (Ritter) Braun & Esteves Pereira ... BR
Platyopuntia saxatilis Ritter
Opuntia saxicola Howell ... EC
Opuntia scheeri Weber ... MX
Opuntia schickendantzii Weber ... AR
Austrocylindropuntia schickendantzii (Weber) Backeberg
Opuntia schottii Engelmann ... US, MX
Corynopuntia schottii (Engelmann) F. Knuth
Opuntia schumannii Weber ex Berger ... CO, VE
Opuntia securigera Borg ... AR
Opuntia shaferi Britton & Rose ... BO, AR
Austrocylindropuntia humahuacana (Backeberg) Backeberg
Cylindropuntia humahuacana Backeberg
Austrocylindropuntia shaferi (Britton & Rose) Backeberg
Austrocylindropuntia steiniana Backeberg
Austrocylindropuntia weingartiana (Backeberg) Backeberg
Cylindropuntia weingartiana (Backeberg) Backeberg
Opuntia weingartiana Backeberg
Opuntia silvestris Backeberg ... BO
Tephrocactus silvestris (Backeberg) Backeberg
Opuntia soederstromiana Britton & Rose ... EC
Opuntia dobbieana Britton & Rose
Opuntia soehrensii Britton & Rose ... PE, BO, AR, CL
Opuntia boliviensis Backeberg
Opuntia cedergreniana Backeberg
Platyopuntia soehrensii (Britton & Rose) Ritter
Opuntia tilcarensis Backeberg
Ⓤ*Opuntia multiareolata* Backeberg
Ⓤ*Opuntia obliqua* Backeberg
Opuntia sphaerica Foerster ... PE, BO, CL
Ⓜ*Cumulopuntia berteri* (Colla) Ritter
Opuntia campestris Britton & Rose
Opuntia dimorpha Foerster
Tephrocactus dimorphus (Foerster) Backeberg
Opuntia kuehnrichiana Werdermann & Backeberg
Tephrocactus kuehnrichianus (Werdermann & Backeberg) Backeberg
Opuntia mira (Rauh & Backeberg) Rowley
Tephrocactus mirus Rauh & Backeberg
Tephrocactus muellerianus Backeberg
Cumulopuntia multiareolata (Ritter) Ritter
Tephrocactus multiareolatus Ritter
Cumulopuntia rauppiana (Schumann) Ritter
Opuntia rauppiana Schumann
Tephrocactus sphaericus (Foerster) Backeberg
Cumulopuntia tubercularis Ritter
Ⓤ*Cumulopuntia kuehnrichiana* (Werdermann & Backeberg) Ritter
Ⓜ*Opuntia berteri* (Colla) A. Hoffmann

Opuntia sphaerocarpa Engelmann & Bigelow . . . . US
Opuntia juniperina Britton & Rose
Opuntia ×spinosibacca Anthony . . . . US
Opuntia spinosior (Engelmann) Toumey ex Bailey . . . . US, MX
Cylindropuntia spinosior (Engelmann) F. Knuth
Opuntia spinosissima Miller . . . . US, KY, JM
Consolea corallicola Small
Consolea spinosissima (Miller) Lemaire
Opuntia spinulifera Salm-Dyck . . . . MX
Opuntia candelabriformis Martius ex Pfeiffer
Opuntia heliabravoana Scheinvar
Opuntia spraguei J.G. Ortega . . . . MX
Opuntia stenarthra Schumann . . . . PY
Opuntia stenopetala Engelmann . . . . MX
Opuntia arrastradillo Backeberg
Opuntia glaucescens Salm-Dyck
Opuntia grandis Hort. Angl. ex Pfeiff.
Opuntia marnieriana Backeberg
®*Opuntia riviereana* Backeberg
Opuntia streptacantha Lemaire . . . . MX
Opuntia stricta (Haworth) Haworth . . . . US, MX, CU
Opuntia bahamana Britton & Rose
Opuntia keyensis Britton ex Small
Opuntia macrarthra Gibbes
Opuntia magnifica Small
Opuntia strigil Engelmann . . . . US
Opuntia subsphaerocarpa Spegazzini . . . . AR
Opuntia subterranea R. Fries . . . . BO, AR
Puna subterranea (R. Fries) Kiesling
Tephrocactus subterraneus (R. Fries) Backeberg
?*Puna bonnieae* D.J. Ferguson & Kiesling
®*Cumulopuntia subterranea* (R. Fries) Ritter
®*Tephrocactus variflorus* Backeberg
Opuntia subulata (Muehlenpfordt) Engelmann . . . . PE, BO
Austrocylindropuntia exaltata (Berger) Backeberg
Opuntia exaltata Berger
Austrocylindropuntia subulata (Muehlenpfordt) Backeberg
Opuntia sulphurea Gillies ex Salm-Dyck . . . . PY, BO, AR
Opuntia brunnescens Britton & Rose
Platyopuntia sulphurea (Gillies ex Salm-Dyck) Ritter
Opuntia vulpina Weber
Opuntia sulphurea ssp. ***brachyacantha*** (Ritter) Braun & Esteves Pereira . . . . BO
Opuntia brachyacantha (Ritter) Crook & Mottram
Platyopuntia brachyacantha Ritter
Opuntia sulphurea ssp. ***spinibarbis*** (Ritter) Braun & Esteves Pereira . . . . BO
Platyopuntia spinibarbis Ritter
Opuntia superbospina Griffiths . . . . US
Opuntia tapona Engelmann ex Coulter . . . . MX
Opuntia comonduensis Britton & Rose
Opuntia tarapacana Philippi CL
Maihueniopsis tarapacana (Philippi) Ritter
Tephrocactus tarapacanus (R. Philippi) Backeberg
Opuntia taylorii Britton & Rose . . . . HT, DO
Opuntia tehuacana S. Arias & L.U. Guzman . . . . MX
Opuntia tehuantepecana (Bravo) Bravo . . . . MX
Opuntia tenuiflora Small . . . . US
Opuntia tesajo Engelmann ex Coulter . . . . MX
Cylindropuntia tesajo (Engelmann ex Coulter) F. Knuth
Opuntia ×tetracantha Toumey . . . . US, MX
Cylindropuntia tetracantha (Toumey) F. Knuth

Opuntia thurberi Engelmann . . . . MX
Cylindropuntia alamosensis (Britton & Rose) Backeberg
Cylindropuntia thurberi (Engelmann) F. Knuth
Opuntia ticnamarensis (Ritter) Hunt CL
Cumulopuntia ticnamarensis Ritter
Opuntia tomentella Berger . . . . MX, GT
Opuntia tomentosa Salm-Dyck . . . . MX
Opuntia hernandezii De Candolle
Opuntia macdougaliana Rose
Opuntia sarca Griffiths ex Scheinvar
Opuntia treleasei Coulter . . . . US
Opuntia triacantha (Willdenow) Sweet . . . . US, CU, PR, VI, WI
Opuntia abjecta Small ex Britton & Rose
Opuntia militaris Britton & Rose
Opuntia trichophora (Engelmann & Bigelow) Britton & Rose . . . . US, MX
Opuntia tumida (Ritter) Hunt . . . . PE
Cumulopuntia tumida Ritter
Opuntia tuna (Linnaeus) Miller . . . . JM, DO
Opuntia tunicata (Lehmann) Link & Otto ex Pfeiffer . US, MX, CU, [EC, PE, AR, CL]
Opuntia puelchana Castellanos
Cylindropuntia tunicata (Lehmann) F. Knuth
Opuntia turbinata Small . . . . US
Opuntia undulata Griffiths . . . . MX
Opuntia unguispina Backeberg . . . . PE
Cumulopuntia unguispina (Backeberg) Ritter
Opuntia urbaniana Werdermann . . . . HT, DO
Opuntia vaginata Engelmann . . . . MX
Opuntia ×vaseyi Britton & Rose . . . . US
Opuntia covillei Britton & Rose
Opuntia velutina Weber ex Roland-Gosselin . . . . MX
Opuntia affinis Griffiths
Opuntia verschaffeltii Cels ex Weber . . . . BO, AR
Austrocylindropuntia haematacantha (Backeberg) Backeberg
Cylindropuntia haematacantha Backeberg
Austrocylindropuntia inarmata Backeberg
Opuntia posnanskyana Cardenas
Opuntia steiniana (Backeberg) Rowley
Austrocylindropuntia verschaffeltii (Cels ex Weber) Backeberg
Opuntia versicolor Engelmann ex Toumey . . . . US, MX
Cylindropuntia versicolor (Engelmann) F. Knuth
Opuntia vestita Salm-Dyck . . . . BO, AR
Austrocylindropuntia chuquisacana (Cardenas) Ritter
Opuntia chuquisacana Cardenas
Austrocylindropuntia teres (Cels ex Weber) Backeberg
Cylindropuntia teres (Cels ex Weber) Backeberg
Austrocylindropuntia vestita (Salm-Dyck) Backeberg
Opuntia vilis Rose . . . . MX
Corynopuntia vilis (Rose) F. Knuth
Opuntia ×viridiflora Britton & Rose . . . . US
Cylindropuntia viridiflora (Britton & Rose) F. Knuth
Opuntia viridirubra (Ritter) Braun & Esteves Pereira . . . . BR
Platyopuntia viridirubra Ritter
Opuntia viridirubra ssp. ***rubrogemmia*** (Ritter) Braun & Esteves Pereira . . . . BR
Platyopuntia rubrogemmia Ritter
Opuntia vitelliniflora (Ritter) Braun & Esteves Pereira . . . . BO
Platyopuntia vitelliniflora Ritter
Opuntia vitelliniflora ssp. ***interjecta*** (Ritter) Braun & Esteves Pereira . . . . BO
Platyopuntia interjecta Ritter
Opuntia ×vivipara Rose . . . . US
Opuntia wagenknechtii (Ritter) Hunt CL
Maihueniopsis wagenknechtii Ritter

Opuntia weberi Spegazzini . . . . AR
Tephrocactus weberi (Spegazzini) Backeberg
?*Opuntia aulacothele* Weber
Opuntia werneri Eggli . . . . BR
Opuntia wetmorei Britton & Rose . . . . AR
Opuntia whipplei Engelmann & Bigelow . . . . US
Cylindropuntia hualpaensis (Hester) Backeberg
Cylindropuntia whipplei (Engelmann & Bigelow) F. Knuth
Opuntia wigginsii L. Benson . . . . US
Cylindropuntia wigginsii (L. Benson) H. Robinson
Opuntia wilcoxii Britton & Rose . . . . MX
Opuntia wolfei (Benson) M.A. Baker . . . . US
Opuntia wootonii Griffiths . . . . US
Opuntia yanganucensis (Rauh & Backeberg) Rowley . . . . PE
Tephrocactus yanganucensis Rauh & Backeberg
Opuntia zehnderi (Rauh & Backeberg) Rowley . . . . PE
Cumulopuntia alboareolata (Ritter) Ritter
Tephrocactus alboareolatus Ritter ex Backeberg
Cumulopuntia zehnderi (Rauh & Backeberg) Ritter
Tephrocactus zehnderi Rauh & Backeberg

Oreocereus celsianus (Salm-Dyck) Riccobono . . . . PE, BO, AR
Borzicactus celsianus (Lemaire ex Salm-Dyck) Kimnach
Borzicactus fossulatus (Labouret) Kimnach
Oreocereus maximus Backeberg
Oreocereus neocelsianus Backeberg
Oreocereus doelzianus (Backeberg) Borg . . . . PE
Morawetzia doelziana Backeberg
Borzicactus doelzianus (Backeberg) Kimnach
Morawetzia sericata Ritter
Oreocereus hempelianus (Guerke) Hunt . . . . PE, CL
Arequipa australis Ritter
Oreocereus australis (Ritter) A.E. Hoffmann
Arequipa erectocylindrica Rauh & Backeberg
Arequipa hempeliana (Guerke) Oehme
Arequipa rettigii (Quehl) Oehme
Oreocereus rettigii (Quehl) Buxbaum
Arequipa soehrensii (Kreuzinger) Backeberg
Arequipa spinosissima Ritter
Arequipa weingartiana Backeberg
Oreocereus leucotrichus (Philippi) Wagenknecht . . . . PE, BO, CL
Borzicactus hendriksenianus (Backeberg) Kimnach
Oreocereus hendriksenianus (Backeberg) Backeberg
Arequipa leucotricha (Philippi) Britton & Rose
Borzicactus leucotrichus (Philippi) Kimnach
Oreocereus pseudofossulatus Hunt . . . . BO
Cleistocactus fossulatus Mottram
⊗*Oreocereus fossulatus* (Labouret) Backeberg
Oreocereus ritteri Cullmann . . . . PE
Oreocereus tacnaensis Ritter . . . . PE
Oreocereus trollii (Kupper) Backeberg . . . . BO, AR
Oreocereus crassiniveus Backeberg
Borzicactus trollii (Kupper) Kimnach
Oreocereus varicolor Backeberg . . . . PE, CL
Morawetzia varicolor (Backeberg) Knize

Oroya borchersii (Boedeker) Backeberg . . . . PE
Oroya peruviana (Schumann) Britton & Rose . . . . PE
Oroya baumannii Knize
Oroya gibbosa Ritter

Oroya laxiareolata Rauh & Backeberg
Oroya neoperuviana Backeberg
Oroya subocculta Rauh & Backeberg

Ortegocactus macdougallii Alexander . . . . MX
Neobesseya macdougallii (Alexander) Kladiwa

×***Pacherocactus orcuttii*** (K. Brandegee) Rowley . . . . MX
Cereus orcuttii K. Brandegee
Pachycereus orcuttii (K. Brandegee) Britton & Rose

Pachycereus fulviceps (Lemaire) Hunt . . . . MX
Carnegiea fulviceps (Schumann) P.V. Heath
Cephalocereus fulviceps (Weber) H.E. Moore
Mitrocereus fulviceps (Weber) Backeberg
Pseudomitrocereus fulviceps (Weber) Bravo & Buxbaum
Pachycereus gatesii (M.E. Jones) Hunt . . . . MX
Lophocereus gatesii M.E. Jones
Pachycereus gaumeri Britton & Rose . . . . MX
Anisocereus foetidus (MacDougall & Miranda) W.T. Marshall
Pachycereus foetidus (Miranda) P.V. Heath
Pterocereus foetidus MacDougall & Miranda
Anisocereus gaumeri (Britton & Rose) Backeberg
Pterocereus gaumeri (Britton & Rose) MacDougall & Miranda
Pachycereus grandis Rose . . . . MX
Pachycereus hollianus (Weber) Buxbaum . . . . MX
Lemaireocereus hollianus (Weber) Britton & Rose
Pachycereus lepidanthus (Eichlam) Britton & Rose . . . . GT, HN
Escontria lepidantha (Eichlam) Buxbaum
Anisocereus lepidanthus (Eichlam) Backeberg
Pachycereus marginatus (De Candolle) Britton & Rose . . . . MX
Marginatocereus marginatus (De Candolle) Backeberg
Stenocereus marginatus (De Candolle) Berger ex Buxbaum
Pachycereus militaris (Audot) Hunt . . . . ♦MX
Pachycereus chrysomallus (Lemaire) Britton & Rose♦
Backebergia militaris (Audot) Sanchez-Mejorada♦
Cephalocereus militaris (Audot) H.E. Moore♦
Mitrocereus militaris (Audot) Bravo♦
Pachycereus pecten-aboriginum (Engelmann) Britton & Rose . . . . MX
Pachycereus pecten-aboriginum ssp. *tehuantepecanus* (MacDougall & Bravo) P.V. Heath
Pachycereus tehuantepecanus MacDougall & Bravo
Pachycereus pringlei (Watson) Britton & Rose . . . . MX
Pachycereus calvus (Engelmann ex Coulter) Britton & Rose
Pachycereus schottii (Engelmann) Hunt . . . . US, MX
Lophocereus mieckleyanus (Weingart) Backeberg
Lophocereus sargentianus (Orcutt) Britton & Rose
Cereus schottii Engelmann
Lophocereus schottii (Engelmann) Britton & Rose
Pachycereus weberi (Coulter) Backeberg . . . . MX
Pachycereus gigas (Backeberg) Backeberg
Lemaireocereus weberi (Coulter) Britton & Rose
Stenocereus weberi (Coulter) Buxbaum

Parodia alacriportana Backeberg & Voll . . . . BR
Parodia alacriportana ssp. **alacriportana** . . . . BR
Brasiliparodia alacriportana (Backeberg & Voll) Ritter
Notocactus alacriportanus (Backeberg & Voll) Buxbaum
Parodia alacriportana ssp. **brevihamata** (W. Haage ex Backeberg) Hofacker & Braun . . . . BR
Brasiliparodia brevihamata (W. Haage ex Backeberg) Ritter

Parodia brevihamata W. Haage ex Backeberg
Notocactus brevihamatus (W. Haage ex Backeberg) Buxbaum
Parodia alacriportana ssp. **buenekeri** (Buining) Hofacker & Braun . . . . BR
Brasiliparodia buenekeri (Buining) Ritter
Notocactus buenekeri (Buining) Krainz
Parodia buenekeri Buining
Parodia alacriportana ssp. **catarinensis** (Ritter) Hofacker & Braun . . . . BR
Brasiliparodia catarinensis Ritter
Notocactus catarinensis (Ritter) Scheinvar
Parodia catarinensis (Ritter) Brandt
Parodia allosiphon (Marchesi) Taylor . . . . UY
Notocactus allosiphon Marchesi
Parodia arnostiana (Lisal & Kolarik) Hofacker . . . . BR
Notocactus arnostianus Lisal & Kolarik
Parodia aureicentra (Backeberg) Backeberg . . . . AR
Parodia muhrii Brandt
Parodia rauschii Backeberg ex Hunt
Parodia varicolor Ritter
Parodia ayopayana Cardenas . . . . BO
Parodia borealis Ritter
Parodia buxbaumiana Brandt
Parodia comosa Ritter
Parodia cotacajensis Brandt
Parodia echinus Ritter
Parodia elata Brandt
Parodia macednosa Brandt
Parodia miguillensis Cardenas
Parodia pseudoayopayana Cardenas
Parodia buiningii (Buxbaum) Taylor . . . . BR, UY
Notocactus buiningii Buxbaum
Parodia carambeiensis (Buining & Brederoo) Hofacker . . . . BR
Notocactus carambeiensis Buining & Brederoo
Notocactus villa-velhensis (Backeberg & Voll) Slaba
Parodia chrysacanthion (Schumann) Backeberg . . . . AR
Parodia columnaris Cardenas . . . . BO
Parodia legitima Brandt
Parodia comarapana Cardenas . . . . BO
Parodia mairanana Cardenas
Parodia neglecta Brandt
Parodia neglectoides Brandt
Parodia commutans Ritter . . . . BO
Parodia maxima Ritter
Parodia obtusa Ritter
Parodia concinna (Monville) Taylor . . . . BR, UY
Notocactus apricus (Arechavaleta) Berger
Frailea caespitosa (Spegazzini) Britton & Rose
Parodia caespitosa (Spegazzini) Taylor
Notocactus caespitosus (Spegazzini) Backeberg
Notocactus concinnioides Prauser
Notocactus eremiticus Ritter
Notocactus gibberulus Prestle
Notocactus joadii (Hooker f.) Herter
Notocactus olimarensis Prestle
Notocactus rubrigemmatus Abraham
Parodia concinna ssp. **concinna** . . . . BR, UY
Notocactus concinnus (Monville) Backeberg
Parodia concinna ssp. **agnetae** (Vliet) Hofacker . . . . BR
Notocactus agnetae Vliet
Parodia concinna ssp. **blauuwiana** (Vliet) Hofacker . . . . BR
Notocactus blaauwianus Vliet
Notocactus multicostatus Buining & Brederoo

Parodia crassigibba (Ritter) Taylor . . . . BR
Notocactus crassigibbus Ritter
Parodia curvispina (Ritter) Hunt . . . . BR
Notocactus curvispinus Ritter
Notocactus rubropedatus Ritter
Parodia erinacea (Haworth) Taylor . . . . BR, UY, AR
Wigginsia acuata (Link & Otto) Ritter
Notocactus acuatus (Link & Otto) Theunissen
Wigginsia erinacea (Haworth) Porter
Notocactus erinaceus (Haworth) Krainz
Notocactus tetracanthus (Lemaire) Gerloff et al.
Parodia erubescens (Osten) Hunt . . . . UY
Notocactus erubescens (Osten) Marchesi
Notocactus schlosseri Vliet
Parodia formosa Ritter . . . . BO
Parodia bellavistana Brandt
Parodia carapariana Brandt
Parodia cardenasii Ritter
Parodia chaetocarpa Ritter
Parodia chirimoyarana Brandt
Parodia pachysa Brandt
Parodia parvula Brandt
Parodia purpureo-aurea Ritter
Parodia pusilla Brandt
Parodia setispina Ritter
Parodia tillii Weskamp
Parodia winbergii Weskamp
Parodia fusca (Ritter) Hofacker & Braun . . . . BR
Notocactus fuscus Ritter
Parodia haselbergii (Haage ex Ruempler) Brandt . . . . BR
Parodia haselbergii ssp. **haselbergii** . . . . BR
Brasilicactus haselbergii (Haage ex Ruempler) Backeberg ex Schoff.
Notocactus haselbergii (Ruempler) Backeberg
Parodia haselbergii ssp. **graessneri** (Schumann) Hofacker & Braun . . . . BR
Parodia elachisantha (Weber) Brandt
Brasilicactus elachisanthus (Weber) Backeberg
Brasilicactus graessneri (Schumann) Backeberg
Notocactus graessneri (Schumann) Berger
Parodia graessneri (Schumann) Brandt
Parodia hausteiniana Rausch . . . . BO
Parodia laui Brandt
Parodia herteri (Werdermann) Taylor . . . . BR, UY
Notocactus herteri (Werdermann) Buining & Kreuzinger
Notocactus pseudoherteri Buining
Notocactus rubriflorus Hort. ex Berger
Parodia horstii (Ritter) Taylor . . . . BR
Notocactus horstii Ritter
Notocactus purpureus Ritter
Parodia langsdorfii (Lehmann) Hunt . . . . BR
Notocactus langsdorfii (Lehmann) Krainz
Wigginsia langsdorfii (Lehmann) Porter
Notocactus leprosorum (Ritter) Havlicek
Wigginsia leprosorum Ritter
Wigginsia longispina Ritter
Notocactus longispinus (Ritter) Havlicek
Wigginsia polyacantha (Link & Otto) Ritter
Notocactus polyacanthus (Link & Otto) Theunissen
Notocactus prolifer (Ritter) Theunissen
Wigginsia prolifera Ritter
Notocactus pulvinatus Vliet

Parodia leninghausii (Schumann) Brandt . . . . BR, PY, UY
Eriocactus leninghausii (Schumann) Backeberg
Notocactus leninghausii (Schumann) Backeberg
Parodia linkii (Lehmann) Kiesling . . . . BR, UY, AR
Notocactus linkii (Lehmann) Herter
Notocactus megapotamicus (Osten) Herter
Parodia maassii (Heese) Berger . . . . BO, AR
Parodia bermejoensis Brandt
Parodia escayachensis (Vaupel) Backeberg
Parodia haageana Brandt
Parodia koehresiana Brandt
Parodia lamprospina Brandt
Parodia mendeziana Brandt
Parodia otaviana Cardenas
Parodia suprema Ritter
Parodia thieleana Brandt
®*Parodia obtusa* ssp. *atochana* Brandt
Parodia magnifica (Ritter) Brandt . . . . BR, UY
Eriocactus magnificus Ritter
Notocactus magnificus (Ritter) Krainz ex Taylor
Parodia mammulosa (Lemaire) Taylor . . . . BR, UY, AR
Notocactus cristatoides Ritter
Notocactus floricomus (Arechavaleta) Backeberg
Notocactus hypocrateriformis (Otto & Dietrich) Herter
Notocactus macambarensis Prestle
Notocactus mueller-moelleri Fric ex Backeberg & F. Knuth
Notocactus paulus Schlosser & Brederoo
Notocactus ritterianus Lisal & Kolarik
Notocactus roseoluteus Vliet
?*Notocactus megalanthus* Schlosser
?*Notocactus orthacanthus* (Link & Otto) Vliet
Parodia mammulosa ssp. **mammulosa** . . . . BR
Notocactus mammulosus (Lemaire) Backeberg
Parodia mammulosa ssp. **brasiliensis** (Havlicek) Hofacker . . . . BR
Parodia mammulosa ssp. **erythracantha**
(Schlosser & Brederoo) Hofacker . . . . BR
Notocactus erythracanthus Schlosser & Brederoo
Parodia mammulosa ssp. **eugeniae** (Vliet) Hofacker . . . . BR
Notocactus eugeniae Vliet
Parodia mammulosa ssp. **submammulosa** (Lemaire) Hofacker . . . . BR, AR
Notocactus pampeanus (Spegazzini) Backeberg & F. Knuth
Parodia submammulosa (Lemaire) Kiesling
Parodia submammulosa ssp. *minor* Kiesling
Notocactus submammulosus (Lemaire) Backeberg
Parodia meonacantha (Prestle) Hofacker . . . . BR
Notocactus meonacanthus Prestle
Parodia microsperma (Weber) Spegazzini . . . . AR
Parodia aconquijaensis Weskamp
Parodia albofuscata Brandt
Parodia argerichiana Weskamp
Parodia aureispina Backeberg
Parodia belenensis Weskamp
Parodia betaniana Ritter
Parodia cabracorralensis Piens
Parodia capillitaensis Brandt
Parodia catamarcensis Backeberg
Parodia cebilarensis Weskamp
Parodia chlorocarpa Ritter
Parodia elegans Fechser ex Backeberg
Parodia erythrantha (Spegazzini) Backeberg & F. Knuth

Parodia fechseri Backeberg
Parodia fuscato-viridis Backeberg
Parodia glischrocarpa Ritter
Parodia grandiflora Veverka
Parodia guachipasana Weskamp
Parodia heyeriana Weskamp
Parodia hummeliana Lau & Weskamp
Parodia lembckei Weskamp
Parodia malyana Rausch
Parodia mercedesiana Weskamp
Parodia mesembrina Brandt
Parodia microthele Backeberg
Parodia minuscula Rausch
Parodia mutabilis Backeberg
Parodia nana Weskamp
Parodia papagayana Brandt
Parodia rigidispina Krainz
Parodia rubellihamata Ritter ex Backeberg
Parodia rubriflora Backeberg
Parodia rubristaminea Ritter
Parodia sanguiniflora Backeberg
Parodia scopaoides Backeberg
Parodia setifera Backeberg
Parodia spanisa Brandt
Parodia spegazziniana Brandt
Parodia tafiensis Backeberg
Parodia tuberculosi-costata Backeberg
Parodia uebelmanniana Ritter
Parodia wagneriana Weskamp
Parodia weskampiana Krasucka & Spanowsky
®*Parodia amblayensis* Brandt
®*Parodia malyana* ssp. *igneiflora* Brandt

Parodia microsperma ssp. **microsperma** . . . . AR
Parodia campestris Brandt
Parodia herzogii Rausch
Parodia macrancistra (Schumann) Backeberg
Parodia talaensis Brandt
Parodia thionantha Brandt
Parodia tucumanensis Weskamp
Parodia weberiana Brandt

Parodia microsperma ssp. **horrida** (Brandt) Kiesling & Ferrari . . . . AR
Parodia atroviridis Backeberg
Parodia cachiana Weskamp
Parodia dextrohamata Backeberg
Parodia dichroacantha Brandt & Weskamp
Parodia heteracantha Ritter ex Weskamp
Parodia horrida Brandt
Parodia kilianana Backeberg
Parodia lohaniana Lau & Weskamp ex Weskamp
Parodia piltziorum Weskamp
Parodia pluricentralis Backeberg ex Brandt
Parodia rigida Backeberg
Parodia superba Brandt
Parodia tolombona Weskamp

Parodia mueller-melchersii (Backeberg) Taylor . . . . BR, UY

Parodia mueller-melchersii ssp. **mueller-melchersii** . . . . UY
Notocactus mueller-melchersii Fric ex Backeberg

Parodia mueller-melchersii ssp. **gutierrezii** (Abraham) Hofacker . . . . BR
Notocactus gutierrezii Abraham

Parodia mueller-melchersii ssp. **winkleri** (Vliet) Hofacker .... UY
Notocactus winkleri Vliet
Parodia muricata (Otto ex Pfeiffer) Hofacker .... BR
Notocactus laetivirens Ritter
Notocactus muricatus (Otto ex Pfeiffer) Backeberg
Parodia neoarechavaletae (Havlicek) Hunt .... UY
Wigginsia arechavaletae (Schumann ex Spegazzini) Porter
Notocactus maldonadensis (Herter) Herter
Notocactus neoarechavaletae Elsner ex Havlicek
Parodia neohorstii (Theunissen) Taylor .... BR
Wigginsia horstii Ritter
Notocactus neohorstii Theunissen
Parodia nigrispina (Schumann) Brandt .... PY
Eriocactus nigrispinus (Schumann) Ritter
Notocactus nigrispinus (Schumann) Buining ex Rowley
Notocactus schumannianus ssp. *nigrispinus* (Schumann) Engel
Parodia nivosa Backeberg .... AR
Parodia faustiana Backeberg
Parodia uhligiana Backeberg
Parodia nothominuscula Hofacker .... BR
Notocactus minusculus Hofacker & Herm
Parodia nothorauschii Hunt .... UY
Notocactus rauschii Vliet
Notocactus spinibarbis Ritter
Parodia ocampoi Cardenas .... BO
Parodia augustinii Weskamp
Parodia compressa Ritter
Parodia copavilquensis Weskamp
Parodia elachista Brandt
Parodia exquisita Brandt
Parodia gibbulosa Ritter
Parodia gibbulosoides Brandt
Parodia minuta Ritter
Parodia punae Cardenas
Parodia zecheri Vasquez
®*Parodia zecheri* ssp. *elachista* (Brandt) Brandt
Parodia ottonis (Lehmann) Taylor .... BR, PY, UY, AR
Notocactus acutus Ritter
Notocactus arechavaletae (Spegazzini) Herter
Parodia paraguayensis Spegazzini
Notocactus tenuispinus (Link & Otto) Herter
Notocactus uruguayus (Arechavaleta) Herter
?*Notocactus ruoffii* Gerloff
®*Notocactus grandiensis* Bergner
Parodia ottonis ssp. **ottonis** .... BR, PY, UY, AR
Notocactus ottonis (Lehmann) Backeberg
Parodia ottonis ssp. **horstii** (Ritter) Hofacker .... BR
Notocactus ottonis ssp. *horstii* Ritter
Parodia oxycostata (Buining & Brederoo) Hofacker .... BR
Notocactus campestrensis Ritter
Notocactus glaucinus Ritter
Notocactus incomptus Gerloff
Notocactus miniatispinus Havlicek
Notocactus oxycostatus Buining & Brederoo
Notocactus securituberculatus Ritter
Parodia oxycostata ssp. **oxycostata** .... BR
Parodia oxycostata ssp. **gracilis** (Ritter) Hofacker .... BR
Notocactus gracilis Gemmrich
Notocactus harmonianus Ritter
Notocactus ibicuiensis Prestle
®*Notocactus eurypleurus* Prestle

Parodia penicillata Fechser & Steeg . . . . AR
Parodia permutata (Ritter) Hofacker . . . . BR
Notocactus permutatus Ritter
Parodia procera Ritter . . . . BO
Parodia andreae Brandt
Parodia andreaeoides Brandt
Parodia challamarcana Brandt
Parodia echinopsoides Brandt
Parodia gracilis Ritter
Parodia lychnosa Brandt
Parodia perplexa Brandt
Parodia prestoensis Brandt
Parodia pseudoprocera Brandt
Parodia riograndensis Brandt
Parodia separata Brandt
Parodia subtilihamata Ritter
Parodia tredecimcostata Ritter
®*Parodia pseudoprocera* ssp. *aurantiaciflora* Brandt
Parodia rechensis (Buining) Brandt . . . . BR
Brasiliparodia rechensis (Buining) Ritter
Notocactus rechensis Buining
Parodia ritteri Buining . . . . BO
Parodia agasta Brandt
Parodia aglaisma Brandt
Parodia belliata Brandt
Parodia camargensis Buining & Ritter
Parodia carrerana Cardenas
Parodia castanea (Ritter) Ritter
Parodia cintiensis Ritter
Parodia fulvispina Ritter
Parodia roseoalba Ritter
Parodia rostrum-sperma Brandt
Parodia rubida Ritter
Parodia splendens Cardenas
Parodia tojoensis Brandt
®*Parodia camblayana* (Ritter) Brandt
Parodia rudibuenekeri (Abraham) Hofacker & Braun . . . . BR
Parodia rudibuenekeri ssp. **rudibuenekeri** . . . . BR
Notocactus rudibuenekeri Abraham
Parodia rudibuenekeri ssp. **glomerata** (Menges ex Gerloff) Hofacker . . . . BR
Notocactus glomeratus Menges ex Gerloff
Parodia rutilans (Daeniker & Krainz) Taylor . . . . BR, UY
Notocactus roseiflorus Schlosser & Brederoo
Parodia rutilans ssp. **rutilans** . . . . BR
Notocactus rutilans Daeniker & Krainz
Parodia rutilans ssp. **veeniana** (Vliet) Hofacker . . . . UY
Notocactus veenianus Vliet
Parodia saint-pieana Backeberg . . . . AR
Parodia schumanniana (Nicolai) Brandt . . . . BR, PY, AR
Parodia ampliocostata (Ritter) Brandt
Eriocactus ampliocostatus Ritter
Notocactus ampliocostatus (Ritter) Theunissen
Parodia grossei (Schumann) Brandt
Parodia schumanniana ssp. **schumanniana** . . . . BR
Eriocactus schumannianus (Nicolai) Backeberg
Notocactus schumannianus (Nicolai) Backeberg
Parodia schumanniana ssp. **claviceps** (Ritter) Hofacker . . . . BR
Eriocactus claviceps Ritter
Notocactus claviceps (Ritter) Krainz
Parodia claviceps (Ritter) Brandt

Parodia schwebsiana (Werdermann) Backeberg . . . . BO
Parodia applanata (Hoffmann & Backeberg) Brandt
Parodia minima Brandt
Parodia salmonea Brandt
Parodia scopa (Sprengel) Taylor . . . . BR, UY
Parodia scopa ssp. **scopa** . . . . BR, UY
Notocactus scopa (Sprengel) Backeberg
Notocactus soldtianus Vliet
Parodia scopa ssp. **marchesii** (Abraham) Hofacker . . . . BR
Notocactus scopa ssp. *marchesii* Abraham
Parodia scopa ssp. **neobuenekeri** (Ritter) Hofacker & Braun . . . . BR
Notocactus neobuenekeri Ritter
Parodia scopa ssp. **succinea** (Ritter) Hofacker & Braun . . . . BR
Parodia succinea (Ritter) Taylor
Notocactus succineus Ritter
Parodia sellowii (Link & Otto) Hunt . . . . BR, UY, AR
Notocactus corynodes (Otto ex Pfeiffer) Krainz
Wigginsia corynodes (Otto ex Pfeiffer) Porter
Wigginsia courantii (Lemaire) Ritter
Notocactus fricii (Arechavaleta) Krainz
Wigginsia fricii (Arechavaleta) Porter
Wigginsia leucocarpa (Arechavaleta) Porter
Notocactus leucocarpus (Arechavaleta) Schaefer
Wigginsia macracantha (Arechavaleta) Porter
Notocactus macracanthus (Arechavaleta) Schaefer
Wigginsia macrogona (Arechavaleta) Porter
Notocactus macrogonus (Arechavaleta) Schaefer
Notocactus pauciareolatus (Arechavaleta) Krainz
Notocactus rubricostatus (Fric ex Fleischer & Schuetz) Schuetz
Notocactus sellowii (Link & Otto) Theunissen
Wigginsia sellowii (Link & Otto) Ritter
Wigginsia sessiliflora (Hooker) Porter
Notocactus sessiliflorus (Hooker) Krainz
Notocactus stegmannii (Backeberg) Krainz
Wigginsia stegmannii (Backeberg) Porter
Wigginsia tephracantha (Link & Otto) Porter
Notocactus tephracanthus (Link & Otto) Krainz
Wigginsia vorwerkiana (Werdermann) Porter
Notocactus vorwerkianus (Werdermann) Krainz
Parodia stockingeri (Prestle) Hofacker & Braun . . . . BR
Notocactus stockingeri Prestle
Parodia stuemeri (Werdermann) Backeberg . . . . AR
Parodia gutekunstiana Backeberg
Parodia rubricentra Backeberg
Parodia rubrispina Koehler
Parodia subterranea Ritter . . . . BO
Parodia culpinensis Brandt
Parodia miranda Brandt
Parodia nigresca Brandt
Parodia occulta Ritter
Parodia pseudosubterranea Brandt
Parodia salitrensis Brandt
Parodia zaletaewana Brandt
Parodia tabularis (Cels ex Ruempler) Hunt . . . . UY
Parodia tabularis ssp. **tabularis** . . . . UY
Notocactus tabularis (Cels ex Ruempler) Backeberg
Parodia tabularis ssp. **bommeljei** (Vliet) Hofacker . . . . UY
Notocactus bommeljei Vliet
Notocactus brederooianus Prestle

Parodia taratensis Cardenas . . . . BO
Parodia bilbaoensis Cardenas
Parodia caineana Brandt
Parodia krahnii Weskamp
Parodia tenuicylindrica (Ritter) Hunt . . . . BR
Notocactus minimus Fric & Kreuzinger ex Berger
Notocactus tenuicylindricus Ritter
Parodia tilcarensis (Werdermann & Backeberg) Backeberg . . . . AR
Parodia carminata Backeberg
Parodia friciana Brandt
Parodia jujuyana Fric ex Subik
Parodia pseudostuemeri Backeberg
Parodia schuetziana Jajo
Parodia scoparia Ritter
Parodia tumbayana Weskamp
®*Parodia setosa* Backeberg
Parodia tuberculata Cardenas . . . . BO
Parodia backebergiana Brandt
Parodia candidata Brandt
Parodia firmisissima Brandt
Parodia idiosa Brandt
Parodia ignorata Brandt
Parodia krasuckana Brandt
Parodia multicostata Ritter & Jelinek
Parodia otuyensis Ritter
Parodia quechua Brandt
Parodia sotomayorensis Ritter
Parodia stereospina Brandt
Parodia sucrensis Brandt
Parodia tarabucina Cardenas
Parodia yamparaezi Cardenas
Parodia turbinata (Arechavaleta) Hofacker . . . . UY
Notocactus calvescens Gerloff & Nilson
Wigginsia schaeferiana Abraham & Theunissen
Notocactus schaeferianus (Abraham & Theunissen) Havlicek
Wigginsia turbinata (Arechavaleta) Porter
Notocactus turbinatus (Arechavaleta) Krainz
Parodia turecekiana Kiesling . . . . AR, UY
Parodia warasii (Ritter) Brandt . . . . BR
Eriocactus warasii Ritter
Notocactus warasii (Ritter) Hewitt & Donald
Parodia werdermanniana (Herter) Taylor . . . . UY
Notocactus ferrugineus Schlosser
Notocactus vanvlietii Rausch
Notocactus werdermannianus Herter
?*Notocactus memorialis* Prestle
Parodia werneri Hofacker . . . . BR
Notocactus arachnites Ritter
Parodia werneri ssp. **werneri** . . . . BR
Notocactus uebelmannianus Buining
Parodia werneri ssp. **pleiocephala** (Gerloff & Koenigs) Hofacker . . . . BR
Notocactus uebelmannianus ssp. *pleiocephalus* Gerloff & Koenigs

Pediocactus bradyi L. Benson . . . . ♦US
Toumeya bradyi (L. Benson) Earle♦
Pediocactus simpsonii ssp. *bradyi* (Benson) Halda♦
Pediocactus bradyi ssp. ***despainii*** (Welsh & Goodrich) Hochstaetter . . . . ♦US
Pediocactus despainii Welsh & Goodrich♦
Pediocactus bradyi ssp. ***winkleri*** (Heil) Hochstaetter . . . . ♦US
Pediocactus winkleri Heil♦

Pediocactus hermannii W.T. Marshall . . . . US
Pediocactus knowltonii L. Benson . . . . ♦US
Pediocactus nigrispinus (Hochstaetter) Hochstaetter . . . . US
Pediocactus nigrispinus ssp. ***beastonii*** (Hochstaetter) Hochstaetter . . . . US
Pediocactus nigrispinus ssp. ***puebloensis*** Hochstaetter . . . . US
Pediocactus paradinei B.W. Benson . . . . ♦US
Pediocactus peeblesianus (Croizat) L. Benson . . . . ♦US
Toumeya fickeisenii (Backeberg) Earle♦
Navajoa peeblesiana Croizat♦
Utahia peeblesiana (Croizat) Kladiwa♦
Navajoa peeblesiana ssp. *fickeisenii* (Hochstaetter) Hochstaetter♦
®*Navajoa fickeisenii* Backeberg♦
Pediocactus sileri (Engelmann ex Coulter) L. Benson . . . . ♦US
Utahia sileri (Engelmann ex Coulter) Britton & Rose♦
Pediocactus simpsonii (Engelmann) Britton & Rose . . . . US
Pediocactus simpsonii ssp. ***bensonii*** (Engelmann) Hochstaetter . . . . US
Pediocactus simpsonii ssp. ***idahoensis*** Hochstaetter . . . . US
Pediocactus simpsonii ssp. ***indranus*** (Hochstaetter) Hochstaetter . . . . US
Pediocactus simpsonii ssp. ***robustior*** (Coulter) Hochstaetter . . . . US
Pediocactus robustior (Coulter) Arp

Pelecyphora aselliformis Ehrenberg . . . . ♦MX
Pelecyphora strobiliformis (Werdermann) Fric & Schelle ex Kreuzinger . . . . . . ♦MX
Encephalocarpus strobiliformis (Werdermann) Berger♦

Peniocereus castellae Sanchez-Mejorada . . . . MX
Peniocereus cuixmalensis Sanchez-Mejorada . . . . MX
Peniocereus fosterianus Cutak . . . . MX
Peniocereus greggii (Engelmann) Britton & Rose . . . . US, MX
Cereus greggii Engelmann
Peniocereus hirschtianus (Schumann) Hunt . . . . GT, NI, SV, CR
Nyctocereus guatemalensis Britton & Rose
Nyctocereus hirschtianus (Schumann) Britton & Rose
Nyctocereus neumannii (Schumann) Britton & Rose
Peniocereus johnstonii Britton & Rose . . . . MX
Peniocereus lazaro-cardenasii (Contreras et al.) Hunt . . . . MX
Neoevansia lazaro-cardenasii Contreras et al.
Wilcoxia lazaro-cardenasii (Contreras et al.) Cartier
Peniocereus macdougallii Cutak . . . . MX
Peniocereus maculatus (Weingart) Cutak . . . . MX
Acanthocereus maculatus (Weingart) F. Knuth
Peniocereus marianus (Gentry) Sanchez-Mejorada . . . . MX
Peniocereus oaxacensis (Britton & Rose) Hunt . . . . MX
Nyctocereus oaxacensis Britton & Rose
Peniocereus occidentalis Bravo . . . . MX
Peniocereus rosei J.G. Ortega . . . . MX
Peniocereus serpentinus (Lagasca & Rodrigues) Taylor . . . . MX
Nyctocereus castellanosii Scheinvar
Nyctocereus serpentinus (Lagasca & Rodriguez) Britton & Rose
Peniocereus striatus (Brandegee) Buxbaum . . . . US, MX
Peniocereus diguetii (Weber) Backeberg
Neoevansia striata (Brandegee) Sanchez-Mejorada
Wilcoxia striata (Brandegee) Britton & Rose
Cereus striatus T. Brandegee
Peniocereus tepalcatepecanus Sanchez-Mejorada . . . . MX
Peniocereus viperinus (Weber) Kreuzinger . . . . MX
Wilcoxia papillosa Britton & Rose
Wilcoxia tomentosa Bravo
Wilcoxia viperina (Weber) Britton & Rose
®*Peniocereus papillosus* (Britton & Rose) hort.

Peniocereus zopilotensis (Meyran) Buxbaum . . . . MX
Neoevansia zopilotensis (Meyran) Sanchez-Mejorada
Wilcoxia zopilotensis Meyran

Pereskia aculeata Miller . . . . US, MX, PA, CU, HT, DO, PR, VI, WI, TT
. . . . GY, GF, SR, BR, PY, VE, EC, PE, UY, AR
Pereskia pereskia (Linnaeus) Karsten
Pereskia aureiflora Ritter . . . . BR
Pereskia bahiensis Guerke . . . . BR
Pereskia bleo (Kunth) De Candolle . . . . PA, CO
Rhodocactus bleo (Kunth) F. Knuth
Rhodocactus corrugatus (Cutak) Backeberg
Pereskia diaz-romeroana Cardenas . . . . BO
Pereskia grandifolia Haworth . . . . BR [widely introduced]
Pereskia grandifolia ssp. **grandifolia** . . . . BR
Rhodocactus grandifolius (Haworth) F. Knuth
Pereskia tampicana Weber
Rhodocactus tampicanus (Weber) Backeberg
⊗*Pereskia bleo* auctt.
Pereskia grandifolia ssp. **violacea** (Leuenberger) Taylor & Zappi . . . . BR
Pereskia guamacho Weber . . . . AN, VE, CO
Pereskia colombiana Britton & Rose
Rhodocactus colombianus (Britton & Rose) F. Knuth
Rhodocactus guamacho (Weber) F. Knuth
Pereskia horrida (Kunth) De Candolle . . . . PE
Pereskia horrida ssp. **horrida** . . . . PE
Pereskia humboldtii Britton & Rose
Pereskia vargasii H. Johnson
Pereskia horrida ssp. **rauhii** (Backeberg) Ostolaza . . . . PE
Pereskia lychnidiflora De Candolle . . . . MX, GT, HN, SV, NI, CR
Pereskia autumnalis Rose
Rhodocactus autumnalis (Eichlam) F. Knuth
Pereskia conzattii Britton & Rose
Rhodocactus lychnidiflorus (De Candolle) F. Knuth
Rhodocactus nicoyanus (Weber) F. Knuth
Pereskiopsis opuntiiflora (De Candolle) Britton & Rose
Pereskia pititache Karwinski ex Pfeiffer
Pereskiopsis pititache (Karwinski ex Pfeiffer) Britton & Rose
Pereskia marcanoi Areces . . . . DO
Pereskia nemorosa Rojas . . . . BR, AR, PY, UY
Pereskia amapola Weber
Pereskia portulacifolia (Linnaeus) Haworth . . . . HT, DO
Rhodocactus portulacifolius (Linnaeus) F. Knuth
Pereskia quisqueyana Liogier . . . . DO
Pereskia sacharosa Grisebach . . . . BR, PY, BO, AR
Pereskia moorei Britton & Rose
Rhodocactus sacharosa (Grisebach) Backeberg
Pereskia saipinensis Cardenas
Rhodocactus saipinensis (Cardenas) Backeberg
Pereskia sparsiflora Ritter
Pereskia stenantha Ritter . . . . BR
Pereskia weberiana Schumann . . . . BO
Rhodocactus antonianus Backeberg
Pereskia zinniiflora De Candolle . . . . CU
Pereskia cubensis Britton & Rose
Rhodocactus cubensis (Britton & Rose) F. Knuth
Rhodocactus zinniiflorus (De Candolle) F. Knuth

Pereskiopsis aquosa (Weber) Britton & Rose . . . . MX
Pereskiopsis blakeana J.G. Ortega . . . . MX

Pereskiopsis diguetii (Weber) Britton & Rose . . . . MX
Pereskiopsis velutina Rose
Pereskiopsis kellermanii Rose . . . . MX, GT, HN
Pereskiopsis scandens Britton & Rose
Pereskiopsis porteri (Brandegee ex Weber) Britton & Rose . . . . MX
?*Pereskiopsis gatesii* Baxter
Pereskiopsis rotundifolia (De Candolle) Britton & Rose . . . . MX
Pereskiopsis chapistle (Weber) Britton & Rose
Pereskiopsis spathulata (Otto ex Pfeiffer) Britton & Rose XC
Pereskia higuerana Cardenas
Rhodocactus higueranus (Cardenas) Backeberg

Pilosocereus albisummus Braun & Esteves . . . . BR
Pilosocereus alensis (Weber ex Gosselin) Byles & Rowley . . . . MX
Cephalocereus alensis (Weber) Britton & Rose
Cephalocereus guerreronis (Backeberg) Buxbaum
Pilosocereus guerreronis (Backeberg) Byles & Rowley
Pilosocereus arrabidae (Lemaire) Byles & Rowley . . . . BR
Cephalocereus arrabidae (Lemaire) Britton & Rose
Pseudopilocereus arrabidae (Lemaire) Buxbaum
Pilosocereus aureispinus (Buining & Brederoo) Ritter . . . . BR
Coleocephalocereus aureispinus Buining & Brederoo
Pilosocereus aurisetus (Werdermann) Byles & Rowley . . . . BR
®*Pilosocereus coerulescens* (Lemaire) Ritter
Pilosocereus aurisetus ssp. **aurisetus** . . . . BR
Pseudopilocereus aurisetus (Werdermann) Buxbaum
Pilosocereus aurisetus ssp. *densilanatus* (Ritter) Braun & Esteves Pereira
Pilosocereus aurisetus ssp. *supthutianus* (Braun) Braun & Esteves Pereira
Pilosocereus aurisetus ssp. *werdermannianus*
(Buining & Brederoo) Braun & Esteves Pereira
Pilosocereus supthutianus Braun
Pilosocereus werdermannianus (Buining & Brederoo) Ritter
Pseudopilocereus werdermannianus Buining & Brederoo
Pilosocereus aurisetus ssp. **aurilanatus** (Ritter) Zappi . . . . BR
Pilosocereus aurilanatus Ritter
Pseudopilocereus aurilanatus (Ritter) P.V. Heath
Pilosocereus azulensis Taylor & Zappi . . . . BR
Pilosocereus brasiliensis (Britton & Rose) Backeberg . . . . BR
Cephalocereus brasiliensis Britton & Rose
?*Pilosocereus sublanatus* (Salm-Dyck) Byles & Rowley
Pilosocereus brasiliensis ssp. **brasiliensis** . . . . BR
Pilosocereus brasiliensis ssp. **ruschianus** (Buining & Brederoo) Zappi . . . . BR
Pilosocereus ruschianus (Buining & Brederoo) Braun
Pseudopilocereus ruschianus Buining & Brederoo
Pilosocereus catingicola (Guerke) Byles & Rowley . . . . BR
Pilosocereus arenicola (Werdermann) Byles & Rowley
Pilosocereus catingicola ssp. *arenicola* (Werdermann) Braun & Esteves Pereira
Pilosocereus catingicola ssp. **catingicola** . . . . BR
Cephalocereus catingicola (Guerke) Britton & Rose
Pseudopilocereus catingicola (Guerke) Buxbaum
Pilosocereus catingicola ssp. *robustus* (Ritter) Braun & Esteves Pereira
Pilosocereus robustus Ritter
Pseudopilocereus robustus (Ritter) P.V. Heath
Pilosocereus catingicola ssp. **salvadorensis** (Werdermann) Zappi . . . . BR
Pilosocereus catingicola ssp. *hapalacanthus* (Werdermann) Braun & Esteves Pereira
Pilosocereus hapalacanthus (Werdermann) Byles & Rowley
Pseudopilocereus hapalacanthus (Werdermann) Buxbaum
Pilosocereus rupicola (Werdermann) Byles & Rowley
Pseudopilocereus rupicola (Werdermann) Buxbaum
Austrocephalocereus salvadorensis (Werdermann) Buxbaum

Pilosocereus salvadorensis (Werdermann) Byles & Rowley
Pseudopilocereus salvadorensis (Werdermann) Buxbaum
Pilosocereus sergipensis (Werdermann) Byles & Rowley
Pseudopilocereus sergipensis (Werdermann) Buxbaum
Pilosocereus chrysacanthus (Weber) Byles & Rowley .... MX
Cephalocereus chrysacanthus (Weber ex Schumann) Britton & Rose
Pilosocereus chrysostele (Vaupel) Byles & Rowley .... BR
Pseudopilocereus chrysostele (Vaupel) Buxbaum
Pilosocereus densiareolatus Ritter .... BR
Pseudopilocereus densiareolatus (Ritter) P.V. Heath
Pilosocereus diersianus (Esteves) Braun .... BR
Pseudopilocereus diersianus Esteves
Pilosocereus flavipulvinatus (Buining & Brederoo) Ritter .... BR
Pilosocereus carolinensis Ritter
Pseudopilocereus carolinensis (Ritter) P.V. Heath
Pseudopilocereus flavipulvinatus Buining & Brederoo
Pilosocereus flavipulvinatus ssp. *carolinensis* (Ritter) Braun & Esteves Pereira
Pilosocereus flexibilispinus Braun & Esteves Pereira .... BR
Pilosocereus floccosus Byles & Rowley .... BR
Pilosocereus floccosus ssp. **floccosus** .... BR
Pseudopilocereus floccosus (Byles & Rowley) Buxbaum
Pilosocereus floccosus ssp. **quadricostatus** (Ritter) Zappi .... BR
Pilosocereus quadricostatus Ritter
Pseudopilocereus quadricostatus (Ritter) P.V. Heath
Pilosocereus fulvilanatus (Buining & Brederoo) Ritter .... BR
Pilosocereus fulvilanatus ssp. **fulvilanatus** .... BR
Pseudopilocereus fulvilanatus Buining & Brederoo
Pilosocereus fulvilanatus ssp. **rosae** (Braun) Zappi .... BR
Pilosocereus rosae Braun
Pilosocereus glaucochrous (Werdermann) Byles & Rowley .... BR
Pseudopilocereus glaucochrous (Werdermann) Buxbaum
Pilosocereus gounellei (Weber) Byles & Rowley .... BR
Pilosocereus gounellei ssp. **gounellei** .... BR
Cephalocereus gounellei (Weber) Britton & Rose
Pseudopilocereus gounellei (Weber) Buxbaum
Pilosocereus gounellei ssp. **zehntneri** (Britton & Rose) Zappi .... BR
Pilosocereus braunii Esteves Pereira
Pilosocereus superfloccosus (Buining & Brederoo) Ritter
Pseudopilocereus superfloccosus Buining & Brederoo
Cephalocereus zehntneri Britton & Rose
Pilosocereus zehntneri (Britton & Rose) Ritter
Pseudopilocereus zehntneri (Britton & Rose) P.V. Heath
Pilosocereus lanuginosus (Linnaeus) Byles & Rowley .... AN, VE, CO, EC, PE
Pilosocereus backebergii (Weingart) Byles & Rowley
Pilosocereus claroviridis (Backeberg) Byles & Rowley
Cephalocereus colombianus Rose
Pilosocereus colombianus (Rose) Byles & Rowley
Pilosocereus gironensis Rauh & Backeberg
Cephalocereus lanuginosus (Linnaeus) Britton & Rose
Cephalocereus moritzianus (Otto ex Pfeiffer) Britton & Rose
Pilosocereus moritzianus (Otto) Byles & Rowley
Pilosocereus tillianus Gruber & Schatzl
Pilosocereus tuberculosus Rauh & Backeberg
Cephalocereus tweedyanus Britton & Rose
Pilosocereus tweedyanus (Britton & Rose) Byles & Rowley
Pilosocereus leucocephalus (Poselger) Byles & Rowley .... MX, GT, HN
?*Cephalocereus cometes* (Scheidweiler) Britton & Rose
?*Pilosocereus cometes* (Scheidweiler) Byles & Rowley
Cephalocereus leucocephalus (Poselger) Britton & Rose
Cephalocereus maxonii Rose

Pilosocereus maxonii (Rose) Byles & Rowley

Cephalocereus palmeri Rose

Pilosocereus palmeri (Rose) Byles & Rowley

Cephalocereus sartorianus Rose

Pilosocereus sartorianus (Rose) Byles & Rowley

Pilosocereus tehuacanus (Weingart) Byles & Rowley

Pilosocereus machrisii (Dawson) Backeberg . . . . BR, PY

Pilosocereus cristalinensis Braun & Esteves Pereira

Pilosocereus densivillosus Braun & Esteves Pereira

Pilosocereus juaruensis (Buining & Brederoo) Braun

Pseudopilocereus juaruensis Buining & Brederoo

Pilosocereus lindanus Braun & Esteves Pereira

Pseudopilocereus machrisii (Dawson) Buxbaum

Pilosocereus paraguayensis Ritter

Pilosocereus parvus (Diers & Esteves Pereira) Braun

Pseudopilocereus parvus Diers & Esteves Pereira

Pilosocereus pusillibaccatus Braun & Esteves Pereira

Pilosocereus saudadensis Ritter

Pseudopilocereus saudadensis (Ritter) P.V. Heath

Ⓤ*Pilosocereus cuyabensis* (Backeberg) Byles & Rowley

Ⓤ*Pseudopilocereus cuyabensis* (Backeberg) Buxbaum

Ⓤ*Pilosocereus lindaianus* Braun & Esteves Pereira

Pilosocereus magnificus (Buining & Brederoo) Ritter . . . . BR

Pseudopilocereus magnificus Buining & Brederoo

Pilosocereus multicostatus Ritter . . . . BR

Pseudopilocereus multicostatus (Ritter) P.V. Heath

Pilosocereus oligolepis (Vaupel) Byles & Rowley . . . . GY, BR

Pilosocereus kanukuensis (Alexander) Leuenberger

Pseudopilocereus oligolepis (Vaupel) Buxbaum

Pilosocereus oligolepis ssp. *kanukuensis* (Alexander) Braun & Esteves Pereira

Pilosocereus pachycladus Ritter . . . . BR

Ⓜ*Pilosocereus glaucescens* (Linke) Byles & Rowley

Pilosocereus pachycladus ssp. **pachycladus** . . . . BR

Pilosocereus atroflavispinus Ritter

Pseudopilocereus atroflavispinus (Ritter) P.V. Heath

Pilosocereus azureus Ritter

Pseudopilocereus azureus Buining & Brederoo

Pilosocereus cenepequei Rizzini & Mattos

Pilosocereus cyaneus Ritter

Pilosocereus oreus Ritter

Pseudopilocereus oreus (Ritter) P.V. Heath

Pseudopilocereus pachycladus (Ritter) P.V. Heath

Pilosocereus schoebelii Braun

Pilosocereus splendidus Ritter

Pseudopilocereus splendidus (Ritter) P.V. Heath

Pilosocereus superbus Ritter

Pseudopilocereus superbus (Ritter) P.V. Heath

Pilosocereus pachycladus ssp. **pernambucoensis** (Ritter) Zappi . . . . BR

Pilosocereus pernambucoensis Ritter

Pseudopilocereus pernambucoensis (Ritter) P.V. Heath

Pilosocereus pentaedrophorus (Cels) Byles & Rowley . . . . BR

Pilosocereus pentaedrophorus ssp. **pentaedrophorus** . . . . BR

Cephalocereus pentaedrophorus (Cels) Britton & Rose

Pseudopilocereus pentaedrophorus (Cels) Buxbaum

Pilosocereus pentaedrophorus ssp. **robustus** Zappi . . . . BR

Pilosocereus piauhyensis (Guerke) Byles & Rowley . . . . BR

Pilosocereus gaturianensis Ritter

Pseudopilocereus gaturianensis (Ritter) P.V. Heath

Pilosocereus mucosiflorus (Buining & Brederoo) Ritter

Pseudopilocereus mucosiflorus Buining & Brederoo
Cephalocereus piauhyensis (Guerke) Britton & Rose
Pseudopilocereus piauhyensis (Guerke) Buxbaum
Pilosocereus piauhyensis ssp. *gaturianensis* (Ritter) Braun & Esteves Pereira
Pilosocereus piauhyensis ssp. *mucosiflorus* (Buining & Brederoo) Braun & Esteves Pereira

Pilosocereus polygonus (Lamarck) Byles & Rowley . . . . US, BS, CU, HT, DO
Cephalocereus bahamensis Britton & Rose
Pilosocereus bahamensis (Britton) Byles & Rowley
Pilosocereus brooksianus (Vaupel) Byles & Rowley
Pilosocereus deeringii (Small) Byles & Rowley
Cephalocereus keyensis Britton & Rose
Pilosocereus keyensis (Britton & Rose) Byles & Rowley
Cephalocereus polygonus (Lamarck) Britton & Rose
Cephalocereus robinii (Lemaire) Britton & Rose
Cereus robinii (Lemaire) L. Benson
Pilosocereus robinii (Lemaire) Byles & Rowley

Pilosocereus purpusii (Britton & Rose) Byles & Rowley . . . . MX
Cephalocereus purpusii Britton & Rose
?*Cephalocereus collinsii* Britton & Rose
?*Pilosocereus collinsii* (Britton & Rose) Byles & Rowley

Pilosocereus quadricentralis (Dawson) Backeberg . . . . MX
Cephalocereus quadricentralis Dawson

Pilosocereus royenii (Linnaeus) Byles & Rowley . . . . MX, BS, CU, KY, JM, DO, PR, VI, WI, TT
Cephalocereus barbadensis Britton & Rose
Pilosocereus barbadensis (Britton & Rose) Byles & Rowley
Cephalocereus brooksianus Britton & Rose
Cephalocereus gaumeri Britton & Rose
Pilosocereus gaumeri (Britton & Rose) Backeberg
Cephalocereus millspaughii Britton
Pilosocereus millspaughii (Britton) Byles & Rowley
Cephalocereus monoclonos (De Candolle) Britton & Rose
Pilosocereus monoclonos (De Candolle) Byles & Rowley
Cephalocereus royenii (Linnaeus) Britton & Rose
Cephalocereus swartzii (Grisebach) Britton & Rose
Pilosocereus swartzii (Grisebach) Byles & Rowley
Pilosocereus urbanianus (Schumann) Byles & Rowley
⊗*Cephalocereus nobilis* (Haworth) Britton & Rose
⊗*Pilosocereus nobilis* (Haworth) Byles & Rowley
⊗*Pseudopilocereus nobilis* (Haworth) Buxbaum

Pilosocereus* ×*subsimilis Rizzini & Mattos . . . . BR

Pilosocereus tuberculatus (Werdermann) Byles & Rowley . . . . BR
Pseudopilocereus tuberculatus (Werdermann) Buxbaum

Pilosocereus ulei (Schumann) Byles & Rowley . . . . BR
Cephalocereus robustus Britton & Rose
Pseudopilocereus ulei (Schumann) Buxbaum

Pilosocereus vilaboensis (Diers & Esteves Pereira) Braun . . . . BR
Pilosocereus rizzoanus Braun & Esteves Pereira
Pseudopilocereus vilaboensis Diers & Esteves Pereira

Polaskia chende (Roland-Gosselin) Gibson & Horak . . . . MX
Heliabravoa chende (Roland-Gosselin) Backeberg
Lemaireocereus chende (Roland-Gosselin) Britton & Rose
Myrtillocactus chende (Roland-Gosselin) P.V. Heath

Polaskia chichipe (Roland-Gosselin) Backeberg . . . . MX
Lemaireocereus chichipe (Roland-Gosselin) Britton & Rose
Myrtillocactus chichipe (Roalnd-Gosselin) P.V. Heath
Lemaireocereus mixtecensis (Purpus) Britton & Rose

Praecereus euchlorus (Weber) Taylor . . . . BR, PY, VE, CO, EC, PE, BO, AR
Praecereus euchlorus ssp. **euchlorus** . . . . BR, PY, BO, AR
Monvillea alticostata Ritter
Cereus alticostatus (Ritter) Braun
Cereus campinensis (Backeberg & Voll) Braun
Monvillea campinensis (Backeberg & Voll) Backeberg
Praecereus campinensis (Backeberg & Voll) Buxbaum
Cereus campinensis ssp. *piedadensis* (Ritter) Braun & Esteves Pereira
Monvillea euchlora (Weber) Backeberg
Cereus euchlorus Weber
Cereus euchlorus ssp. *alticostatus* (Ritter) Braun & Esteves Pereira
Cereus euchlorus ssp. *leucanthus* (Ritter) Braun & Esteves Pereira
Cereus lauterbachii Schumann ex Chodat & Hassler
Monvillea lauterbachii (Schumann ex Chodat & Hassler) Borg
Monvillea leucantha Ritter
Monvillea piedadensis Ritter
?®*Monvillea cavendishii* (Monville) Britton & Rose
Praecereus euchlorus ssp. **amazonicus** (Schumann ex Vaupel) Taylor . . . . PE, BO
Monvillea amazonica (Schumann ex Vaupel) Britton & Rose
Cereus amazonicus Schumann ex Vaupel
Praecereus amazonicus (Schumann ex Vaupel) Buxbaum
Cereus apoloensis (Cardenas) Braun & Esteves Pereira
Monvillea apoloensis Cardenas
Praecereus apoloensis (Cardenas) Buxbaum
Cereus ballivianii (Cardenas) Braun & Esteves Pereira
Monvillea ballivianii Cardenas
Praecereus euchlorus ssp. **diffusus** (Britton & Rose) Taylor . . . . EC, PE
Monvillea diffusa Britton & Rose
Cereus diffusus (Britton & Rose) Werdermann ex Backeberg
Monvillea maritima Britton & Rose
Praecereus maritimus (Britton & Rose) Buxbaum
Monvillea pugionifera Ritter
Praecereus euchlorus ssp. **jaenensis** (Rauh & Backeberg) Ostolaza . . . . PE
Monvillea jaenensis Rauh & Backeberg
Praecereus jaenensis (Rauh & Backeberg) Buxbaum
Praecereus euchlorus ssp. **smithianus** (Britton & Rose) Taylor . . . . VE, CO
Monvillea smithiana (Britton & Rose) Backeberg
Cephalocereus smithianus Britton & Rose
Cereus smithianus (Britton & Rose) Werdermann ex Backeberg
Praecereus smithianus (Britton & Rose) Buxbaum
Praecereus saxicola (Morong) Taylor . . . . PY, BO, AR
Monvillea chacoana Ritter
Cereus euchlorus ssp. *rhodoleucanthus* (Schumann) Braun & Esteves Pereira
Monvillea parapetiensis Ritter
Monvillea rhodoleucantha (Schumann) Berger
Cereus rhodoleucanthus Schumann
Cereus ritteri Braun & Esteves Pereira
Cereus ritteri ssp. *parapetiensis* (Ritter) Braun & Esteves Pereira
Cereus saxicola Morong
Monvillea saxicola (Morong) Berger

Pseudoacanthocereus brasiliensis (Britton & Rose) Ritter . . . . BR
Pseudoacanthocereus boreominarum Rizzini & Mattos
Acanthocereus brasiliensis Britton & Rose
Pseudoacanthocereus sicariguensis (Croizat & Tamayo) Taylor . . . . VE, CO
Acanthocereus sicariguensis Croizat & Tamayo

Pseudorhipsalis acuminata Cufodontis . . . . CR
Disocactus acuminatus (Cufodontis) Kimnach

Pseudorhipsalis alata (Swartz) Britton & Rose . . . . JM
Disocactus alatus (Swartz) Kimnach
Pseudorhipsalis himantoclada (Roland-Gosselin) Britton & Rose . . . . CR, PA
Wittia himantoclada (Roland-Goss.) R.E. Woodson
Disocactus himantocladus (Roland-Gosselin) Kimnach
Pseudorhipsalis horichii (Kimnach) Barthlott . . . . CR
Disocactus horichii Kimnach
Pseudorhipsalis lankesteri (Kimnach) Barthlott . . . . CR
Disocactus lankesteri Kimnach
Pseudorhipsalis ramulosa (Salm-Dyck) Barthlott
. . . . MX, GT, BZ, HN, NI, SV, CR, JM, HT, BR, CO, VE, EC, PE, BO
Rhipsalis angustissima Weber
Rhipsalis coriacea Polakowsky
Rhipsalis jamaicensis Britton & Harris
Rhipsalis leiophloea Vaupel
Rhipsalis purpusii Weingart
Rhipsalis ramulosa (Salm-Dyck) Pfeiffer
Disocactus ramulosus (Salm-Dyck) Kimnach

Pterocactus araucanus Castellanos . . . . AR
Pterocactus australis (Weber) Backeberg . . . . AR
Pterocactus fischeri Britton & Rose . . . . AR
Pterocactus gonjianii Kiesling . . . . AR
Pterocactus hickenii Britton & Rose . . . . AR
Opuntia skottsbergii Skottsberg ex Britton & Rose
Pterocactus skottsbergii (Britton & Rose) Backeberg
Pterocactus megliolii Kiesling . . . . AR
Pterocactus reticulatus Kiesling . . . . AR
Pterocactus tuberosus (Pfeiffer) Britton & Rose . . . . AR
Pterocactus decipiens Guerke
Pterocactus kuntzei Schumann
Pterocactus valentinii Spegazzini . . . . AR
Pterocactus pumilus Britton & Rose

Pygmaeocereus bieblii Diers . . . . PE
Pygmaeocereus bylesianus Andreae & Backeberg . . . . PE
Pygmaeocereus akersii Johnson ex Backeberg
Arthrocereus rowleyanus (Backeberg) Buxbaum
Pygmaeocereus rowleyanus Backeberg
Pygmaeocereus familiaris Ritter . . . . PE

Quiabentia verticillata (Vaupel) Vaupel . . . . PY, BO, AR
Quiabentia chacoensis Backeberg
Quiabentia pereziensis Backeberg
Pereskia pflanzii Vaupel
Quiabentia pflanzii (Vaupel) Vaupel
Pereskia verticillata Vaupel
Quiabentia zehntneri (Britton & Rose) Britton & Rose . . . . BR
Pereskia zehntneri Britton & Rose

Rauhocereus riosaniensis Backeberg . . . . PE
Rauhocereus riosaniensis ssp. **riosaniensis** . . . . PE
Browningia riosaniensis (Backeberg) Rowley
Rauhocereus riosaniensis ssp. **jaenensis** (Rauh ex Backeberg) Ostolaza . . . . PE

Rebutia albiflora Ritter & Buining . . . . BO
Aylostera albiflora (Ritter & Buining) Backeberg
Rebutia albopectinata Rausch . . . . BO
Lobivia albipectinata (Rausch) Neirinck
Rebutia schatzliana Rausch
Rebutia supthutiana Rausch

Rebutia arenacea Cardenas . . . . BO

Sulcorebutia arenacea (Cardenas) Ritter

Weingartia arenacea (Cardenas) Brandt

Rebutia candiae Cardenas

Sulcorebutia candiae (Cardenas) Buining & Donald

Weingartia candiae (Cardenas) Brandt

Rebutia glomeriseta Cardenas

Sulcorebutia glomeriseta (Cardenas) Ritter

Weingartia glomeriseta (Cardenas) Brandt

Rebutia menesesii Cardenas

Sulcorebutia menesesii (Cardenas) Buining & Donald

Weingartia menesesii (Cardenas) Brandt

Sulcorebutia muschii Vasquez

Weingartia muschii (Vasquez) Brandt

Ⓤ*Sulcorebutia xanthoantha* Backeberg

Rebutia aureiflora Backeberg . . . . AR

Mediolobivia aureiflora (Backeberg) Backeberg

Rebutia aureiflora ssp. *elegans* (Backeberg) Donald

Mediolobivia elegans Backeberg

Rebutia oculata Werdermann

Rebutia sarothroides Werdermann

Rebutia tilcarensis (Rausch) Sida

Ⓤ*Lobivia euanthema* Backeberg

Ⓤ*Mediolobivia euanthema* (Backeberg) Krainz

Ⓤ*Rebutia euanthema* (Backeberg) Buining & Donald

Rebutia brunescens Rausch . . . . BO

Mediolobivia brunescens (Rausch) W. Haage

Rebutia caineana Cardenas . . . . BO

Sulcorebutia caineana (Cardenas) Donald

Weingartia caineana (Cardenas) Brandt

Rebutia haseltonii Cardenas

Sulcorebutia haseltonii (Cardenas) Donald

Weingartia haseltonii (Cardenas) Brandt

Ⓤ*Sulcorebutia breviflora* Backeberg

Ⓤ*Weingartia breviflora* (Backeberg) Brandt

Rebutia canigueralii Cardenas . . . . BO

Sulcorebutia alba Rausch

Weingartia alba (Rausch) Brandt

Sulcorebutia albaoides (Brandt) Pilbeam

Weingartia albaoides Brandt

Weingartia albaoides ssp. *subfusca* Brandt

Weingartia aureispina (Rausch) Brandt

Sulcorebutia brevispina (Brandt) Pilbeam

Weingartia brevispina Brandt

Sulcorebutia callecallensis (Brandt) Pilbeam

Sulcorebutia canigueralii (Cardenas) Buining & Donald

Weingartia canigueralii (Cardenas) Brandt

Rebutia caracarensis Cardenas

Sulcorebutia caracarensis (Cardenas) Donald

Weingartia caracarensis (Cardenas) Brandt

Sulcorebutia fischeriana Augustin

Sulcorebutia frankiana Rausch

Weingartia frankiana (Rausch) Brandt

Rebutia inflexiseta Cardenas

Sulcorebutia inflexiseta (Cardenas) Donald

Weingartia inflexiseta (Cardenas) Brandt

Sulcorebutia losenickyana Rausch

Weingartia losenickyana (Rausch) Brandt

Sulcorebutia pasopayana (Brandt) W. Gertel

Weingartia pasopayana Brandt
Sulcorebutia perplexiflora (Brandt) Pilbeam
Weingartia perplexiflora Brandt
Sulcorebutia rauschii Frank
Weingartia rauschii (Frank) Brandt
Sulcorebutia ritteri (Brandt) Ritter
Weingartia ritteri Brandt
Weingartia rubro-aurea Brandt
Sulcorebutia rubroaurea (Brandt) Pilbeam
Weingartia saxatilis Brandt
Weingartia tarabucina Brandt
Sulcorebutia tarabucoensis Rausch
Weingartia tarabucoensis (Rausch) Brandt
Sulcorebutia vasqueziana Rausch
Aylostera zavaletae Cardenas
Sulcorebutia zavaletae (Cardenas) Backeberg
Weingartia zavaletae (Cardenas) Brandt
®*Rebutia rauschii* (G. Frank) Hunt
Rebutia canigueralii ssp. **crispata** (Rausch) Donald . . . . BO
Sulcorebutia crispata Rausch
Weingartia crispata (Rausch) Brandt
Rebutia canigueralii ssp. **pulchra** (Cardenas) Donald ex Hunt . . . . BO
Rebutia pulchra Cardenas
Sulcorebutia pulchra (Cardenas) Donald
Weingartia pulchra (Cardenas) Brandt
Rebutia cardenasiana (R. Vasquez) G. Navarro . . . . BO
Sulcorebutia cardenasiana Vasquez
Weingartia cardenasiana (Vasquez) Brandt
Rebutia cylindrica (Donald & Lau) Hunt . . . . BO
Sulcorebutia cylindrica Donald & Lau
Weingartia cylindrica (Donald) Brandt
Rebutia deminuta (Weber) Britton & Rose . . . . AR
Aylostera deminuta (Weber) Backeberg
Aylostera pseudominuscula (Spegazzini) Spegazzini
Rebutia pseudominuscula (Spegazzini) Britton & Rose
Rebutia einsteinii Fric . . . . BO, AR
Lobivia auranitida Wessner
Mediolobivia auranitida (Wessner) Krainz
Rebutia auranitida (Wessner) Buining & Donald
Mediolobivia conoidea (Wessner) Krainz
Lobivia einsteinii (Fric ex Kreuzinger & Buining) Rausch
Mediolobivia schmiedcheniana (Kohler) Krainz
?*Mediolobivia neopygmaea* Backeberg
Rebutia fabrisii Rausch . . . . AR
Rebutia fidaiana (Backeberg) Hunt . . . . BO
Sulcorebutia lecoriensis (Cardenas) Brandt
Weingartia lecoriensis Cardenas
Sulcorebutia vilcayensis (Cardenas) Brandt
Weingartia vilcayensis Cardenas
Sulcorebutia westii (Hutchison) Brandt
Weingartia westii (Hutchison) Donald
Rebutia fidaiana ssp. **fidaiana** . . . . BO
Spegazzinia fidaiana Backeberg
Sulcorebutia fidaiana (Backeberg) Brandt
Weingartia fidaiana (Backeberg) Werdermann
Rebutia fidaiana ssp. **cintiensis** (Cardenas) Hunt . . . . BO
Sulcorebutia cintiensis (Cardenas) Brandt
Weingartia cintiensis Cardenas
Weingartia fidaiana ssp. *cintiensis* (Cardenas) Donald

Rebutia fiebrigii (Guerke) Britton & Rose ex Bailey .... BO, AR
Aylostera albipilosa (Ritter) Backeberg
Rebutia albipilosa Ritter
Rebutia cajasensis Ritter
Rebutia cintiensis Ritter
Rebutia donaldiana Lau & Rowley
Aylostera fiebrigii (Guerke) Backeberg
Mediolobivia ithyacantha Cardenas
Rebutia ithyacantha (Cardenas) Diers & Esteves Pereira
Aylostera jujuyana (Rausch) W. Haage
Rebutia jujuyana Rausch
Rebutia kieslingii Rausch
Aylostera muscula (Ritter & Thiele) Backeberg
Rebutia muscula Ritter & Thiele
Aylostera pulchella (Rausch) W. Haage
Rebutia pulchella Rausch
Rebutia tamboensis Ritter
Rebutia vallegrandensis Cardenas
Rebutia flavistyla Ritter .... BO
Rebutia fulviseta Rausch .... BO
Aylostera fulviseta (Rausch) W. Haage
Rebutia gonjianii Kiesling .... AR
Rebutia heliosa Rausch .... BO
Aylostera heliosa (Rausch) W. Haage
Rebutia huasiensis Rausch .... BO
Rebutia leucanthema Rausch .... BO
Rebutia marsoneri Werdermann .... AR
Rebutia krainziana Kesselring
®*Rebutia hyalacantha* (Backeberg) Backeberg
Rebutia mentosa (Ritter) Donald .... BO
Sulcorebutia albissima (Brandt) Pilbeam
Weingartia albissima Brandt
Sulcorebutia augustinii Hentzschel
Sulcorebutia flavida (Brandt) Pilbeam
Sulcorebutia flavissima Rausch
Weingartia flavissima (Rausch) Brandt
Sulcorebutia formosa (Brandt) Pilbeam
Weingartia formosa Brandt
Sulcorebutia markusii Rausch
Weingartia markusii (Rausch) Brandt
Sulcorebutia santiaginensis Rausch
Sulcorebutia swobodae Augustin
Sulcorebutia torotorensis (Cardenas) Brandt
Sulcorebutia torotorensis (Cardenas) Brederoo & Donald
Weingartia torotorensis Cardenas
Sulcorebutia unguispina Rausch
Rebutia mentosa ssp. **mentosa** .... BO
Sulcorebutia mentosa Ritter
Weingartia mentosa (Ritter) Brandt
Rebutia mentosa ssp. **purpurea** (Donald & Lau) Donald ex Hunt .... BO
Sulcorebutia purpurea (Donald & Lau) Brederoo & Donald
Weingartia purpurea Donald & Lau
Rebutia minuscula Schumann .... AR
Rebutia carminea Buining
Rebutia chrysacantha Backeberg
Rebutia graciliflora Backeberg
Rebutia grandiflora Backeberg
Rebutia grandilacea hort. ex Pickoff
Rebutia kariusiana Wessner

Rebutia minuscula ssp. *grandiflora* (Backeberg) Donald
Rebutia minuscula ssp. *violaciflora* (Backeberg) Donald
Rebutia senilis Backeberg
Rebutia senilis ssp. *chrysacantha* (Backeberg) Donald
Rebutia violaciflora Backeberg

Rebutia narvaecensis (Cardenas) Donald . . . . BO
Aylostera narvaecensis Cardenas

Rebutia neocumingii (Backeberg) Hunt . . . . PE, BO
Weingartia attenuata Brandt
Weingartia brachygraphisa Brandt
Weingartia buiningiana Ritter
Weingartia callecallensis Brandt
Weingartia columnaris Brandt
Rebutia corroana Cardenas
Sulcorebutia corroana (Cardenas) Brederoo & Donald
Weingartia corroana (Cardenas) Cardenas
Sulcorebutia erinacea (Ritter) Brandt
Weingartia erinacea Ritter
Weingartia flavida Brandt
Weingartia gracilispina Ritter
Sulcorebutia hediniana (Backeberg) Brandt
Weingartia hediniana Backeberg
Weingartia knizei Brandt
Weingartia mairanana Brandt
Weingartia mataralensis Brandt
Weingartia miranda Brandt
Sulcorebutia multispina (Ritter) Brandt
Weingartia multispina Ritter
Weingartia neglecta Brandt
Sulcorebutia neocorroana Brandt
Sulcorebutia platygona (Cardenas) Brandt
Weingartia platygona Cardenas
Weingartia saetosa Brandt
⊗*Weingartia ambigua* (Hildmann ex Schumann) Backeberg

Rebutia neocumingii ssp. **neocumingii** (Backeberg) Hunt . . . . BO
Lobivia cumingii (Salm-Dyck) Britton & Rose
Sulcorebutia neocumingii (Backeberg) Brandt
Weingartia neocumingii Backeberg
Weingartia neocumingii ssp. *pulquinensis* (Cardenas) Donald
Weingartia neocumingii ssp. *sucrensis* (Ritter) Donald
Sulcorebutia pulquinensis (Cardenas) Brandt
Weingartia pulquinensis Cardenas
Sulcorebutia sucrensis (Ritter) Brandt
Weingartia sucrensis Ritter

Rebutia neocumingii ssp. **riograndensis** (Ritter) Hunt . . . . BO
Sulcorebutia lanata (Ritter) Brandt
Weingartia lanata Ritter
Weingartia lanata ssp. *riograndensis* (Ritter) Donald
Sulcorebutia riograndensis (Ritter) Brandt
Weingartia riograndensis Ritter

Rebutia neocumingii ssp. **pilcomayensis** (Cardenas) Hunt . . . . BO
Weingartia lanata ssp. *longigibba* (Ritter) Donald
Weingartia lanata ssp. *pilcomayensis* (Cardenas) Donald
Sulcorebutia longigibba (Ritter) Brandt
Weingartia longigibba Ritter
Sulcorebutia pilcomayensis (Cardenas) Brandt
Weingartia pilcomayensis Cardenas

Rebutia neocumingii ssp. **saipinensis** (Brandt) Hunt . . . . BO
Weingartia saipinensis Brandt

Rebutia neocumingii ssp. **trollii** (Oeser) Hunt . . . . BO
Weingartia trollii Oeser
Rebutia neumanniana (Werdermann) Hunt . . . . AR
Weingartia kargliana Rausch
Sulcorebutia neumanniana (Backeberg) Brandt
Weingartia neumanniana Werdermann
Gymnocalycium neumannianum (Backeberg) Hutchison
Weingartia pygmaea Ritter
Rebutia nigricans (Wessner) Hunt . . . . AR
Rebutia albispina (Backeberg) Sida
Rebutia carmeniana Rausch
Lobivia nigricans Wessner
Mediolobivia nigricans (Wessner) Krainz
Rebutia oligacantha (Brandt) Donald ex Hunt . . . . BO
Sulcorebutia oligacantha (Brandt) Pilbeam
Weingartia oligacantha Brandt
Weingartia sanguineo-tarijensis Brandt
Sulcorebutia sanguineotarijensis (Brandt) Pilbeam
Sulcorebutia tarijensis Ritter
Weingartia tarijensis Brandt
Rebutia padcayensis Rausch . . . . BO
Rebutia margarethae Rausch
Weingartia margarethae (Rausch) Brandt
Aylostera padcayensis (Rausch) W. Haage
Rebutia singularis Ritter
Rebutia perplexa Donald . . . . BO
Rebutia pseudodeminuta Backeberg . . . . BO, AR
Rebutia albiareolata Ritter
Rebutia buiningiana Rausch
Rebutia minutissima Ritter
Rebutia nitida Ritter
Rebutia nogalesensis Ritter
Aylostera pseudodeminuta (Backeberg) Backeberg
Rebutia robustispina Ritter
Rebutia sanguinea Ritter
Rebutia wahliana Rausch
?*Aylostera kupperiana* (Boedeker) Backeberg
?*Rebutia kupperiana* Boedeker
Rebutia pulvinosa Ritter & Buining . . . . BO
Aylostera pulvinosa (Ritter & Buining) Backeberg
Rebutia pygmaea (R. Fries) Britton & Rose . . . . BO, AR
Rebutia canacruzensis Rausch
Rebutia colorea Ritter
Rebutia crassa (Rausch) Sida
Rebutia diersiana Rausch
Rebutia elegantula (Rausch) Sida
Mediolobivia eos (Rausch) W. Haage
Rebutia eos Rausch
Rebutia friedrichiana Rausch
Rebutia gracilispina Ritter
Lobivia haagei (Fric & Schelle) Wessner
Rebutia haagei Fric & Schelle
Mediolobivia haefneriana Cullmann
Rebutia haefneriana (Cullmann) Sida
Rebutia iridescens Ritter
Rebutia iscayachensis Rausch
Rebutia knizei (Rausch) Sida
Rebutia lanosiflora Ritter
Rebutia mixta Ritter

Rebutia mixticolor Ritter
Rebutia mudanensis Rausch
Rebutia nazarenoensis (Rausch) B. Fearn & L. Pearcy
Lobivia neohaageana Backeberg
Rebutia odontopetala Ritter
Lobivia orurensis Backeberg
Rebutia orurensis (Backeberg) Ritter ex Sida
Rebutia pallida Rausch
Rebutia pauciareolata Ritter
Rebutia paucicostata Ritter
Lobivia pectinata Backeberg
Mediolobivia pectinata (Backeberg) Backeberg ex Krainz
Rebutia pelzliana (Rausch) Sida
Rebutia polypetala (Rausch) Sida
Rebutia pseudoritteri (Rausch) Sida
Mediolobivia pygmaea (Fries) Krainz
Rebutia rosalbiflora Ritter
Rebutia rutiliflora Ritter
Rebutia salpingantha Ritter
Rebutia torquata Ritter & Buining
Rebutia tropaeolipicta Ritter
Rebutia villazonensis Brandt
Rebutia violacistaminea (Rausch) Sida
Rebutia violascens Ritter
Rebutia yuncharasensis (Rausch) Sida
Rebutia yuquinensis Rausch
?*Lobivia atrovirens* Backeberg
?*Rebutia atrovirens* (Backeberg) Sida

Rebutia ritteri (Wessner) Buining & Donald . . . . BO
Rebutia raulii Rausch
Lobivia ritteri Wessner
Mediolobivia ritteri (Wessner) Krainz

Rebutia simoniana Rausch . . . . BO

Rebutia spegazziniana Backeberg . . . . BO, AR
Rebutia froehlichiana Rausch
Rebutia fusca Ritter
Rebutia mamillosa Rausch
Rebutia patericalyx Ritter
Aylostera rubiginosa (Ritter) Backeberg
Rebutia rubiginosa Ritter
Aylostera spegazziniana (Backeberg) Backeberg
Rebutia sumayana Rausch
Rebutia tarijensis Rausch
Rebutia tarvitaensis Ritter
Aylostera tuberosa (Ritter) Backeberg
Rebutia tuberosa Ritter
Rebutia vulpina Ritter
Rebutia zecheri Rausch

Rebutia spinosissima Backeberg . . . . BO, AR
Rebutia archibuiningiana Ritter
Rebutia hoffmannii Diers & Rausch
Aylostera spinosissima (Backeberg) Backeberg
Rebutia walteri Diers & Esteves Pereira

Rebutia steinbachii Werdermann . . . . BO
Weingartia aglaia Brandt
Weingartia ansaldoensis Brandt
Weingartia backebergiana Brandt
Sulcorebutia clavata (Brandt) Pilbeam
Weingartia clavata Brandt

Sulcorebutia cochabambina Rausch
Sulcorebutia croceareolata (Brandt) Pilbeam
Weingartia croceareolata Brandt
Weingartia electracantha (Backeberg) Brandt
Rebutia glomerispina Cardenas
Sulcorebutia glomerispina (Cardenas) Buining & Donald
Weingartia glomerispina (Cardenas) Brandt
Sulcorebutia krahnii Rausch
Weingartia krahnii (Rausch) Brandt
Sulcorebutia lepida Ritter
Weingartia lepida (Ritter) Brandt
Sulcorebutia mariana Swoboda
Weingartia minima (Rausch) Brandt
Sulcorebutia mizquensis Rausch
Weingartia mizquensis (Rausch) Brandt
Weingartia nigro-fuscata Brandt
Sulcorebutia nigrofuscata (Brandt) Pilbeam
Sulcorebutia oenantha Rausch
Weingartia oenantha (Rausch) Brandt
Sulcorebutia pampagrandensis Rausch
Weingartia pampagrandensis (Rausch) Brandt
Rebutia polymorpha Cardenas
Sulcorebutia polymorpha (Cardenas) Backeberg
Weingartia polymorpha (Cardenas) Brandt
Sulcorebutia steinbachii ssp. *australis* (Rausch) W. Gertel
Rebutia taratensis Cardenas
Sulcorebutia taratensis (Cardenas) Buining & Donald
Weingartia taratensis (Cardenas) Brandt
Sulcorebutia totoralensis (Brandt) Pilbeam
Weingartia totoralensis Brandt
Sulcorebutia totorensis (Cardenas) Ritter
Weingartia totorensis (Cardenas) Brandt
Rebutia tuberculato-chrysantha Cardenas
Sulcorebutia tuberculato-chrysantha (Cardenas) Brederoo & Donald
Weingartia tuberculato-chrysantha (Cardenas) Brandt
Rebutia tunariensis Cardenas
Sulcorebutia tunariensis (Cardenas) Buining & Donald
Weingartia tunariensis (Cardenas) Brandt
Rebutia vizcarrae Cardenas
Sulcorebutia vizcarrae (Cardenas) Donald
Weingartia vizcarrae (Cardenas) Brandt
®*Lobivia hoffmanniana* Backeberg
®*Sulcorebutia hoffmanniana* (Backeberg) Backeberg
®*Weingartia hoffmanniana* (Backeberg) Brandt

Rebutia steinbachii ssp. **steinbachii** . . . . BO
Sulcorebutia steinbachii (Werdermann) Backeberg
Weingartia steinbachii (Werdermann) Brandt

Rebutia steinbachii ssp. **kruegeri** (Cardenas) Hunt . . . . BO
Rebutia kruegeri (Cardenas) Backeberg
Sulcorebutia kruegeri (Cardenas) Ritter
Weingartia kruegeri (Cardenas) Brandt
®*Sulcorebutia steinbachii* ssp. *kruegeri* (Cardenas) Gertel

Rebutia steinbachii ssp. **tiraquensis** (Cardenas) Hunt . . . . BO
Sulcorebutia steinbachii ssp. *tiraquensis* (Cardenas) Gertel
Rebutia tiraquensis Cardenas
Sulcorebutia tiraquensis (Cardenas) Ritter
Weingartia tiraquensis (Cardenas) Brandt

Rebutia steinbachii ssp. **verticillacantha** (Ritter) Donald ex Hunt . . . . BO
Sulcorebutia steinbachii ssp. *verticillacantha* (Ritter) Gertel

Sulcorebutia verticillacantha Ritter
Weingartia verticillacantha (Ritter) Brandt
Rebutia steinmannii (Solms-Laubach) Britton & Rose . . . . BO
Rebutia applanata (Rausch) Sida
Lobivia brachyantha Wessner
Mediolobivia brachyantha (Wessner) Krainz
Rebutia brachyantha (Wessner) Buining & Donald
Rebutia brunneoradicata Ritter
Rebutia camargoensis Rausch
Rebutia christinae Rausch
Rebutia cincinnata Rausch
Mediolobivia costata (Werdermann) Krainz
Rebutia costata Werdermann
Lobivia eucaliptana Backeberg
Mediolobivia eucaliptana (Backeberg) Krainz
Rebutia eucaliptana (Backeberg) Ritter
Rebutia leucacantha (Rausch) Sida
Rebutia major (Rausch) Sida
Rebutia melanocentra (Rausch) Sida
Rebutia parvula (Rausch) Sida
Rebutia poecilantha Ritter
Rebutia potosina Ritter
Rebutia rauschii Zecher
Aylostera steinmannii (Solms-Laubach) Backeberg
Lobivia steinmannii ssp. *melanocentra* Rausch
Rebutia tuberculata (Rausch) Sida
Rebutia wessneriana Bewerunge . . . . AR
Rebutia calliantha Bewerunge
Rebutia wessneriana ssp. *beryllioides* (Buining & Donald) Donald
Rebutia xanthocarpa Backeberg . . . . AR

Rhipsalis baccifera (J.S. Mueller) Stearn . . . . US, MX, GT, BZ, HN, NI, CR, PA, CU,
. . . . JM, HT, DO, PR, WI, TT, AN, GY, GF, SR, BR, PY,
. . . . CO, VE, EC, PE, BO, AR, OW
Rhipsalis baccifera ssp. *fasciculata* (Willdenow) Suepplie
Rhipsalis baccifera ssp. *rhodocarpa* (Weber) Suepplie
Rhipsalis bartlettii Clover
Rhipsalis cassutha Gaertner
Rhipsalis cassuthopsis Backeberg
Rhipsalis cassytha auctt
Rhipsalis cassythoides G. Don
Rhipsalis fasciculata (Willdenow) Haworth
Rhipsalis heptagona Rauh & Backeberg
Rhipsalis hylaea Ritter
Rhipsalis minutiflora Schumann
Rhipsalis undulata Pfeiffer
®*Rhipsalis quellebambensis* Johnson ex Backeberg
Rhipsalis baccifera ssp. **baccifera** . . . . US, MX, GT, BZ, HN, NI, CR, PA, CU,
. . . . JM, HT, DO, PR, WI, TT, AN, GY, GF, SR, BR, CO, VE, EC, PE
Rhipsalis baccifera ssp. **erythrocarpa** (Schumann) Barthlott . . . . OW
Rhipsalis erythrocarpa Schumann
Rhipsalis baccifera ssp. **hileiabaiana** Taylor & Barthlott . . . . BR
Rhipsalis baccifera ssp. **horrida** (Baker) Barthlott . . . . OW
Rhipsalis coralloides Rauh
Rhipsalis horrida Baker
Rhipsalis madagascariensis Weber
Rhipsalis pilosa Weber ex Schumann
Rhipsalis baccifera ssp. **mauritiana** (De Candolle) Barthlott . . . . OW
Rhipsalis baccifera ssp. *fortdauphinensis* F. Suepplie
Rhipsalis baccifera ssp. **shaferi** (Britton & Rose) Barthlott & Taylor . BR, PY, BO, AR
Rhipsalis shaferi Britton & Rose

Rhipsalis burchellii Britton & Rose . . . . BR
Erythrorhipsalis burchellii (Britton & Rose) Volgin
Rhipsalis campos-portoana Loefgren . . . . BR
Erythrorhipsalis campos-portoana (Loefgren) Volgin
Rhipsalis cereoides (Backeberg & Voll) Backeberg . . . . BR
Rhipsalis cereuscula Haworth . . . . BR, PY, BO, UY, AR
Erythrorhipsalis cereuscula (Haworth) Volgin
Rhipsalis penduliflora N.E. Brown
Rhipsalis simmleri Beauverd
Rhipsalis clavata Weber . . . . BR
Rhipsalis crispata (Haworth) Pfeiffer . . . . BR
Rhipsalis cuneata Britton & Rose . . . . BO
Rhipsalis dissimilis (Lindberg) Schumann . . . . BR
?*Rhipsalis chrysantha* Loefgren
Lepismium dissimile Lindberg
Lepismium epiphyllanthoides (Backeberg) Backeberg
Rhipsalis epiphyllanthoides Backeberg
Rhipsalis rigida Loefgren
Lepismium saxatile A. Friedrich & Redecker
Rhipsalis saxatilis (A. Friedrich & Redecker) A. Friedrich & Redecker
Rhipsalis spinescens J.A. Lombardi
?*Lepismium chrysanthum* (Loefgren) Backeberg
Rhipsalis elliptica Lindberg ex Schumann . . . . BR
Rhipsalis chloroptera Weber
Rhipsalis ewaldiana Barthlott & Taylor . . . . BR
Rhipsalis floccosa Salm-Dyck ex Pfeiffer . . . . BR, PY, VE, PE, BO, AR
Rhipsalis floccosa ssp. **floccosa** . . . . BR
Lepismium floccosum (Salm-Dyck) Backeberg
Rhipsalis floccosa ssp. **hohenauensis** (Ritter) Barthlott & Taylor . . . . PY, AR
Rhipsalis hohenauensis Ritter
Rhipsalis floccosa ssp. **oreophila** Taylor & Zappi . . . . BR
Rhipsalis monteazulensis Ritter
Rhipsalis floccosa ssp. **pittieri** (Britton & Rose) Barthlott & Taylor . . . . VE
Lepismium pittieri (Britton & Rose) Backeberg
Rhipsalis pittieri Britton & Rose
Rhipsalis floccosa ssp. **pulvinigera** (Lindberg) Barthlott & Taylor . . . . BR
Rhipsalis flosculosa Ritter
Rhipsalis gibberula Weber
Lepismium gibberulum (Weber) Backeberg
Rhipsalis pulvinigera Lindberg
Lepismium pulvinigerum (Lindberg) Backeberg
Rhipsalis floccosa ssp. **tucumanensis** (Weber) Barthlott & Taylor . . . . . . AR, BO, PE
Lepismium tucumanense (Weber) Backeberg
Rhipsalis tucumanensis Weber
Rhipsalis goebeliana Backeberg . . . . BO
Rhipsalis grandiflora Haworth . . . . BR
Rhipsalis fastigiata Hjelmquist
Lepismium grandiflorum (Haworth) Backeberg
Rhipsalis hadrosoma Lindberg
Rhipsalis hoelleri Barthlott & Taylor . . . . BR
Rhipsalis juengeri Barthlott & Taylor . . . . BR
Rhipsalis lindbergiana Schumann . . . . BR
Rhipsalis densiareolata Loefgren
Rhipsalis mesembryanthemoides Haworth . . . . BR
®*Rhipsalis mesembryanthoides* Haworth
Rhipsalis micrantha (Kunth) De Candolle . . . . GT, CR, CO, EC, PE
Rhipsalis kirbergii Barthlott
Rhipsalis rauhiorum Barthlott
Rhipsalis roseana Berger
Rhipsalis tonduzii Weber
Rhipsalis wercklei Berger

Rhipsalis neves-armondii Schumann . . . . BR

Rhipsalis megalantha Loefgren

Lepismium megalanthum (Loefgren) Backeberg

Lepismium neves-armondii (Schumann) Backeberg

Rhipsalis oblonga Loefgren . . . . BR

Rhipsalis crispimarginata Loefgren

Rhipsalis occidentalis Barthlott . . . . SR, EC, PE

Rhipsalis olivifera Taylor & Zappi . . . . BR

Rhipsalis ormindoi Taylor & Zappi . . . . BR

Rhipsalis pacheco-leonis Loefgren . . . . BR

Rhipsalis pacheco-leonis ssp. **pacheco-leonis** . . . . BR

Lepismium pacheco-leonii (Loefgren) Backeberg

Rhipsalis pacheco-leonis ssp. **catenulata** (Kimnach) Barthlott & Taylor . . . . BR

Rhipsalis pachyptera Pfeiffer . . . . BR

Rhipsalis dusenii Hjelmquist

Rhipsalis robusta Lemaire

Rhipsalis paradoxa (Salm-Dyck ex Pfeiffer) Salm-Dyck . . . . BR

Rhipsalis paradoxa ssp. **paradoxa** . . . . BR

Lepismium paradoxum Salm-Dyck ex Pfeiffer

Rhipsalis paradoxa ssp. **septentrionalis** Taylor & Barthlott . . . . BR

Rhipsalis pentaptera Pfeiffer . . . . BR

Rhipsalis pilocarpa Loefgren . . . . BR

Erythrorhipsalis pilocarpa (Loefgren) Berger

Rhipsalis pulchra Loefgren . . . . BR

Rhipsalis puniceodiscus Lindberg . . . . BR

Rhipsalis chrysocarpa Loefgren

Lepismium puniceodiscus (Lindberg) Backeberg

Rhipsalis russellii Britton & Rose . . . . BR

Rhipsalis sulcata Weber . . . . BR

Rhipsalis teres (Vellozo) Steudel . . . . BR

Rhipsalis alboareolata Ritter

Rhipsalis capilliformis Weber

Rhipsalis clavellina Ritter

Rhipsalis heteroclada Britton & Rose

Rhipsalis maricaensis Scheinvar

Rhipsalis prismatica (Lemaire) Ruempler

Rhipsalis virgata Weber

Ⓜ*Erythrorhipsalis cribrata* (Lemaire) Volgin

Ⓜ*Rhipsalis cribrata* (Lemaire) Pfersdorf ex N.E. Brown

Rhipsalis trigona Pfeiffer . . . . BR

Lepismium trigonum (Pfieiffer) Backeberg

Samaipaticereus corroanus Cardenas . . . . BO

Schlumbergera ×buckleyi (T. Moore) Tjaden XC

Schlumbergera ×exotica Barthlott & Rauh XC

Schlumbergera kautskyi (Horobin & McMillan) Taylor . . . . BR

Schlumbergera truncata ssp. *kautskyi* Horobin & McMillan

Schlumbergera microsphaerica (Schumann) Hoevel . . . . BR

Epiphyllanthus microsphaericus (Schumann) Britton & Rose

Schlumbergera obtusangula (Schumann) Hunt

Epiphyllanthus obtusangulus (Schumann) Berger

Arthrocereus microsphaericus (Schumann) Berger

Schlumbergera microsphaerica ssp. ***candida*** (Loefgren) Hunt . . . . BR

Schlumbergera candida (Loefgren) Hoevel

Epiphyllanthus candidus (Loefgren) Britton & Rose

Schlumbergera opuntioides (Loefgren & Dusen) Hunt . . . . BR

Ⓜ*Epiphyllanthus obovatus* (Engelmann ex Schumann) Britton & Rose

Schlumbergera orssichiana Barthlott & McMillan . . . . BR

Schlumbergera ×reginae XC
Schlumbergera russelliana (Hooker) Britton & Rose . . . . BR
   ⊛*Schlumbergera bridgesii* auctt
Schlumbergera truncata (Haworth) Moran . . . . BR
   Epiphyllum bridgesii Lemaire
   Zygocactus truncatus (Haworth) Schumann

Sclerocactus brevihamatus (Engelmann) Hunt . . . . US
Sclerocactus brevihamatus ssp. **brevihamatus** . . . . US
   Ancistrocactus brevihamatus (Engelmann) Britton & Rose
   Pediocactus brevihamatus (Engelmann) Halda
Sclerocactus brevihamatus ssp. **tobuschii** (Marshall) Taylor . . . . US
   Pediocactus brevihamatus ssp. *tobuschii* (Marshall) Halda
   Ancistrocactus tobuschii (W.T. Marshall) Backeberg ex Benson♦
   Ferocactus tobuschii (W.T. Marshall) Taylor♦
Sclerocactus erectocentrus (Coulter) Taylor . . . . ♦US, MX
   Echinomastus acunensis W.T. Marshall♦
   Neolloydia erectocentra (Coulter) Benson♦
   Echinomastus erectocentrus (Coulter) Britton & Rose♦
   Pediocactus erectocentrus (Coulter) Halda♦
   Echinomastus krausei (Hildmann) Borg♦
   ⊛*Gymnocactus beguinii* (Weber ex Schumann) Backeberg♦
   ⊛*Neolloydia beguinii* (Weber ex Schumann) Britton & Rose♦
Sclerocactus glaucus (Schumann) Benson . . . . ♦US
   Sclerocactus brevispinus Heil & Porter♦
   Ferocactus glaucus (Schumann) Taylor♦
   Pediocactus glaucus (Schumann) Arp♦
   Pediocactus wetlandicus (Hochstaetter) Halda♦
   Sclerocactus wetlandicus Hochstaetter♦
   Sclerocactus wetlandicus ssp. *ilseae* (Hochstaetter) Hochstaetter♦
Sclerocactus intertextus (Engelmann) Taylor . . . . US, MX
   Echinomastus dasyacanthus (Engelmann) Britton & Rose
   Neolloydia intertexta (Engelmann) Benson
   Echinomastus intertextus (Engelmann) Britton & Rose
   Pediocactus intertextus (Engelmann) Halda
Sclerocactus johnsonii (Engelmann) Taylor . . . . US
   Echinomastus johnsonii (Engelmann) Baxter
   Ferocactus johnsonii (Engelmann) Britton & Rose
   Neolloydia johnsonii (Engelmann) L. Benson
   Pediocactus johnsonii (Engelmann) Halda
Sclerocactus mariposensis (Hester) Taylor . . . . ♦US, MX
   Echinomastus mariposensis Hester♦
   Neolloydia mariposensis (Hester) Benson♦
   Pediocactus mariposensis (Hester) Halda♦
Sclerocactus mesae-verdae (Boissevain & Davidson ex Marshall & Bock) Benson
. . . . ♦US
   Coloradoa mesae-verdae Boissevain & Davidson ex Marshall & Bock♦
   Ferocactus mesae-verdae (Boissevain & Davidson ex Marshall & Bock) Taylor♦
   Pediocactus mesae-verdae (Boissevain & Davidson ex Marshall & Bock) Arp♦
Sclerocactus nyensis Hochstaetter . . . . US
   Pediocactus nyensis (Hochstaetter) Halda
Sclerocactus papyracanthus (Engelmann) Taylor . . . . ♦US, MX
   Toumeya papyracantha (Engelmann) Britton & Rose♦
   Pediocactus papyracanthus (Engelmann) Benson♦
Sclerocactus parviflorus Clover & Jotter . . . . US
   Pediocactus cloverae (Heil & Porter) Halda
   Sclerocactus cloverae Heil & Porter
   Pediocactus cloverae ssp. *brackii* (Heil & Porter) Halda
   Sclerocactus cloverae ssp. *brackii* Heil & Porter

Sclerocactus parviflorus ssp. **parviflorus** . . . . US
Sclerocactus contortus Heil
Ferocactus parviflorus (Clover & Jotter) Taylor
Pediocactus parviflorus (Clover & Jotter) Halda
Sclerocactus parviflorus ssp. ***havasupaiensis*** (Clover) Hochstaetter . . . . US
Sclerocactus havasupaiensis Clover
Pediocactus parviflorus ssp. *havasupaiensis* (Clover) Halda
Sclerocactus parviflorus ssp. ***intermedius*** (Peebles) Heil & Porter . . . . US
Sclerocactus intermedius Peebles
Pediocactus parviflorus ssp. *intermedius* (Peebles) Halda
Sclerocactus parviflorus ssp. ***terrae-canyonae*** (Heil) Heil & Porter . . . . US
Pediocactus parviflorus ssp. *terrae-canyonae* (Heil) Halda
Sclerocactus terrae-canyonae Heil
Sclerocactus polyancistrus (Engelmann & Bigelow) Britton & Rose . . . . US
Ferocactus polyancistrus (Engelmann & Bigelow) Taylor
Pediocactus polyancistrus (Engelmann & Bigelow) Arp
Sclerocactus pubispinus (Engelmann) Benson . . . . ♦US
Ferocactus pubispinus (Engelmann) Taylor♦
Pediocactus pubispinus (Engelmann) Halda♦
Sclerocactus scheeri (Salm-Dyck) Taylor . . . . US, MX
Ancistrocactus megarhizus (Rose) Britton & Rose
Ancistrocactus scheeri (Salm-Dyck) Britton & Rose
Ferocactus scheeri (Salm-Dyck) Taylor
Pediocactus scheeri (Salm-Dyck) Halda
Sclerocactus sileri (Benson) Heil & Porter . . . . US
Pediocactus pubispinus ssp. *sileri* (Benson) Halda
Sclerocactus spinosior (Engelmann) Woodruff & Benson . . . . US
Ferocactus spinosior (Engelmann) Taylor
Pediocactus spinosior (Engelmann) Halda
Sclerocactus spinosior ssp. ***blainei*** (Welsh & Thorne) Hochstaetter . . . . US
Sclerocactus blainei Welsh & Thorne
Sclerocactus schlesseri Heil & Welsh
Pediocactus spinosior ssp. *blainei* (Welsh & Thorne) Halda
Sclerocactus uncinatus (Galeotti) Taylor . . . . US, MX
Sclerocactus uncinatus ssp. **uncinatus** . . . . MX
Ancistrocactus uncinatus (Galeotti) Benson
Ferocactus uncinatus (Galeotti) Britton & Rose
Glandulicactus uncinatus (Galeotti) Backeberg
Hamatocactus uncinatus (Galeotti) Orcutt
Pediocactus uncinatus (Galeotti) Halda
Sclerocactus uncinatus ssp. **crassihamatus** (Weber) Taylor . . . . US
Ancistrocactus crassihamatus (Weber) Benson
Ferocactus crassihamatus (Weber) Britton & Rose
Glandulicactus crassihamatus (Weber) Backeberg
Hamatocactus crassihamatus (Weber) Buxbaum
Ferocactus mathssonii (Berge ex Schumann) Taylor
Glandulicactus mathssonii (Berge ex Schumann) D.J. Ferguson
Sclerocactus uncinatus ssp. **wrightii** (Engelmann) Taylor . . . . US, MX
Glandulicactus wrightii (Engelmann) D.J. Ferguson
Sclerocactus unguispinus (Engelmann) Taylor . . . . MX
Echinomastus durangensis (Runge) Britton & Rose
Neolloydia durangensis (Runge) Benson
Thelocactus durangensis (Runge) Rowley
Echinomastus laui G. Frank & Zecher
Echinomastus mapimiensis Backeberg
Neolloydia unguispina (Engelmann) Benson
Echinomastus unguispinus (Engelmann) Britton & Rose
Pediocactus unguispinus (Engelmann) Halda
Thelocactus unguispinus (Engelmann) Rowley
Echinomastus unguispinus ssp. *laui* (Frank & Zecher) Glass

Sclerocactus warnockii (Benson) Taylor . . . . US
Echinomastus warnockii (Benson) Glass & Foster
Neolloydia warnockii Benson
Pediocactus warnockii (Benson) Halda
?*Echinomastus gautii* (Benson) Mosco & Zanovello
?*Neolloydia gautii* Benson
?*Pediocactus gautii* (Benson) Halda
?*Turbinicarpus gautii* (Benson) A. Zimmerman
Sclerocactus whipplei (Engelmann & Bigelow) Britton & Rose . . . . US
Ferocactus whipplei (Engelmann & Bigelow) Taylor
Pediocactus whipplei (Engelmann & Bigelow) Arp
Sclerocactus whipplei ssp. ***busekii*** Hochstaetter . . . . US
Pediocactus whipplei ssp. *busekii* (Hochstaetter) Halda
Sclerocactus wrightiae Benson . . . . ♦US
Ferocactus wrightiae (Benson) Taylor♦
Pediocactus wrightiae (Benson) Arp♦

Selenicereus anthonyanus (Alexander) Hunt . . . . MX
Cryptocereus anthonyanus Alexander
Selenicereus atropilosus Kimnach . . . . MX
Selenicereus boeckmannii (Otto ex Salm-Dyck) Britton & Rose . . . . . . MX, CU, KY
Cereus boeckmannii Otto ex Salm-Dyck
Selenicereus vaupelii (Weingart) Berger
Selenicereus brevispinus Britton & Rose . . . . CU
Selenicereus chontalensis (Alexander) Kimnach . . . . MX
Nyctocereus chontalensis Alexander
Selenicereus chrysocardium (Alexander) Kimnach . . . . MX
Epiphyllum chrysocardium Alexander
Marniera chrysocardium (Alexander) Backeberg
Selenicereus coniflorus (Weingart) Britton & Rose . . . . MX
Selenicereus pringlei Rose
Selenicereus donkelaarii (Salm-Dyck) Britton & Rose ex Bailey . . . . MX
Selenicereus grandiflorus (Linnaeus) Britton & Rose
. . . . MX, BZ, NI, CR, CU, KY, JM, HT, DO, PR, VI, WI
Cereus grandiflorus (Linnaeus) Miller
Selenicereus hallensis Weingart ex Borg
Selenicereus kunthianus (Otto ex Salm-Dyck) Britton & Rose
Selenicereus hamatus (Scheidweiler) Britton & Rose . . . . MX
Selenicereus hondurensis (Schumann) Britton & Rose . . . . GT, HN
Selenicereus inermis (Otto) Britton & Rose . . . . CO, VE
Epiphyllum steyermarkii Croizat
Selenicereus innesii Kimnach . . . . WI
Selenicereus macdonaldiae (Hooker) Britton & Rose . . . . HN
Selenicereus rothii (Weingart) Berger
Selenicereus megalanthus (Schumann ex Vaupel) Moran . . . . CO, EC, PE
Mediocactus megalanthus (Schumann ex Vaupel) Britton & Rose
Selenicereus murrillii Britton & Rose . . . . MX
Selenicereus nelsonii (Weingart) Britton & Rose . . . . MX
Selenicereus pteranthus (Link & Otto) Britton & Rose . . . . MX
Selenicereus nycticalus (Link) W.T. Marshall
Cereus pteranthus Link & Otto
Selenicereus setaceus (Salm-Dyck ex De Candolle) Werdermann
. . . . GY, GF, SR, BR, BO, PY, AR
Cereus hassleri Schumann
Mediocactus hassleri (Schumann) Backeberg
Cereus lindbergianus Weber ex Schumann
Cereus lindmanii Weber ex Schumann
Mediocactus lindmanii (Weber) Backeberg
Selenicereus rizzinii Scheinvar
¶?*Cereus extensus* Salm-Dyck ex De Candolle
®*Mediocactus coccineus* (Salm-Dyck ex De Candolle) Britton & Rose

Selenicereus spinulosus (De Candolle) Britton & Rose . . . . US, MX
Selenicereus pseudospinulosus Weingart
Cereus spinulosus De Candolle
Selenicereus testudo (Karwinski ex Zuccarini) Buxbaum MX, GT, BZ, HN, NI, CR, CO
Deamia diabolica Clover
Deamia testudo (Karwinski ex Zuccarini) Britton & Rose
Selenicereus tricae Hunt . . . . BZ, MX
Selenicereus urbanianus (Guerke ex Weingart) Britton & Rose . . . . CU, HT, DO
Selenicereus maxonii Rose
Selenicereus vagans (K. Brandegee) Britton & Rose . . . . MX
Selenicereus validus Arias & Guzman . . . . MX
Selenicereus wercklei (Weber) Britton & Rose . . . . CR
Selenicereus wittii (Schumann) Rowley . . . . BR, CO, VE, EC, PE
Strophocactus wittii (Schumann) Britton & Rose

Stenocactus coptonogonus (Lemaire) Berger ex Hill . . . . MX
Echinofossulocactus coptonogonus (Lemaire) Lawrence
Ferocactus coptonogonus (Lemaire) Taylor
Stenocactus crispatus (De Candolle) Berger ex Hill . . . . MX
Echinofossulocactus arrigens (Link ex Dietrich) Britton & Rose
Stenocactus arrigens (Link ex Dietrich) Berger ex Hill
Echinofossulocactus confusus Britton & Rose
Echinofossulocactus crispatus (De Candolle) Lawrence
Ferocactus crispatus (De Candolle) Taylor
Echinofossulocactus dichroacanthus (Martius ex Pfeiffer) Britton & Rose
Stenocactus dichroacanthus (Martius ex Pfeiffer) Berger ex Backeberg & F. Knuth
Echinofossulocactus flexispinus (Salm-Dyck) Bravo
Echinofossulocactus guerraianus Backeberg
Echinofossulocactus kellerianus Krainz
Echinofossulocactus lamellosus (A. Dietrich) Britton & Rose
Stenocactus lamellosus (A. Dietrich) Berger ex Hill
Echinofossulocactus lancifer (A. Dietrich) Britton & Rose
Stenocactus lancifer (A. Dietrich) Berger ex Backeberg & F. Knuth
Echinofossulocactus multiareolatus Bravo
Echinofossulocactus violaciflorus (Quehl) Britton & Rose
Stenocactus hastatus (Hoppfer ex Schumann) Berger ex Hill . . . . MX
Echinofossulocactus hastatus (Hopffer ex Schumann) Britton & Rose
Stenocactus multicostatus (Hildmann ex Schumann) Berger ex Hill . . . . MX
Echinofossulocactus lloydii Britton & Rose
Stenocactus lloydii (Britton & Rose) Berger ex Hill
Echinofossulocactus multicostatus (Hildmann ex Schumann) Britton & Rose
Echinofossulocactus zacatecasensis Britton & Rose
Stenocactus zacatecasensis (Britton & Rose) Berger ex Hill
®*Echinofossulocactus erectocentrus* Backeberg
Stenocactus obvallatus (De Candolle) Berger ex Hill . . . . MX
Echinofossulocactus caespitosus Backeberg
Echinofossulocactus obvallatus (De Candolle) Lawrence
Echinofossulocactus pentacanthus (Lemaire) Britton & Rose
Stenocactus pentacanthus (Lemaire) Berger ex Hill
Stenocactus ochoterenanus Tiegel . . . . MX
Stenocactus bustamantei Bravo ex Miranda
Echinofossulocactus densispinus Tiegel ex Pechanek
Echinofossulocactus lexarzae (Bravo ex Miranda) Croizat
Echinofossulocactus parksianus C. Schmoll ex Pechanek
Echinofossulocactus rosasianus Whitmore ex Pechanek
?*Echinofossulocactus heteracanthus* (Muehlenpfordt) Britton & Rose
?*Stenocactus heteracanthus* (Muehlenpfordt) Berger ex Hill
®*Echinofossulocactus ochoterenaus* (Tiegel) Whitmore
Stenocactus phyllacanthus (Dietrich & Otto) Berger ex Hill . . . . MX
Echinofossulocactus phyllacanthus (Dietrich & Otto) Lawrence

Ferocactus phyllacanthus (Dietrich & Otto) Taylor
Echinofossulocactus tricuspidatus (Scheidweiler) Britton & Rose
Stenocactus rectispinus Schmoll . . . . MX
Stenocactus sulphureus (A. Dietrich) Bravo . . . . MX
Echinofossulocactus sulphureus (A. Dietrich) J. Meyran
Stenocactus vaupelianus (Werdermann) F. Knuth . . . . MX
Echinofossulocactus vaupelianus (Werdermann) Oehme
Ferocactus vaupelianus (Werdermann) Taylor
?*Echinofossulocactus albatus* (A. Dietrich) Britton & Rose
⊗*Stenocactus albatus* (A. Dietrich) F. Knuth

Stenocereus alamosensis (J. Coulter) Gibson & Horak . . . . MX
Rathbunia alamosensis (J. Coulter) Britton & Rose
Rathbunia neosonorensis Backeberg
Rathbunia sonorensis (Runge) Britton & Rose
Stenocereus aragonii (Weber) Buxbaum . . . . CR
Lemaireocereus aragonii (Weber) Britton & Rose
Marshallocereus aragonii (Weber) Backeberg
Pachycereus aragonii (Weber) P.V. Heath
Stenocereus beneckei (Ehrenberg) Buxbaum . . . . MX
Hertrichocereus beneckei (Ehrenberg) Backeberg
Lemaireocereus beneckei (Ehrenberg) Britton & Rose
Rathbunia beneckei (Ehrenberg) P.V. Heath
Stenocereus chacalapensis (Bravo & MacDougall) Buxbaum . . . . MX
Rathbunia chacalapensis (Bravo & MacDougall) P.V. Heath
Stenocereus chrysocarpus Sanchez-Mejorada . . . . MX
Rathbunia chrysocarpa (Sanchez-Mejorada) P.V. Heath
Stenocereus dumortieri (Scheidweiler) Buxbaum . . . . MX
Isolatocereus dumortieri (Scheidweiler) Backeberg
Lemaireocereus dumortieri (Scheidweiler) Britton & Rose
Rathbunia dumortieri (Scheidweiler) P.V. Heath
Stenocereus eichlamii (Britton & Rose) Buxbaum . . . . MX, GT, HN, NI, SV
Lemaireocereus eichlamii Britton & Rose
Rathbunia eichlamii (Britton & Rose) P.V. Heath
Ritterocereus eichlamii (Britton & Rose) Backeberg
Lemaireocereus longispinus Britton & Rose
Rathbunia longispinus (Britton & Rose) P.V. Heath
Stenocereus longispinus (Britton & Rose) Buxbaum
Stenocereus eruca (T. Brandegee) Gibson & Horak . . . . MX
Machaerocereus eruca (T. Brandegee) Britton & Rose
Rathbunia eruca (Brandegee) P.V. Heath
Stenocereus fimbriatus (Lamarck) Lourteig . . . . CU, HT, DO, JM, PR
Rathbunia fimbriata (Lamarck) P.V. Heath
Ritterocereus fimbriatus (Lamarck) Backeberg
⊗*Lemaireocereus hystrix* (Haworth) Britton & Rose
⊗*Stenocereus hystrix* (Haworth) Buxbaum
⊗*Stenocereus peruvianus* (Linnaeus) Kiesling
Stenocereus fricii Sanchez-Mejorada . . . . MX
Rathbunia fricii (Sanchez-Mejorada) P.V. Heath
Stenocereus griseus (Haworth) Buxbaum . . . . MX, TT, AN, CO, VE
Lemaireocereus deficiens (Otto & Dietrich) Britton & Rose
Rathbunia deficiens (Otto & Dietrich) P.V. Heath
Ritterocereus deficiens (Otto & Dietrich) Backeberg
Stenocereus deficiens (Otto & Dietrich) Buxbaum
Rathbunia grisea (Haworth) P.V. Heath
Lemaireocereus griseus (Haworth) Britton & Rose
Ritterocereus griseus (Haworth) Backeberg
Stenocereus gummosus (Engelmann) Gibson & Horak . . . . MX
Rathbunia gummosa (Brandegee) P.V. Heath
Machaerocereus gummosus (Engelmann) Britton & Rose

Stenocereus kerberi (Schumann) Gibson & Horak . . . . MX
Rathbunia kerberi (Schumann) Britton & Rose
Stenocereus laevigatus (Salm-Dyck) Buxbaum . . . . MX
Rathbunia laevigata (Salm-Dyck) P.V. Heath
Ritterocereus laevigatus (Salm-Dyck) Backeberg
Stenocereus martinezii (J.G. Ortega) Buxbaum . . . . MX
Lemaireocereus martinezii J.G. Ortega
Rathbunia martinezii (J.G. Ortega) P.V. Heath
Stenocereus montanus (Britton & Rose) Buxbaum . . . . MX
Rathbunia montana (Britton & Rose) P.V. Heath
Lemaireocereus montanus Britton & Rose
Stenocereus pruinosus (Otto ex Pfeiffer) Buxbaum . . . . MX
Rathbunia pruinosa (Otto ex Pfeiffer) P.V. Heath
Lemaireocereus pruinosus (Otto ex Pfeiffer) Britton & Rose
Ritterocereus pruinosus (Otto ex Pfeiffer) Backeberg
Stenocereus queretaroensis (Weber) Buxbaum . . . . MX
Lemaireocereus queretaroensis (Weber) Safford
Rathbunia queretaroensis (Weber) P.V. Heath
Ritterocereus queretaroensis (Weber) Backeberg
Stenocereus quevedonis (J.G. Ortega) Buxbaum . . . . MX
Lemaireocereus quevedonis J.G. Ortega
Rathbunia quevedonis (J.G. Ortega) P.V. Heath
Stenocereus standleyi (J.G. Ortega) Buxbaum . . . . MX
Rathbunia standleyi (J.G. Ortega) P.V. Heath
Ritterocereus standleyi (J.G. Ortega) Backeberg
Stenocereus stellatus (Pfeiffer) Riccobono . . . . MX
Rathbunia stellata (Pfeiffer) P.V. Heath
Lemaireocereus stellatus (Pfeiffer) Britton & Rose
Stenocereus thurberi (Engelmann) Buxbaum . . . . US, MX
Stenocereus thurberi ssp. **thurberi** . . . . US, MX
Cereus thurberi Engelmann
Lemaireocereus thurberi (Engelmann) Britton & Rose
Marshallocereus thurberi (Engelmann) Backeberg
Rathbunia thurberi (Engelmann) P.V. Heath
Stenocereus thurberi ssp. **littoralis** (K. Brandegee) Taylor . . . . MX
Stenocereus littoralis (K. Brandegee) L.W. Lenz
Stenocereus treleasei (Britton & Rose) Backeberg . . . . MX
Lemaireocereus treleasei Britton & Rose
Rathbunia treleasei (Vaupel) P.V. Heath
Stenocereus yunckeri (Standley) Bravo & Sanchez-Mejorada . . . . GT, HN
Cereus yunckeri Standley
Rathbunia yunckeri (Standley) P.V. Heath

Stephanocereus leucostele (Guerke) Berger . . . . BR
Cephalocereus leucostele (Guerke) Britton & Rose
Stephanocereus luetzelburgii (Vaupel) Taylor & Eggli . . . . BR
Coleocephalocereus luetzelburgii (Vaupel) Buxbaum
Pilosocereus luetzelburgii (Vaupel) Byles & Rowley
Pseudopilocereus luetzelburgii (Vaupel) Buxbaum

Stetsonia coryne (Foerster) Britton & Rose . . . . PY, BO, AR
Cereus chacoanus Vaupel

Strombocactus disciformis (De Candolle) Britton & Rose . . . . ♦MX
?*Strombocactus jarmilae* Halda♦
Strombocactus disciformis ssp. **disciformis** . . . . ♦MX
Strombocactus disciformis ssp. **esperanzae** Glass & Arias . . . . ♦MX
Strombocactus pulcherrimus Halda♦

Tacinga braunii Esteves Pereira . . . . BR

Tacinga funalis Britton & Rose . . . . BR

Tacinga atropurpurea Werdermann

Tacinga funalis ssp. *atropurpurea* (Werdermann) Braun & Esteves Pereira

Thelocactus bicolor (Galeotti ex Pfeiffer) Britton & Rose . . . . US, MX

Thelocactus schottii (Engelmann) Kladiwa & Fittkau

Thelocactus wagnerianus Berger

Thelocactus bicolor ssp. **bicolor** . . . . US, MX

Ferocactus bicolor (Galeotti ex Pfeiffer) Taylor

Thelocactus bicolor ssp. **flavidispinus** (Backeberg) Taylor . . . . US

Thelocactus flavidispinus (Backeberg) Backeberg

Thelocactus bicolor ssp. **schwarzii** (Backeberg) Taylor . . . . MX

Thelocactus schwarzii Backeberg

Thelocactus conothelos (Regel & Klein) F. Knuth . . . . MX

Thelocactus saussieri (Weber) Berger

Thelocactus conothelos ssp. **conothelos** . . . . MX

Gymnocactus conothelos (Regel & Klein) Backeberg

Torreyocactus conothelos (Regel & Klein) Doweld

¶?*Pediocactus smithii* (Muehlenpfordt) Halda

¶?*Neolloydia smithii* (Muehlenpfordt) Kladiwa & Fittkau

Thelocactus conothelos ssp. **argenteus** (Glass & Foster) Glass . . . . MX

Thelocactus conothelos ssp. **aurantiacus** (Glass & Foster) Glass . . . . MX

Thelocactus garciae Glass & M. Mendoza . . . . MX

Thelocactus hastifer (Werdermann & Boedeker) F. Knuth . . . . MX

Ferocactus hastifer (Werdermann & Boedeker) Taylor

Thelocactus heterochromus (Weber) Oosten . . . . MX

Ferocactus heterochromus (Weber) Taylor

⊗*Thelocactus pottsii* (Salm-Dyck) Britton & Rose

Thelocactus hexaedrophorus (Lemaire) Britton & Rose . . . . MX

Thelocactus fossulatus (Scheidweiler) Britton & Rose

Thelocactus hexaedrophorus ssp. **hexaedrophorus** . . . . MX

Thelocactus hexaedrophorus ssp. **lloydii** (Britton & Rose) Taylor . . . . MX

Thelocactus lloydii Britton & Rose

Thelocactus lausseri Riha & Busek . . . . MX

Thelocactus leucacanthus (Pfeiffer) Britton & Rose . . . . MX

Thelocactus ehrenbergii (Pfeiffer) F. Knuth

Ferocactus leucacanthus (Zuccarini) Taylor

Thelocactus macdowellii (Rebut ex Quehl) W. Marshall . . . . MX

Echinomastus macdowellii (Quehl) Britton & Rose

Neolloydia macdowellii (Rebut ex Quehl) H.E. Moore

Thelocactus rinconensis (Poselger) Britton & Rose . . . . MX

Thelocactus lophothele (Salm-Dyck) Britton & Rose

Thelocactus nidulans (Quehl) Britton & Rose

Thelocactus phymatothelos (Ruempler) Britton & Rose

Thelocactus rinconensis ssp. *nidulans* (Quehl) Glass

Thelocactus rinconensis ssp. *phymatothelos* (Poselger) Glass

?*Thelocactus multicephalus* Halda & Panarotto

Thelocactus rinconensis ssp. **rinconensis** . . . . MX

Thelocactus rinconensis ssp. **hintonii** J.M. Luethy . . . . MX

Thelocactus setispinus (Engelmann) Anderson . . . . US, MX

Ferocactus setispinus (Engelmann) Benson

Hamatocactus setispinus (Engelmann) Britton & Rose

Thelocactus tulensis (Poselger) Britton & Rose . . . . MX

?*Thelocactus krainzianus* Oehme

Thelocactus tulensis ssp. **tulensis** . . . . MX

Thelocactus tulensis ssp. **buekii** (Klein) Taylor . . . . MX

Thelocactus buekii (Klein) Britton & Rose

Thelocactus tulensis ssp. **matudae** (Sanchez-Mejorada & Lau) Taylor . . . . MX

Thelocactus matudae Sanchez-Mejorada & Lau

Turbinicarpus alonsoi Glass & Arias . . . . ♦MX
Pediocactus alonsoi (Glass & Arias) Halda♦
Turbinicarpus bonatzii G. Frank . . . . ♦MX
Pediocactus bonatzii (G. Frank) Halda♦
Turbinicarpus booleanus G.S. Hinton . . . . ♦MX
Turbinicarpus beguinii (Taylor) Mosco & Zanovello . . . . ♦MX
Thelocactus beguinii Taylor♦
ⓂⓊ*Gymnocactus beguinii* Backeberg♦
ⓂⓊ*Neolloydia beguinii* Britton & Rose♦
Ⓜ*Neolloydia smithii* Kladiwa & Fittkau♦
Turbinicarpus gielsdorfianus (Werdermann) John & Riha . . . . ♦MX
Neolloydia gielsdorfiana (Werdermann) F. Knuth♦
Gymnocactus gielsdorfianus (Werdermann) Backeberg♦
Pediocactus gielsdorfianus (Werdermann) Halda♦
Thelocactus gielsdorfianus (Werdermann) Borg♦
Turbinicarpus hoferi Luethy & Lau . . . . ♦MX
Neolloydia hoferi (Luethy & Lau) A.T. Powell♦
Pediocactus hoferi (Luethy & Lau) Halda♦
Turbinicarpus horripilus (Lemaire) John & Riha . . . . ♦MX
Gymnocactus goldii (Bravo) Ito♦
Neolloydia horripila (Lemaire) Britton & Rose♦
Gymnocactus horripilus (Lemaire) Backeberg♦
Pediocactus horripilus (Lemaire) Halda♦
Thelocactus horripilus (Lemaire) Kladiwa♦
Turbinicarpus jauernigii G. Frank . . . . ♦MX
Turbinicarpus lophophoroides ssp. *jauernigii* (Frank) Battaia & Zanovello♦
Turbinicarpus knuthianus (Boedeker) John & Riha . . . . ♦MX
Neolloydia knuthiana (Boedeker) F. Knuth♦
Gymnocactus knuthianus (Boedeker) Backeberg♦
Pediocactus knuthianus (Boedeker) Halda♦
Thelocactus knuthianus (Boedeker) Borg♦
Turbinicarpus laui Glass & Foster . . . . ♦MX
Neolloydia laui (Glass & Foster) Anderson♦
Pediocactus laui (Glass & Foster) Halda♦
Strombocactus laui (Glass & Foster) Mays♦
Ⓜ*Neolloydia pilispina* (Purpus) Britton & Rose♦
Turbinicarpus lophophoroides (Werdermann) Buxbaum & Backeberg . . . . ♦MX
Neolloydia lophophoroides (Werdermann) Anderson♦
Pediocactus lophophoroides (Werdermann) Halda♦
Turbinicarpus mandragora (Fric ex Berger) A. Zimmerman . . . . ♦MX
Gymnocactus mandragora (Fric ex Berger) Backeberg♦
Neolloydia mandragora (Fric ex Berger) Anderson♦
Pediocactus mandragora (Fric ex Berger) Halda♦
Thelocactus mandragora (Fric ex Berger) Berger♦
Turbinicarpus ×mombergeri Riha (T. laui × T. pseudopectinatus) . . . . ♦MX
Turbinicarpus pseudomacrochele (Backeberg) Buxbaum & Backeberg . . . . ♦MX
Neolloydia krainziana (G. Frank) A.T. Powell♦
Toumeya krainziana G. Frank♦
Turbinicarpus krainzianus (G. Frank) Backeberg♦
Turbinicarpus pseudomacrochele ssp. *krainzianus* (G. Frank) Glass
Turbinicarpus pseudomacrochele ssp. **pseudomacrochele** . . . . ♦MX
Kadenicarpus pseudomacrochele (Backeberg) Doweld
Neolloydia pseudomacrochele (Backeberg) Anderson♦
Pediocactus pseudomacrochele (Backeberg) Halda♦
Turbinicarpus pseudomacrochele ssp. **lausseri** (Diers & G. Frank) Glass . . . . . ♦MX
Turbinicarpus pseudopectinatus (Backeberg) Glass & Foster . . . . ♦MX
Neolloydia pseudopectinata (Backeberg) Anderson♦
Normanbokea pseudopectinata (Backeberg) Kladiwa & Buxbaum♦
Pelecyphora pseudopectinata Backeberg♦
Pediocactus pseudopectinatus (Backeberg) Halda♦

Thelocactus pseudopectinatus (Backeberg) Anderson & Boke♦
Pelecyphora pulcherrima Sabatini♦
Turbinicarpus rioverdensis G. Frank . . . . ♦MX
Pediocactus rioverdensis (G. Frank) Halda♦
Turbinicarpus saueri (Boedeker) John & Riha . . . . ♦MX
Gymnocactus saueri (Boedeker) Backeberg♦
Neolloydia saueri (Boedeker) F. Knuth♦
Pediocactus saueri (Boedeker) Halda♦
Thelocactus saueri (Boedeker) Borg♦
Turbinicarpus schmiedickeanus (Boedeker) Buxbaum & Backeberg . . . . ♦MX
Pediocactus schmiedickeanus (Boedeker) Halda♦
Turbinicarpus schmiedickeanus ssp. **schmiedickeanus** . . . . ♦MX
Neolloydia schmiedickeana (Boedeker) Anderson♦
Turbinicarpus schmiedickeanus ssp. **dickisoniae** (Glass & Foster) Taylor . . . . . ♦MX
Turbinicarpus dickisoniae (Glass & Foster) Glass & Hofer
Turbinicarpus schmiedickeanus ssp. **flaviflorus** (G. Frank & Lau) Glass . . . . . . ♦MX
Turbinicarpus flaviflorus G. Frank & Lau♦
Turbinicarpus schmiedickeanus ssp. **gracilis** (Glass & Foster) Glass . . . . ♦MX
Turbinicarpus gracilis Glass & Foster♦
Turbinicarpus schmiedickeanus ssp. **klinkerianus** (Backeberg & H.J. Jacobsen) Taylor
. . . . ♦MX
Turbinicarpus klinkerianus Backeberg & H.J. Jacobsen♦
Turbinicarpus schmiedickeanus ssp. **macrochele** (Werdermann) Taylor . . . . . . ♦MX
Turbinicarpus macrochele (Werdermann) Buxbaum & Backeberg♦
Turbinicarpus schmiedickeanus ssp. **schwarzii** (Shurly) Taylor . . . . ♦MX
Strombocactus polaskii (Backeberg) Hewitt♦
Turbinicarpus schwarzii (Shurly) Backeberg♦
®*Turbinicarpus polaskii* Backeberg♦
Turbinicarpus subterraneus (Backeberg) A. Zimmerman . . . . ♦MX
Neolloydia subterranea (Backeberg) H.E. Moore♦
Gymnocactus subterraneus (Backeberg) Schwarz♦
Pediocactus subterraneus (Backeberg) Halda♦
Thelocactus subterraneus (Backeberg) Backeberg♦
Turbinicarpus swobodae Diers & Esteves Pereira . . . . ♦MX
Pediocactus swobodae (Diers & Esteves Pereira) Halda♦
Turbinicarpus valdezianus (Moeller) Glass & Foster . . . . ♦MX
Neolloydia valdeziana (Moeller) Anderson♦
Normanbokea valdeziana (Moeller) Kladiwa & Buxbaum♦
Gymnocactus valdezianus (Moeller) Backeberg♦
Pediocactus valdezianus (Moeller) Halda♦
Turbinicarpus viereckii (Werdermann) John & Riha . . . . ♦MX
Turbinicarpus viereckii ssp. **viereckii** . . . . ♦MX
Gymnocactus viereckii (Werdermann) Backeberg♦
Neolloydia viereckii (Werdermann) F. Knuth♦
Pediocactus viereckii (Werdermann) Halda♦
Thelocactus viereckii (Werdermann) Bravo♦
Turbinicarpus viereckii ssp. **major** (Glass & Foster) Glass . . . . ♦MX
Turbinicarpus ysabelae (Schlange) John & Riha . . . . ♦MX
Gymnocactus ysabelae (Schlange) Backeberg♦
Pediocactus ysabelae (Schlange) Halda♦
Thelocactus ysabelae Schlange ex Croizat♦
Turbinicarpus zaragozae (Glass & Foster) Glass & Hofer ex Glass . . . . ♦MX

Uebelmannia buiningii Donald . . . . ♦BR
Uebelmannia gummifera (Backeberg & Voll) Buining . . . . ♦BR
Parodia gummifera Backeberg & Voll♦
Uebelmannia gummifera ssp. *meninensis* (Buining) Braun & Esteves Pereira♦
Uebelmannia meninensis Buining♦
Uebelmannia pectinifera Buining . . . . ♦BR
Uebelmannia pectinifera ssp. **pectinifera** . . . . ♦BR

Uebelmannia pectinifera ssp. **flavispina**
(Buining & Brederoo) Braun & Esteves Pereira . . . . ♦BR
Uebelmannia flavispina Buining & Brederoo♦
Uebelmannia pectinifera ssp. **horrida** (Braun) Braun & Esteves Pereira . . . . ♦BR

Weberbauerocereus churinensis Ritter . . . . PE
Haageocereus churinensis (Ritter) P.V. Heath
Weberbauerocereus cuzcoensis Knize . . . . PE
Haageocereus cuzcoensis (Knize) P.V. Heath
Weberbauerocereus johnsonii Ritter . . . . PE
Haageocereus johnsonii (Ritter) P.V. Heath
Weberbauerocereus longicomus Ritter . . . . PE
Weberbauerocereus albus Ritter
Haageocereus longicomus (Ritter) P.V. Heath
Weberbauerocereus rauhii Backeberg . . . . PE
Haageocereus rauhii (Backeberg) P.V. Heath
Weberbauerocereus torataensis Ritter . . . . PE
Haageocereus torataensis (Ritter) P.V. Heath
Weberbauerocereus weberbaueri (Schumann ex Vaupel) Backeberg . . . . PE, CL
Trichocereus fascicularis (Meyen) Britton & Rose
Weberbauerocereus horridispinus Rauh & Backeberg
Weberbauerocereus seyboldianus Rauh & Backeberg
Haageocereus weberbaueri (Schumann ex Vaupel) Hunt
Weberbauerocereus winterianus Ritter . . . . PE
Haageocereus winterianus (Ritter) P.V. Heath

Weberocereus biolleyi (Weber) Britton & Rose . . . . NI, CR
Weberocereus bradei (Britton & Rose) Rowley . . . . CR
Eccremocactus bradei Britton & Rose
Weberocereus glaber (Eichlam) Rowley . . . . MX, GT, CR
Werckleocereus glaber (Eichlam) Britton & Rose
Selenicereus mirandae Bravo
Weberocereus imitans (Kimnach & Hutchison) Buxbaum . . . . CR
Cryptocereus imitans (Kimnach & Hutchison) Backeberg
Eccremocactus imitans (Kimnach & Hutchison) Kimnach
Werckleocereus imitans Kimnach & Hutchison
Weberocereus panamensis Britton & Rose . . . . PA
Weberocereus rosei (Kimnach) Buxbaum . . . . EC
Cryptocereus rosei (Kimnach) Backeberg
Eccremocactus rosei Kimnach
Weberocereus tonduzii (Weber) Rowley . . . . CR
Werckleocereus tonduzii (Weber) Britton & Rose
Weberocereus trichophorus H. Johnson & Kimnach . . . . CR
Weberocereus tunilla (Weber) Britton & Rose . . . . CR

Yungasocereus inquisivensis (Cardenas) Ritter . . . . BO
Samaipaticereus inquisivensis Cardenas
®*Yungasocereus microcarpus* Ritter ex Krainz

Part III. Country/Area Checklists

ARGENTINA (AR)

Acanthocalycium
ferrarii
klimpelianum
spiniflorum
Austrocactus
bertinii
coxii
patagonicus
Blossfeldia
liliputana
Cereus
aethiops
argentinensis
hankeanus
hildmannianus ssp. **uruguayanus**
roseiflorus
spegazzinii
stenogonus
validus
Cleistocactus
baumannii ssp. **baumannii**
ferrarii
hyalacanthus
smaragdiflorus
Denmoza
rhodacantha
Echinopsis
ancistrophora
angelesiae
arboricola
atacamensis ssp. **pasacana**
aurea
baldiana
bruchii
×*cabrerae*
candicans
chamaecereus
chrysantha
chrysochete
crassicaulis
densispina
eyriesii
fabrisii
famatinensis
ferox
formosa
friedrichii
glaucina
haematantha
huascha
korethroides
lamprochlora
leucantha
mamillosa
marsoneri
(*Echinopsis*)
minuana
mirabilis
molesta
nigra
oxygona
pugionacantha
rhodotricha
saltensis
sanguiniflora
schickendantzii
schreiteri
silvestrii
smrziana
spachiana
strigosa
tarijensis
terscheckii
thelegona
thelegonoides
thionantha
tiegeliana
tubiflora
vatteri
walteri
Epiphyllum
phyllanthus
Eriosyce
andreaeana
bulbocalyx
strausiana
umadeave
vertongenii
villicumensis
Frailea
castanea
grahliana
mammifera
pumila
pygmaea ssp. **pygmaea**
schilinzkyana
Gymnocalycium
ambatoense
amerhauseri
andreae
andreae ssp. *carolinense*
angelae
baldianum
bayrianum
berchtii
bodenbenderianum
bodenbenderianum ssp. *intertextum*
borthii
bruchii
calochlorum

(Gymnocalycium)
capillaense
carminanthum
castellanosii
catamarcense
catamarcense ssp. ***acinacispinum***
catamarcense ssp. ***schmidianum***
deeszianum
delaetii
erinaceum
gibbosum
gibbosum ssp. ***ferox***
hossei
hybopleurum
kieslingii
leeanum
leptanthum
mackieanum
marsoneri
mesopotamicum
mihanovichii
monvillei
monvillei ssp. ***achirasense***
monvillei ssp. ***brachyanthum***
monvillei ssp. ***horridispinum***
mostii
mucidum
neuhuberi
obductum
ochoterenaessp. ***herbsthoferianum***
ochoterenae ssp. **ochoterenae**
ochoterenae ssp. **vatteri**
oenanthemum
parvulum
pflanzii
platense
pugionacanthum
quehlianum
ragonesei
riojense
riojense ssp. ***kozelskyanum***
riojense ssp. ***paucispinum***
riojense ssp. ***piltziorum***
ritterianum
rosae
saglionis
saglionis ssp. ***tilcarense***
schickendantzii
schroederianum
schroederianum ssp. ***bayense***
schroederianum ssp. ***paucicostatum***
spegazzinii
stellatum
stellatum ssp. ***occultum***
striglianum
stuckertii
taningaense
terweemeanum
tillianum
uebelmannianum

Harrisia
balansae
martinii
pomanensis
pomanensis ssp. ***regelii***
tortuosa

Lepismium
aculeatum
cruciforme
houlletianum
ianthothele
lorentzianum
lumbricoides
monacanthum
warmingianum

Maihuenia
patagonica
poeppigii

Neowerdermannia
vorwerkii

Opuntia
alexanderi
anacantha
aoracantha
armata
articulata
atrovirens
aurantiaca
boliviana
brasiliensis
chakensis
chichensis
clavarioides
colubrina
corrugata
darwinii
delaetiana
discolor
erectoclada
glomerata
halophila
ianthinantha
longispina
microdisca
minuta
molinensis
monacantha
nigrispina
ovata
paraguayensis
penicilligera
pentlandii
picardoi
pituitosa
prasina
quimilo
rossiana
salagria
salmiana
schickendantzii

(Opuntia)
securigera
shaferi
soehrensii
subsphaerocarpa
subterranea
sulphurea
tunicata
verschaffeltii
vestita
weberi
wetmorei
Oreocereus
celsianus
trollii
Parodia
aureicentra
chrysacanthion
erinacea
linkii
maassii
mammulosa ssp. submammulosa
microsperma ssp. microsperma
microsperma ssp. horrida
nivosa
ottonis ssp. ottonis
penicillata
saint-pieana
schumanniana ssp. schumanniana
sellowii
stuemeri
tilcarensis
turecekiana
Pereskia
aculeata
nemorosa
sacharosa
Praecereus
euchlorus ssp. euchlorus
saxicola
Pterocactus
araucanus
australis
fischeri
gonjianii
hickenii
megliolii
reticulatus
tuberosus
valentinii
Quiabentia
verticillata
Rebutia
aureiflora
deminuta
einsteinii
fabrisii
fiebrigii
gonjianii
marsoneri
minuscula
neumanniana
nigricans
pseudodeminuta
pygmaea
spegazziniana
spinosissima
wessneriana
xanthocarpa
Rhipsalis
baccifera ssp. shaferi
cereuscula
floccosa ssp. hohenauensis
floccosa ssp. tucumanensis
Selenicereus
setaceus
Stetsonia
coryne

BAHAMAS (BS)

Harrisia
brookii
Mammillaria
nivosa
Melocactus
intortus ssp. intortus
Opuntia
darrahiana
(Opuntia)
dillenii
×*lucayana*
millspaughii
nashii
Pilosocereus
polygonus
royenii

BELIZE (BZ)

Acanthocereus
tetragonus
Epiphyllum
crenatum
hookeri
pumilum
Hylocereus
minutiflorus
Pseudorhipsalis
ramulosa
Rhipsalis
baccifera ssp. **baccifera**
Selenicereus
grandiflorus
testudo
tricae

BOLIVIA (BO)

Blossfeldia
liliputana
Browningia
caineana
Cephalocleistocactus
chrysocephalus
Cereus
braunii
cochabambensis
comarapanus
hankeanus
huilunchu
kroenleinii
lamprospermus
spegazzinii
stenogonus
tacuaralensis
Cintia
knizei
Cleistocactus
baumannii ssp. **baumannii**
baumannii ssp. *chacoanus*
baumannii ssp. *santacruzensis*
brookeae
buchtienii
candelilla
dependens
hildegardiae
laniceps
luribayensis
micropetalus
muyurinensis
orthogonus
palhuayensis
parapetiensis
parviflorus
piraymirensis
reae
ritteri
roezlii
samaipatanus
smaragdiflorus
strausii
tarijensis
tominensis
(Cleistocactus)
tupizensis
varispinus
vulpis-cauda
winteri
Corryocactus
ayopayanus
charazanensis
melanotrichus
otuyensis
perezianus
pulquinensis
tarijensis
Discocactus
♦*ferricola*
♦**heptacanthus** ssp. **heptacanthus**
Echinopsis
ancistrophora
ancistrophora ssp. *arachnacantha*
ancistrophora ssp. *cardenasiana*
ancistrophora ssp. *pojoensis*
antezanae
arebaloi
atacamensis ssp. **pasacana**
backebergii
bertramiana
boyuibensis
bridgesii
bridgesii ssp. *yungasensis*
caineana
cajasensis
callianthＯlilacina
callichroma
camarguensis
cerdana
cinnabarina
clavatus
cochabambensis
comarapana
conaconensis
coronata
cotacajesii
escayachensis
ferox
graciliflora

(Echinopsis)
haematantha
hammerschmidii
hertrichiana
huotii
huotii ssp. ***vallegrandensis***
hystrichoides
ibicuatensis
kladiwana
klingleriana
lageniformis
lateritia
macrogona
mamillosa
mamillosa ssp. ***silvatica***
marsoneri
mataranensis
maximiliana
mieckleyi
obrepanda
obrepanda ssp. ***calorubra***
obrepanda ssp. ***tapecuana***
pentlandii
pojoensis
pseudomamillosa
pugionacantha
pugionacantha ssp. ***rossii***
quadratiumbonatus
riviere-de-caraltii
schieliana
scopulicola
spachiana
subdenudata
sucrensis
tacaquirensis
tacaquirensis ssp. ***taquimbalensis***
taratensis
tarijensis
tarijensis ssp. ***herzogianus***
tarijensis ssp. ***totorensis***
terscheckii
tiegeliana
trichosa
tunariensis
uyupampensis
vasquezii
volliana
yuquina
Epiphyllum
phyllanthus
Espostoa
guentheri
Frailea
cataphracta ssp. **duchii**
chiquitana
Gymnocalycium
anisitsii
chiquitanum
eytianum
marsoneri

(Gymnocalycium)
marsoneri ssp. ***matoense***
pflanzii
spegazzinii ssp. **cardenasianum**
stenopleurum
Harrisia
pomanensis
tetracantha
Lepismium
bolivianum
crenatum
ianthothele
incachacanum
lorentzianum
lumbricoides
miyagawae
monacanthum
paranganiense
Neoraimondia
herzogiana
Neowerdermannia
vorwerkii
Opuntia
albisaetacens
alko-tuna
anacantha
arcei
backebergii
boliviana
brasiliensis
chichensis
cochabambensis
conjungens
dactylifera
flexuosa
floccosa
frigida
glomerata
heteromorpha
lagopus
microdisca
minuscula
nigrispina
orurensis
pentlandii
pubescens
pyrrhantha
quimilo
roborensis
rossiana
salmiana
shaferi
silvestris
soehrensii
sphaerica
subterranea
subulata
sulphurea
sulphurea ssp. ***brachyacantha***
sulphurea ssp. ***spinibarbis***

(Opuntia)
verschaffeltii
vestita
vitelliniflora
vitelliniflora ssp. *interjecta*
Oreocereus
celsianus
leucotrichus
pseudofossulatus
trollii
Parodia
ayopayana
columnaris
comarapana
commutans
formosa
hausteiniana
maassii
ocampoi
procera
ritteri
schwebsiana
subterranea
taratensis
tuberculata
Pereskia
diaz-romeroana
sacharosa
weberiana
Praecereus
euchlorus ssp. **amazonicus**
euchlorus ssp. **euchlorus**
saxicola
Pseudorhipsalis
ramulosa
Quiabentia
verticillata
Rebutia
albiflora
albopectinata
arenacea
brunescens
caineana
canigueralii ssp. **crispata**
canigueralii ssp. **pulchra**
cardenasiana
cylindrica

(Rebutia)
einsteinii
fidaiana ssp. **cintiensis**
fidaiana ssp. **fidaiana**
fiebrigii
flavistyla
fulviseta
heliosa
huasiensis
leucanthema
mentosa ssp. **mentosa**
mentosa ssp. **purpurea**
narvaecensis
neocumingii ssp. **neocumingii**
neocumingii ssp. **pilcomayensis**
neocumingii ssp. **riograndensis**
neocumingii ssp. **saipinensis**
neocumingii ssp. **trollii**
oligacantha
padcayensis
perplexa
pseudodeminuta
pulvinosa
pygmaea
ritteri
simoniana
spegazziniana
spinosissima
steinbachii ssp. **kruegeri**
steinbachii ssp. **steinbachii**
steinbachii ssp. **tiraquensis**
steinbachii ssp. **verticillacantha**
steinmannii
Rhipsalis
baccifera ssp. **shaferi**
cereuscula
cuneata
floccosa ssp. **tucumanensis**
goebeliana
Samaipaticereus
corroanus
Selenicereus
setaceus
Stetsonia
coryne
Yungasocereus
inquisivensis

BRAZIL (BR)

Arrojadoa
×*albiflora*
bahiensis
dinae ssp. **dinae**
dinae ssp. **eriocaulis**
penicillata
rhodantha

Arthrocereus
glaziovii
melanurus ssp. **magnus**
melanurus ssp. **melanurus**
melanurus ssp. **odorus**
rondonianus
spinosissimus

Brasilicereus
- markgrafii
- phaeacanthus

Cereus
- adelmarii
- aethiops
- albicaulis
- *bicolor*
- fernambucensis ssp. fernambucensis
- fernambucensis ssp. sericifer
- hexagonus
- hildmannianus ssp. hildmannianus
- hildmannianus ssp. uruguayanus
- insularis
- jamacaru ssp. calcirupicola
- *jamacaru* ssp. *goiasensis*
- jamacaru ssp. jamacaru
- kroenleinii
- mirabella
- *ridleii*
- saddianus
- spegazzinii

Cipocereus
- bradei
- crassisepalus
- laniflorus
- minensis ssp. minensis
- minensis ssp. pleurocarpus
- pusilliflorus

Cleistocactus
- baumannii ssp. horstii

Coleocephalocereus
- aureus
- buxbaumianus ssp. buxbaumianus
- buxbaumianus ssp. flavisetus
- fluminensis ssp. decumbens
- fluminensis ssp. fluminensis
- goebelianus
- pluricostatus
- purpureus

Discocactus
- ♦bahiensis
- ♦*ferricola*
- ♦heptacanthus ssp. catingicola
- ♦heptacanthus ssp. heptacanthus
- ♦heptacanthus ssp. magnimammus
- ♦horstii
- ♦placentiformis
- ♦pseudoinsignis
- ♦zehntneri ssp. boomianus
- ♦zehntneri ssp. zehntneri

Disocactus
- amazonicus

Echinopsis
- *brasiliensis*
- calochlora
- *calochlora* ssp. *glaetzleana*
- eyriesii
- *hammerschmidii*
- oxygona

(Echinopsis)
- rhodotricha

Epiphyllum
- phyllanthus

Espostoopsis
- dybowskii

Facheiroa
- cephaliomelana ssp. cephaliomelana
- cephaliomelana ssp. estevesii
- squamosa
- ulei

Frailea
- *buenekeri*
- *buenekeri* ssp. *densispina*
- castanea
- *castanea* ssp. *harmoniana*
- cataphracta ssp. cataphracta
- cataphracta ssp. duchii
- curvispina
- gracillima ssp. gracillima
- gracillima ssp. horstii
- mammifera
- *perumbilicata*
- phaeodisca
- pumila
- *pumila* ssp. *deminuta*
- pygmaea ssp. albicolumnaris
- *pygmaea* ssp. *fulviseta*
- pygmaea ssp. pygmaea

Gymnocalycium
- anisitsii
- *anisitsii* ssp. *multiproliferum*
- *buenekeri*
- denudatum
- horstii
- marsoneri
- *marsoneri* ssp. *matoense*
- uruguayense

Harrisia
- adscendens
- balansae

Hatiora
- epiphylloides ssp. bradei
- epiphylloides ssp. epiphylloides
- gaertneri
- herminiae
- rosea
- salicornioides

Leocereus
- bahiensis

Lepismium
- cruciforme
- houlletianum
- lumbricoides
- warmingianum

Melocactus
- ×*albicephalus*
- azureus ssp. azureus
- azureus ssp. ferreophilus
- bahiensis ssp. amethystinus

(Melocactus)
bahiensis ssp. **bahiensis**
concinnus
♦**conoideus**
♦**deinacanthus**
ernestii ssp. **ernestii**
ernestii ssp. **longicarpus**
estevesii
♦**glaucescens**
×***horridus***
lanssensianus
levitestatus
neryi
oreas ssp. **cremnophilus**
oreas ssp. **oreas**
pachyacanthus ssp. **pachyacanthus**
pachyacanthus ssp. **viridis**
♦**paucispinus**
salvadorensis
smithii
violaceus ssp. **margaritaceus**
violaceus ssp. **ritteri**
violaceus ssp. **violaceus**
zehntneri
Micranthocereus
albicephalus
auriazureus
dolichospermaticus
estevesii
flaviflorus
polyanthus
purpureus
streckeri
violaciflorus
Opuntia
brasiliensis
estevesii
inamoena
monacantha
palmadora
×***quipa***
salmiana
saxatilis
viridirubra
viridirubra ssp. ***rubrogemmia***
werneri
Parodia
alacriportana ssp. **alacriportana**
alacriportana ssp. **brevihamata**
alacriportana ssp. **buenekeri**
alacriportana ssp. **catarinensis**
arnostiana
buiningii
carambeiensis
concinna ssp. **agnetae**
concinna ssp. **blauuwiana**
concinna ssp. **concinna**
crassigibba
curvispina
erinacea

(Parodia)
fusca
haselbergii ssp. **graessneri**
haselbergii ssp. **haselbergii**
herteri
horstii
langsdorfii
leninghausii
linkii
magnifica
mammulosa ssp. **brasiliensis**
mammulosa ssp. **erythracantha**
mammulosa ssp. **eugeniae**
mammulosa ssp. **mammulosa**
mammulosa ssp. **submammulosa**
meonacantha
mueller-melchersii ssp. **gutierrezii**
mueller-melchersii ssp. **winkleri**
muricata
neohorstii
nothominuscula
ottonis ssp. **horstii**
ottonis ssp. **ottonis**
oxycostata ssp. **gracilis**
oxycostata ssp. **oxycostata**
permutata
rechensis
rudibuenekeri ssp. **glomerata**
rudibuenekeri ssp. **rudibuenekeri**
rutilans ssp. **rutilans**
schumanniana ssp. **claviceps**
schumanniana ssp. **schumanniana**
scopa ssp. **marchesii**
scopa ssp. **neobuenekeri**
scopa ssp. **scopa**
scopa ssp. **succinea**
sellowii
stockingeri
tenuicylindrica
warasii
werneri ssp. **pleiocephala**
werneri ssp. **werneri**
Pereskia
aculeata
aureiflora
bahiensis
grandifolia ssp. **grandifolia**
grandifolia ssp. **violacea**
nemorosa
sacharosa
stenantha
Pilosocereus
albisummus
arrabidae
aureispinus
aurisetus ssp. **aurilanatus**
aurisetus ssp. **aurisetus**
azulensis
brasiliensis ssp. **brasiliensis**
brasiliensis ssp. **ruschianus**

(Pilosocereus)
catingicola ssp. **catingicola**
catingicola ssp. **salvadorensis**
chrysostele
densiareolatus
diersianus
flavipulvinatus
flexibilispinus
floccosus ssp. **floccosus**
floccosus ssp. **quadricostatus**
fulvilanatus ssp. **fulvilanatus**
fulvilanatus ssp. **rosae**
glaucochrous
gounellei ssp. **gounellei**
gounellei ssp. **zehntneri**
machrisii
magnificus
multicostatus
oligolepis
pachycladus ssp. **pachycladus**
pachycladus ssp. **pernambucoensis**
pentaedrophorus
ssp. **pentaedrophorus**
pentaedrophorus ssp. **robustus**
piauhyensis
×*subsimilis*
tuberculatus
ulei
vilaboensis
Praecereus
euchlorus ssp. **euchlorus**
Pseudoacanthocereus
brasiliensis
Pseudorhipsalis
ramulosa
Quiabentia
zehntneri
Rhipsalis
baccifera ssp. **baccifera**
baccifera ssp. **hileiabaiana**
baccifera ssp. **shaferi**
burchellii
campos-portoana
cereoides
cereuscula
clavata
crispata
dissimilis
elliptica
ewaldiana

(Rhipsalis)
floccosa ssp. **floccosa**
floccosa ssp. **oreophila**
floccosa ssp. **pulvinigera**
grandiflora
hoelleri
juengeri
lindbergiana
mesembryanthemoides
neves-armondii
oblonga
olivifera
ormindoi
pacheco-leonis ssp. **catenulata**
pacheco-leonis ssp. **pacheco-leonis**
pachyptera
paradoxa ssp. **paradoxa**
paradoxa ssp. **septentrionalis**
pentaptera
pilocarpa
pulchra
puniceodiscus
russellii
sulcata
teres
trigona
Schlumbergera
kautskyi
microsphaerica
microsphaerica ssp. ***candida***
opuntioides
orssichiana
russelliana
truncata
Selenicereus
setaceus
wittii
Stephanocereus
leucostele
luetzelburgii
Tacinga
braunii
funalis
Uebelmannia
♦**buiningii**
♦**gummifera**
♦**pectinifera** ssp. **flavispina**
♦**pectinifera** ssp. **horrida**
♦**pectinifera** ssp. **pectinifera**

CANADA (CA)

Escobaria
vivipara
Opuntia
×*columbiana*

(Opuntia
fragilis
polyacantha

CAYMAN ISLANDS (KY)

Harrisia
  gracilis
Leptocereus
  leonii
Opuntia
  dillenii
  spinosissima
Pilosocereus
  royenii
Selenicereus
  boeckmannii
  grandiflorus

CHILE (CL)

Austrocactus
  philippii
  spiniflorus
Browningia
  candelaris
Copiapoa
  bridgesii
  calderana ssp. **calderana**
  calderana ssp. **longistaminea**
  chanaralensis
  cinerascens
  cinerea ssp. **cinerea**
  cinerea ssp. **haseltoniana**
  cinerea ssp. **krainziana**
  copiapensis
  coquimbana
  desertorum
  echinoides
  fiedleriana
  hornilloensis
  humilis
  hypogaea
  laui
  macracantha
  malletiana
  marginata
  megarhiza
  montana ssp. **grandiflora**
  montana ssp. **montana**
  rupestris
  serpentisulcata
  solaris
  tenuissima
  tocopillana
  varispinata
Corryocactus
  brevistylus ssp. **brevistylus**
Echinopsis
  atacamensis ssp. **atacamensis**
  chiloensis
  coquimbana
  deserticola
  formosa
  glauca
  litoralis
  skottsbergii
(Echinopsis)
  spinibarbis
Eriosyce
  aerocarpa
  aspillagae
  aurata
  chilensis
  confinis
  crispa ssp. **atroviridis**
  crispa ssp. **crispa**
  curvispina
  engleri
  esmeraldana
  garaventae
  heinrichiana ssp. **heinrichiana**
  heinrichiana ssp. **intermedia**
  heinrichiana ssp. **simulans**
  islayensis
  krausii
  kunzei
  laui
  limariensis
  marksiana
  napina ssp. **lembckei**
  napina ssp. **napina**
  occulta
  odieri ssp. **fulva**
  odieri ssp. **glabrescens**
  odieri ssp. **odieri**
  recondita ssp. **iquiquensis**
  recondita ssp. **recondita**
  rodentiophila
  senilis ssp. **coimasensis**
  senilis ssp. **elquiensis**
  senilis ssp. **senilis**
  sociabilis
  subgibbosa ssp. **clavata**
  subgibbosa ssp. **subgibbosa**
  taltalensis ssp. **echinus**
  taltalensis ssp. **paucicostata**
  taltalensis ssp. **pilispina**
  taltalensis ssp. **taltalensis**
  tenebrica
  villosa
Eulychnia
  acida

(Eulychnia)
aricensis
breviflora
castanea
iquiquensis
procumbens
Haageocereus
australis
fascicularis
Maihuenia
poeppigii
Neowerdermannia
chilensis ssp. **chilensis**
Opuntia
alcerrecensis
archiconoidea
atacamensis
boliviana
camachoi
colorea
conoidea
crassipina
domeykoensis
(Opuntia)
glomerata
ignescens
leucophaea
llanos-de-huanta
miquelii
ovata
rahmeri
rossiana
sanctae-barbarae
soehrensii
sphaerica
tarapacana
ticnamarensis
tunicata
wagenknechtii
Oreocereus
hempelianus
leucotrichus
varicolor
Weberbauerocereus
weberbaueri

COLOMBIA (CO)

Acanthocereus
colombianus
Armatocereus
humilis
Cereus
horrispinus
Disocactus
amazonicus
Epiphyllum
columbiense
phyllanthus
rubrocoronatum
trimetrale
Hylocereus
microcladus
monacanthus
polyrhizus
Mammillaria
columbiana ssp. **columbiana**
Melocactus
andinus
curvispinus ssp. **caesius**
curvispinus ssp. **curvispinus**
mazelianus
schatzlii
Opuntia
bella
(Opuntia)
curassavica
elatior
pennellii
pittieri
schumannii
Pereskia
bleo
guamacho
Pilosocereus
lanuginosus
Praecereus
euchlorus ssp. **smithianus**
Pseudoacanthocereus
sicariguensis
Pseudorhipsalis
ramulosa
Rhipsalis
baccifera ssp. **baccifera**
micrantha
Selenicereus
inermis
megalanthus
testudo
wittii
Stenocereus
griseus

COSTA RICA (CR)

Acanthocereus
- **tetragonus**

Disocactus
- **amazonicus**
- **kimnachii**

Epiphyllum
- **cartagense**
- *columbiense*
- *costaricense*
- **grandilobum**
- *hookeri*
- **lepidocarpum**
- **oxypetalum**
- *pittieri*
- **thomasianum**

Hylocereus
- **calcaratus**
- **costaricensis**
- *monacanthus*
- *polyrhizus*
- **stenopterus**

Melocactus
- **curvispinus** ssp. **curvispinus**

Opuntia
- **elatior**
- *guatemalensis*

Peniocereus
- **hirschtianus**

Pereskia
- **lychnidiflora**

Pseudorhipsalis
- **acuminata**
- **himantoclada**
- *horichii*
- *lankesteri*
- **ramulosa**

Rhipsalis
- **baccifera** ssp. **baccifera**
- **micrantha**

Selenicereus
- **grandiflorus**
- **testudo**
- **wercklei**

Stenocereus
- **aragonii**

Weberocereus
- **biolleyi**
- **bradei**
- **glaber**
- **imitans**
- **tonduzii**
- **trichophorus**
- **tunilla**

CUBA (CU)

Acanthocereus
- *baxaniensis*
- **tetragonus**

Dendrocereus
- *nudiflorus*

Epiphyllum
- *hookeri*

Escobaria
- **cubensis**

Harrisia
- **earlei**
- **eriophora**
- **fernowii**
- *taetra*
- **taylori**

Hylocereus
- **triangularis**

Leptocereus
- *arboreus*
- **assurgens**
- *carinatus*
- *ekmanii*
- *leonii*
- *maxonii*
- *prostratus*
- *santamarinae*

(Leptocereus)
- *scopulophilus*
- *sylvestris*
- *wrightii*

Mammillaria
- **prolifera** ssp. **prolifera**

Melocactus
- **curvispinus** ssp. **curvispinus**
- **harlowii**
- **matanzanus**
- *perezassoi*

Opuntia
- **auberi**
- ×*cubensis*
- **dejecta**
- **dillenii**
- *elata*
- **humifusa**
- *macracantha*
- *millspaughii*
- **moniliformis**
- *nashii*
- **stricta**
- **triacantha**
- **tunicata**

Pereskia
  aculeata
  zinniiflora
Pilosocereus
  polygonus
  royenii
Rhipsalis
  baccifera ssp. baccifera
Selenicereus
  boeckmannii
  brevispinus
  grandiflorus
  urbanianus
Stenocereus
  fimbriatus

DOMINICAN REPUBLIC (DO)

Harrisia
  divaricata
  hurstii
  nashii
Hylocereus
  triangularis
Leptocereus
  weingartianus
Mammillaria
  prolifera ssp. haitiensis
Melocactus
  intortus ssp. domingensis
  lemairei
Opuntia
  antillana
  caribaea
  dillenii
  falcata
  moniliformis
  nashii
(*Opuntia*)
  taylorii
  tuna
  urbaniana
Pereskia
  aculeata
  marcanoi
  portulacifolia
  quisqueyana
Pilosocereus
  polygonus
  royenii
Rhipsalis
  baccifera ssp. baccifera
Selenicereus
  grandiflorus
  urbanianus
Stenocereus
  fimbriatus

ECUADOR (EC)

Armatocereus
  brevispinus
  cartwrightianus
  godingianus
  matucanensis
Brachycereus
  nesioticus
Cleistocactus
  icosagonus
  neoroezlii
  sepium
Disocactus
  amazonicus
Echinopsis
  pachanoi
Epiphyllum
  columbiense
  phyllanthus
  rubrocoronatum
  thomasianum
Espostoa
  frutescens
(*Espostoa*)
  lanata
Hylocereus
  polyrhizus
Jasminocereus
  thouarsii
Melocactus
  bellavistensis ssp. bellavistensis
  peruvianus
Opuntia
  ×*aequatorialis*
  bonplandii
  cylindrica
  dillenii
  echios
  galapageia
  helleri
  insularis
  megasperma
  pubescens
  quitensis
  saxicola

(Opuntia)
soederstromiana
tunicata
Pereskia
aculeata
Pilosocereus
lanuginosus
Praecereus
euchlorus ssp. diffusus
Pseudorhipsalis
ramulosa
Rhipsalis
baccifera ssp. baccifera
micrantha
occidentalis
Selenicereus
megalanthus
wittii
Weberocereus
rosei

EL SALVADOR (SV)

Acanthocereus
horridus
tetragonus
Disocactus
cinnabarinus
Epiphyllum
hookeri
oxypetalum
Hylocereus
guatemalensis
Opuntia
salvadorensis
Peniocereus
hirschtianus
Pereskia
lychnidiflora
Pseudorhipsalis
ramulosa
Stenocereus
eichlamii

FRENCH GUIANA (GF)

Cereus
hexagonus
Epiphyllum
phyllanthus
Hylocereus
lemairei
Pereskia
aculeata
Rhipsalis
baccifera ssp. baccifera
Selenicereus
setaceus

GUATEMALA (GT)

Acanthocereus
horridus
tetragonus
Disocactus
biformis
cinnabarinus
eichlamii
nelsonii
quezaltecus
Epiphyllum
crenatum
guatemalense
hookeri
oxypetalum
pumilum
thomasianum
Hylocereus
escuintlensis
(Hylocereus)
guatemalensis
minutiflorus
Mammillaria
columbiana ssp. yucatanensis
voburnensis ssp. eichlamii
voburnensis ssp. voburnensis
Melocactus
curvispinus ssp. curvispinus
Myrtillocactus
eichlamii
Opuntia
deamii
decumbens
eichlamii
guatemalensis
guilanchi
lutea

(Opuntia)
puberula
pubescens
tomentella
Pachycereus
lepidanthus
Peniocereus
hirschtianus
Pereskia
lychnidiflora
Pereskiopsis
kellermanii
Pilosocereus
leucocephalus
Pseudorhipsalis
ramulosa
Rhipsalis
baccifera ssp. **baccifera**
micrantha
Selenicereus
hondurensis
testudo
Stenocereus
eichlamii
yunckeri
Weberocereus
glaber

GUYANA (GY)

Cereus
hexagonus
Epiphyllum
phyllanthus
Hylocereus
scandens
Melocactus
neryi
smithii
Opuntia
cochenillifera
Pereskia
aculeata
Pilosocereus
oligolepis
Rhipsalis
baccifera ssp. **baccifera**
Selenicereus
setaceus

HAITI (HT)

Dendrocereus
undulosus
Harrisia
divaricata
nashii
Leptocereus
paniculatus
weingartianus
Mammillaria
ekmanii
prolifera ssp. **haitiensis**
Melocactus
lemairei
Opuntia
acaulis
caribaea
ekmanii
falcata
(Opuntia)
moniliformis
taylorii
urbaniana
Pereskia
aculeata
portulacifolia
Pilosocereus
polygonus
Pseudorhipsalis
ramulosa
Rhipsalis
baccifera ssp. **baccifera**
Selenicereus
grandiflorus
urbanianus
Stenocereus
fimbriatus

HONDURAS (HN)

Acanthocereus
tetragonus
Disocactus
aurantiacus
(Disocactus)
biformis
cinnabarinus
nelsonii

Epiphyllum
- **crenatum**
- *guatemalense*
- *hookeri*
- **oxypetalum**
- **thomasianum**

Hylocereus
- **minutiflorus**

Mammillaria
- **columbiana** ssp. **yucatanensis**
- **voburnensis** ssp. **eichlamii**

Melocactus
- **curvispinus** ssp. **curvispinus**

Opuntia
- *deamii*
- **decumbens**
- *guatemalensis*
- *hondurensis*
- **lutea**

Pachycereus
- **lepidanthus**

Pereskia
- **lychnidiflora**

Pereskiopsis
- **kellermanii**

Pilosocereus
- **leucocephalus**

Pseudorhipsalis
- **ramulosa**

Rhipsalis
- **baccifera** ssp. **baccifera**

Selenicereus
- *hondurensis*
- *macdonaldiae*
- **testudo**

Stenocereus
- **eichlamii**
- **yunckeri**

JAMAICA (JM)

Harrisia
- **gracilis**

Hylocereus
- **triangularis**

Mammillaria
- **columbiana** ssp. **yucatanensis**

Melocactus
- **caroli-linnaei**

Opuntia
- **dillenii**
- *jamaicensis*
- *sanguinea*
- **spinosissima**

(Opuntia)
- **tuna**

Pilosocereus
- **royenii**

Pseudorhipsalis
- **alata**
- **ramulosa**

Rhipsalis
- **baccifera** ssp. **baccifera**

Selenicereus
- **grandiflorus**

Stenocereus
- **fimbriatus**

LESSER ANTILLES (LA)

Acanthocereus
- **tetragonus**

Hylocereus
- **trigonus**

Mammillaria
- **mammillaris**
- **nivosa**

Melocactus
- *broadwayi*
- **intortus** ssp. **intortus**

Opuntia
- *antillana*
- *caracassana*
- **curassavica**

(Opuntia)
- **dillenii**
- **elatior**
- **rubescens**
- **triacantha**

Pereskia
- **aculeata**

Pilosocereus
- **royenii**

Rhipsalis
- **baccifera** ssp. **baccifera**

Selenicereus
- **grandiflorus**
- *innesii*

MEXICO (MX)

Acanthocereus
horridus
occidentalis
subinermis
tetragonus
Ariocarpus
♦**agavoides**
♦**bravoanus** ssp. **bravoanus**
♦**bravoanus** ssp. **hintonii**
♦**fissuratus**
♦**kotschoubeyanus**
♦**retusus** ssp. **retusus**
♦**retusus** ssp. **trigonus**
♦**scaphirostris**
Astrophytum
♦**asterias**
capricorne
myriostigma
ornatum
Aztekium
hintonii
♦**ritteri**
Bergerocactus
emoryi
Carnegiea
gigantea
Cephalocereus
apicicephalium
columna-trajani
nizandensis
senilis
totolapensis
Coryphantha
calipensis
calochlora
clavata
compacta
cornifera
cornuta
delaetiana
difficilis
durangensis
echinoidea
echinus
elephantidens
erecta
georgii
glanduligera
gracilis
grata
guerkeana
indensis
jalpanensis
jaumavei
longicornis
macromeris ssp. **macromeris**
maiz-tablasensis
maliterrarum
(Coryphantha)
melleospina
neglecta
nickelsiae
octacantha
odorata
ottonis
pallida
poselgeriana
potosiana
pseudoechinus
pseudoradians
pulleineana
pusilliflora
pycnacantha
radians
recurvata
reduncispina
retusa
robustispina ssp. **robustispina**
robustispina ssp. **scheeri**
salinensis
sulcolanata
tripugionacantha
unicornis
vaupeliana
vogtherriana
♦**werdermannii**
wohlschlageri
Disocactus
ackermannii
aurantiacus
cinnabarinus
flagelliformis
♦**macdougallii**
macranthus
martianus
nelsonii
phyllanthoides
schrankii
speciosus
Echinocactus
grusonii
horizonthalonius
parryi
platyacanthus
polycephalus ssp. **polycephalus**
texensis
Echinocereus
adustus ssp. **adustus**
adustus ssp. **bonatzii**
adustus ssp. **schwarzii**
barthelowanus
berlandieri
brandegeei
bristolii
chisoensis
cinerascens ssp. **cinerascens**

(Echinocereus)
cinerascens ssp. **septentrionalis**
cinerascens ssp. **tulensis**
coccineus ssp. **coccineus**
dasyacanthus
dasyacanthus ssp. ***rectispinus***
engelmannii
enneacanthus ssp. **brevispinus**
enneacanthus ssp. **enneacanthus**
fasciculatus
fendleri ssp. **fendleri**
fendleri ssp. ***hempelii***
fendleri ssp. **rectispinus**
ferreirianus ssp. **ferreirianus**
♦**ferreirianus** ssp. **lindsayi**
freudenbergeri
grandis
klapperi
knippelianus
laui
leucanthus
longisetus ssp. **delaetii**
longisetus ssp. **longisetus**
mapimiensis
maritimus
maritimus ssp. ***hancockii***
mojavensis
nicholii
nicholii ssp. ***llanuraensis***
nivosus
ortegae
palmeri
pamanesiorum
papillosus
parkeri ssp. ***arteagensis***
parkeri ssp. **gonzalezii**
parkeri ssp. **mazapilensis**
parkeri ssp. **parkeri**
pectinatus
pensilis
pentalophus ssp. **leonensis**
pentalophus ssp. **pentalophus**
pentalophus ssp. **procumbens**
polyacanthus ssp. **acifer**
polyacanthus ssp. **huitcholensis**
polyacanthus ssp. **pacificus**
polyacanthus ssp. **polyacanthus**
poselgeri
primolanatus
pseudopectinatus
pulchellus ssp. **acanthosetus**
pulchellus ssp. **pulchellus**
pulchellus ssp. **sharpii**
pulchellus ssp. **weinbergii**
rayonesensis
reichenbachii ssp. **armatus**
reichenbachii ssp. **fitchii**
reichenbachii ssp. **reichenbachii**
rigidissimus ssp. **rigidissimus**
rigidissimus ssp. **rubispinus**

(Echinocereus)
russanthus
scheeri ssp. **gentryi**
scheeri ssp. **scheeri**
schereri
♦**schmollii**
sciurus ssp. **floresii**
sciurus ssp. **sciurus**
scopulorum
spinigemmatus
stoloniferus ssp. **stoloniferus**
stoloniferus ssp. **tayopensis**
stramineus ssp. **occidentalis**
stramineus ssp. **stramineus**
subinermis ssp. **ochoterenae**
subinermis ssp. **subinermis**
triglochidiatus
viereckii ssp. **morricalii**
viereckii ssp. **viereckii**
viridiflorus ssp. **chloranthus**
websterianus

Epiphyllum
anguliger
caudatum
crenatum
guatemalense
hookeri
laui
oxypetalum
pumilum
thomasianum

Epithelantha
bokei
micromeris ssp. **greggii**
micromeris ssp. **micromeris**
micromeris ssp. **pachyrhiza**
micromeris ssp. **polycephala**
micromeris ssp. **unguispina**

Escobaria
aguirreana
chihuahuensis ssp. **chihuahuensis**
chihuahuensis ssp. **henricksonii**
dasyacantha ssp. **chaffeyi**
dasyacantha ssp. **dasyacantha**
emskoetteriana
laredoi
lloydii
missouriensis ssp. **asperispina**
robbinsorum
roseana
roseana ssp. ***galeanensis***
tuberculosa
vivipara
zilziana

Escontria
chiotilla

Ferocactus
alamosanus ssp. **alamosanus**
alamosanus ssp. **reppenhagenii**
chrysacanthus ssp. **chrysacanthus**

(Ferocactus)
chrysacanthus ssp. **grandiflorus**
cylindraceus ssp. **cylindraceus**
cylindraceus ssp. **lecontei**
cylindraceus ssp. **tortulispinus**
diguetii
echidne
emoryi ssp. **emoryi**
emoryi ssp. **rectispinus**
flavovirens
fordii
glaucescens
gracilis ssp. **coloratus**
gracilis ssp. **gatesii**
gracilis ssp. **gracilis**
haematacanthus
hamatacanthus ssp. **hamatacanthus**
hamatacanthus ssp. **sinuatus**
herrerae
histrix
johnstonianus
latispinus ssp. **latispinus**
latispinus ssp. **spiralis**
lindsayi
macrodiscus ssp. **macrodiscus**
macrodiscus ssp. **septentrionalis**
peninsulae
pilosus
pottsii
robustus
santa-maria
schwarzii
tiburonensis
townsendianus
viridescens
wislizeni
Geohintonia
mexicana
Hylocereus
ocamponis
purpusii
Leuchtenbergia
principis
Lophophora
diffusa
williamsii
Mammillaria
albicans ssp. **albicans**
albicans ssp. **fraileana**
albicoma
albiflora
albilanata ssp. **albilanata**
albilanata ssp. **oaxacana**
albilanata ssp. **reppenhagenii**
albilanata ssp. **tegelbergiana**
amajacensis
anniana
armillata ssp. **armillata**
armillata ssp. **cerralboa**
aureilanata

(Mammillaria)
backebergiana ssp. **backebergiana**
backebergiana ssp. **ernestii**
barbata
baumii
beneckei
berkiana
blossfeldiana
bocasana ssp. **bocasana**
bocasana ssp. **eschauzieri**
bocensis
boelderliana
bombycina ssp. **bombycina**
bombycina ssp. **perezdelarosae**
boolii
brandegeei ssp. **brandegeei**
brandegeei ssp. **gabbii**
brandegeei ssp. **glareosa**
brandegeei ssp. **lewisiana**
canelensis
capensis
carmenae
carnea
carretii
coahuilensis ssp. **albiarmata**
coahuilensis ssp. **coahuilensis**
columbiana ssp. **yucatanensis**
compressa ssp. **centralifera**
compressa ssp. **compressa**
craigii
crinita ssp. **crinita**
crinita ssp. **leucantha**
crinita ssp. **wildii**
crucigera ssp. **crucigera**
crucigera ssp. **tlalocii**
decipiens ssp. **albescens**
decipiens ssp. **camptotricha**
decipiens ssp. **decipiens**
deherdtiana ssp. **deherdtiana**
deherdtiana ssp. **dodsonii**
densispina
dioica ssp. **angelensis**
dioica ssp. **dioica**
dioica ssp. **estebanensis**
discolor ssp. **discolor**
discolor ssp. **esperanzaensis**
dixanthocentron
duoformis
duwei
elongata ssp. **echinaria**
elongata ssp. **elongata**
eriacantha
erythrosperma
evermanniana
fittkaui ssp. **fittkaui**
fittkaui ssp. **limonensis**
flavicentra
formosa ssp. **chionocephala**
formosa ssp. **formosa**
formosa ssp. **microthele**

(*Mammillaria*)
formosa ssp. **pseudocrucigera**
geminispina ssp. **geminispina**
geminispina ssp. **leucocentra**
gigantea
glassii ssp. **ascensionis**
glassii ssp. **glassii**
glochidiata
goodridgii
grusonii
guelzowiana
guerreronis
guillauminiana
haageana ssp. **acultzingensis**
haageana ssp. **conspicua**
haageana ssp. **elegans**
haageana ssp. **haageana**
haageana ssp. **san-angelensis**
haageana ssp. **schmollii**
hahniana ssp. **bravoae**
hahniana ssp. **hahniana**
hahniana ssp. **mendeliana**
hahniana ssp. **woodsii**
halbingeri
halei
hamata
heidiae
hernandezii
herrerae
hertrichiana
heyderi ssp. **gaumeri**
heyderi ssp. **gummifera**
heyderi ssp. **hemisphaerica**
heyderi ssp. **heyderi**
heyderi ssp. **meiacantha**
huitzilopochtli
humboldtii
hutchisoniana ssp. **hutchisoniana**
hutchisoniana ssp. **louisae**
insularis
jaliscana ssp. **jaliscana**
jaliscana ssp. **zacatecasensis**
johnstonii
karwinskiana ssp. **beiselii**
karwinskiana ssp. **collinsii**
karwinskiana ssp. **karwinskiana**
karwinskiana ssp. **nejapensis**
klissingiana
knippeliana
kraehenbuehlii
lasiacantha ssp. **egregia**
lasiacantha ssp. **hyalina**
lasiacantha ssp. **lasiacantha**
lasiacantha ssp. **magallanii**
laui ssp. **dasyacantha**
laui ssp. **laui**
laui ssp. **subducta**
lenta
lindsayi
lloydii

(*Mammillaria*)
longiflora ssp. **longiflora**
longiflora ssp. **stampferi**
longimamma
luethyi
magnifica
magnimamma
mainiae
marcosii
marksiana
mathildae
matudae
mazatlanensis ssp. **mazatlanensis**
mazatlanensis ssp. **patonii**
melaleuca
melanocentra ssp. **linaresensis**
melanocentra ssp. **melanocentra**
melanocentra ssp. **rubrograndis**
mercadensis
meyranii
microhelia
miegiana
mieheana
moelleriana
morganiana
muehlenpfordtii
multidigitata
mystax
nana
napina
neopalmeri
nunezii ssp. **bella**
nunezii ssp. **nunezii**
oteroi
painteri
parkinsonii
♦**pectinifera**
peninsularis
pennispinosa ssp. **nazasensis**
pennispinosa ssp. **pennispinosa**
perbella
petrophila ssp. **arida**
petrophila ssp. **baxteriana**
petrophila ssp. **petrophila**
petterssonii
phitauiana
picta ssp. **picta**
picta ssp. **viereckii**
pilispina
plumosa
polyedra
polythele ssp. **durispina**
polythele ssp. **obconella**
polythele ssp. **polythele**
pondii ssp. **maritima**
pondii ssp. **pondii**
pondii ssp. **setispina**
poselgeri
pottsii
prolifera ssp. **arachnoidea**

(Mammillaria)
prolifera ssp. **texana**
prolifera ssp. **zublerae**
rekoi ssp. **aureispina**
rekoi ssp. **leptacantha**
rekoi ssp. **rekoi**
rettigiana
rhodantha ssp. **aureiceps**
rhodantha ssp. **fera-rubra**
rhodantha ssp. **mccartenii**
rhodantha ssp. **mollendorffiana**
rhodantha ssp. **pringlei**
rhodantha ssp. **rhodantha**
roseoalba
saboae ssp. **goldii**
saboae ssp. **haudeana**
saboae ssp. **saboae**
sanchez-mejoradae
sartorii
scheinvariana
schiedeana ssp. **dumetorum**
schiedeana ssp. **giselae**
schiedeana ssp. **schiedeana**
schumannii
schwarzii
scrippsiana
sempervivi
senilis
sheldonii
sinistrohamata
♦solisioides
sonorensis
sphacelata ssp. **sphacelata**
sphacelata ssp. **viperina**
sphaerica
spinosissima ssp. **pilcayensis**
spinosissima ssp. **spinosissima**
spinosissima ssp. **tepoxtlana**
standleyi
stella-de-tacubaya
supertexta
surculosa
tayloriorum
tepexicensis
theresae
thornberi ssp. **thornberi**
thornberi ssp. **yaquensis**
tonalensis
uncinata
vari(e)aculeata
vetula ssp. **gracilis**
vetula ssp. **vetula**
voburnensis ssp. **voburnensis**
wagneriana
weingartiana
wiesingeri ssp. **apamensis**
wiesingeri ssp. **wiesingeri**
winterae ssp. **aramberri**
winterae ssp. **winterae**
wrightii ssp. **wilcoxii**

(Mammillaria)
wrightii ssp. **wrightii**
xaltianguensis ssp. **bambusiphila**
xaltianguensis ssp. **xaltianguensis**
zeilmanniana
zephyranthoides
Mammilloydia
candida
Melocactus
curvispinus ssp. **curvispinus**
curvispinus ssp. **dawsonii**
×*Myrtgerocactus*
lindsayi
Myrtillocactus
cochal
geometrizans
schenckii
Neobuxbaumia
euphorbioides
laui
macrocephala
mezcalaensis
multiareolata
polylopha
scoparia
squamulosa
tetetzo
Neolloydia
conoidea
matehualensis
Obregonia
♦denegrii
Opuntia
acanthocarpa
agglomerata
alcahes
amyclaea
anteojoensis
arbuscula
atrispina
atropes
auberi
azurea
basilaris
bensonii
bigelovii
bradtiana
bravoana
bulbispina
burrageana
californica
cardenche
chaffeyi
chavena
chihuahuensis
chlorotica
cholla
cineracea
cochenillifera
crassa

(Opuntia)
crystalenia
deamii
decumbens
depressa
dillenii
dumetorum
durangensis
echinocarpa
eichlamii
elizondoana
emoryi
engelmannii
excelsa
feracantha
ficus-indica
fulgida
fuliginosa
grahamii
guilanchi
hitchcockii
howeyi
huajuapensis
humifusa
hyptiacantha
imbricata
inaperta
invicta
jaliscana
joconostle
karwinskiana
kleiniae
kunzei
lagunae
larreyi
lasiacantha
leptocaulis
leucotricha
lindsayi
littoralis
lloydii
lutea
macrocentra
macrorhiza
mamillata
marenae
megacantha
megarhiza
microdasys
moelleri
molesta
nejapensis
neochrysacantha
nuda
orbiculata
oricola
pachona
pailana
parryi
parviclada

(Opuntia)
phaeacantha
pilifera
plumbea
prolifera
puberula
pubescens
pumila
pycnantha
pyriformis
ramosissima
rastrera
reflexispina
rileyi
ritteri
robinsonii
robusta
rosarica
rosea
rufida
santamaria
scheeri
schottii
spinosior
spinulifera
spraguei
stenopetala
streptacantha
stricta
tapona
tehuacana
tehuantepecana
tesajo
×*tetracantha*
thurberi
tomentella
tomentosa
trichophora
tunicata
undulata
vaginata
velutina
versicolor
vilis
wilcoxii
Ortegocactus
macdougallii
×*Pacherocactus*
orcuttii
Pachycereus
fulviceps
gatesii
gaumeri
grandis
hollianus
marginatus
♦militaris
pecten-aboriginum
pringlei
schottii

(Pachycereus)
weberi
Pelecyphora
♦aselliformis
♦strobiliformis
Peniocereus
castellae
cuixmalensis
fosterianus
greggii
johnstonii
lazaro-cardenasii
macdougallii
maculatus
marianus
oaxacensis
occidentalis
rosei
serpentinus
striatus
tepalcatepecanus
viperinus
zopilotensis
Pereskia
aculeata
lychnidiflora
Pereskiopsis
aquosa
blakeana
diguetii
kellermanii
porteri
rotundifolia
Pilosocereus
alensis
chrysacanthus
leucocephalus
purpusii
quadricentralis
royenii
Polaskia
chende
chichipe
Pseudorhipsalis
ramulosa
Rhipsalis
baccifera ssp. **baccifera**
Sclerocactus
♦erectocentrus
intertextus
♦mariposensis
♦papyracanthus
scheeri
uncinatus ssp. **uncinatus**
uncinatus ssp. **wrightii**
unguispinus
Selenicereus
anthonyanus
atropilosus
boeckmannii

(Selenicereus)
chontalensis
chrysocardium
coniflorus
donkelaarii
grandiflorus
hamatus
murrillii
nelsonii
pteranthus
spinulosus
testudo
tricae
vagans
validus
Stenocactus
coptonogonus
crispatus
hastatus
multicostatus
obvallatus
ochoterenanus
phyllacanthus
rectispinus
sulphureus
vaupelianus
Stenocereus
alamosensis
beneckei
chacalapensis
chrysocarpus
dumortieri
eichlamii
eruca
fricii
griseus
gummosus
kerberi
laevigatus
martinezii
montanus
pruinosus
queretaroensis
quevedonis
standleyi
stellatus
thurberi ssp. **littoralis**
thurberi ssp. **thurberi**
treleasei
Strombocactus
♦disciformis ssp. **disciformis**
♦disciformis ssp. **esperanzae**
Thelocactus
bicolor ssp. **bicolor**
bicolor ssp. **schwarzii**
conothelos ssp. **argenteus**
conothelos ssp. **aurantiacus**
conothelos ssp. **conothelos**
garciae
hastifer

(Thelocactus)
- ***heterochromus***
- **hexaedrophorus**
  - ssp. **hexaedrophorus**
- **hexaedrophorus** ssp. **lloydii**
- **lausseri**
- **leucacanthus**
- **macdowellii**
- **rinconensis** ssp. **hintonii**
- **rinconensis** ssp. **rinconensis**
- **setispinus**
- **tulensis** ssp. **buekii**
- **tulensis** ssp. **matudae**
- **tulensis** ssp. **tulensis**

Turbinicarpus
- ♦**alonsoi**
- ♦**beguinii**
- ♦***bonatzii***
- ♦***booleanus***
- ♦**gielsdorfianus**
- ♦***hoferi***
- ♦**horripilus**
- ♦***jauernigii***
- ♦**knuthianus**
- ♦**laui**
- ♦**lophophoroides**

(Turbinicarpus)
- ♦**mandragora**
- ♦×***mombergeri***
- ♦**pseudomacrochele** ssp. **lausseri**
- ♦**pseudomacrochele**
  - ssp. **pseudomacrochele**
- ♦**pseudopectinatus**
- ♦***rioverdensis***
- ♦**saueri**
- ♦**schmiedickeanus** ssp. **dickisoniae**
- ♦**schmiedickeanus** ssp. **flaviflorus**
- ♦**schmiedickeanus** ssp. **gracilis**
- ♦**schmiedickeanus** ssp. **klinkerianus**
- ♦**schmiedickeanus** ssp. **macrochele**
- ♦**schmiedickeanus**
  - ssp. **schmiedickeanus**
- ♦**schmiedickeanus** ssp. **schwarzii**
- ♦**subterraneus**
- ♦***swobodae***
- ♦**valdezianus**
- ♦**viereckii** ssp. **viereckii**
- ♦**viereckii** ssp. **major**
- ♦**ysabelae**
- ♦***zaragozae***

Weberocereus
- **glaber**

NETHERLANDS ANTILLES (AN)

Acanthocereus
- **tetragonus**

Cereus
- **repandus**

Hylocereus
- **lemairei**

Mammillaria
- **mammillaris**

Melocactus
- **macracanthos**

Opuntia
- **boldinghii**
- ***caracassana***

(Opuntia)
- ***caribaea***
- **curassavica**
- **dillenii**
- **elatior**

Pereskia
- **guamacho**

Pilosocereus
- **lanuginosus**

Rhipsalis
- **baccifera** ssp. **baccifera**

Stenocereus
- **griseus**

NICARAGUA (NI)

Acanthocereus
- **tetragonus**

Disocactus
- **aurantiacus**

Epiphyllum
- ***hookeri***
- **oxypetalum**
- ***pittieri***
- **thomasianum**

Hylocereus
- **costaricensis**
- ***polyrhizus***

Mammillaria
- **voburnensis** ssp. **eichlamii**

Opuntia
- **decumbens**
- ***guatemalensis***
- **lutea**

- Peniocereus
  - hirschtianus
- Pereskia
  - lychnidiflora
- Pseudorhipsalis
  - ramulosa
- Rhipsalis
  - baccifera ssp. baccifera
- Selenicereus
  - grandiflorus
  - testudo
- Stenocereus
  - eichlamii
- Weberocereus
  - biolleyi

PANAMA (PA)

- Acanthocereus
  - tetragonus
- Disocactus
  - amazonicus
- Epiphyllum
  - cartagense
  - *columbiense*
  - *costaricense*
  - grandilobum
  - lepidocarpum
  - phyllanthus
  - *pittieri*
  - *rubrocoronatum*
- Hylocereus
  - costaricensis
  - *monacanthus*
  - *polyrhizus*
- *(Hylocereus)*
  - stenopterus
- Melocactus
  - curvispinus ssp. curvispinus
- Opuntia
  - decumbens
  - elatior
- Pereskia
  - aculeata
  - bleo
- Pseudorhipsalis
  - himantoclada
- Rhipsalis
  - baccifera ssp. baccifera
- Weberocereus
  - panamensis

PARAGUAY (PY)

- Browningia
  - caineana
- Cereus
  - haageanus
  - hildmannianus ssp. hildmannianus
  - kroenleinii
  - lamprospermus
  - lanosus
  - *pachyrrhizus*
  - phatnospermus
  - spegazzinii
  - stenogonus
- Cleistocactus
  - *baumannii* ssp. *anguinus*
  - baumannii ssp. baumannii
  - *baumannii* ssp. *croceiflorus*
  - *grossei*
  - *paraguariensis*
- Discocactus
  - ♦heptacanthus ssp. magnimammus
- Echinopsis
  - *adolfofriedrichii*
  - *derenbergii*
  - hahniana
  - *meyeri*
- *(Echinopsis)*
  - oxygona
  - rhodotricha
  - *rhodotricha* ssp. *chacoana*
  - *werdermannii*
- Epiphyllum
  - phyllanthus
- Frailea
  - cataphracta ssp. cataphracta
  - cataphracta ssp. duchii
  - *friedrichii*
  - gracillima ssp. gracillima
  - grahliana
  - *grahliana* ssp. *moseriana*
  - *knippeliana*
  - pumila
  - schilinzkyana
- Gymnocalycium
  - anisitsii
  - eurypleurum
  - *fleischerianum*
  - marsoneri
  - mihanovichii
  - paraguayense
  - pediophilum

(Gymnocalycium)
pflanzii
stenopleurum
Harrisia
balansae
martinii
pomanensis
tortuosa
Lepismium
cruciforme
lumbricoides
warmingianum
Opuntia
anacantha
assumptionis
aurantiaca
brasiliensis
cardiosperma
chakensis
cognata
delaetiana
elata
grosseana
limitata
mieckleyi
monacantha
paraguayensis
pubescens
quimilo

(Opuntia)
salagria
salmiana
stenarthra
sulphurea
Parodia
leninghausii
nigrispina
ottonis ssp. ottonis
schumanniana ssp. schumanniana
Pereskia
aculeata
nemorosa
sacharosa
Pilosocereus
machrisii
Praecereus
euchlorus ssp. euchlorus
saxicola
Quiabentia
verticillata
Rhipsalis
baccifera ssp. shaferi
cereuscula
floccosa ssp. hohenauensis
Selenicereus
setaceus
Stetsonia
coryne

PERU (PE)

Armatocereus
arduus
cartwrightianus
laetus
mataranus ssp. ancashensis
mataranus ssp. mataranus
matucanensis
oligogonus
procerus
rauhii ssp. balsasensis
rauhii ssp. rauhii
riomajensis
rupicola
Browningia
albiceps
altissima
amstutziae
candelaris
chlorocarpa
columnaris
hertlingiana
microsperma
pilleifera
viridis
Calymmanthium
substerile

Cereus
trigonodendron
vargasianus
Cleistocactus
acanthurus ssp. acanthurus
acanthurus ssp. faustianus
acanthurus ssp. pullatus
chotaensis
clavispinus
×*crassiserpens*
fieldianus ssp. fieldianus
fieldianus ssp. samnensis
fieldianus ssp. tessellatus
hystrix
icosagonus
morawetzianus
neoroezlii
pachycladus
peculiaris
plagiostoma
pungens
roezlii
serpens
sextonianus
tenuiserpens
xylorhizus

×*Cleistocana*
mirabilis
Corryocactus
acervatus
apiciflorus
aureus
ayacuchoensis
brachycladus
brachypetalus
brevispinus
brevistylus ssp. brevistylus
brevistylus ssp. puquiensis
chachapoyensis
chavinilloensis
cuajonesensis
erectus
gracilis
heteracanthus
huincoensis
megarhizus
melaleucus
odoratus
pilispinus
prostratus
pyroporphyranthus
quadrangularis
quivillanus
serpens
solitarius
squarrosus
tenuiculus
Disocactus
amazonicus
Echinopsis
backebergii
cephalomacrostibas
chalaensis
cuzcoensis
glauca
hertrichiana
knuthiana
maximiliana
pachanoi
pampana
pentlandii
peruviana ssp. peruviana
peruviana ssp. puquiensis
santaensis
schieliana
schoenii
tarmaensis
tegeleriana
tulhuayacensis
uyupampensis
Epiphyllum
floribundum
phyllanthus
Eriosyce
islayensis
omasensis
Espostoa
baumannii
blossfeldiorum
calva
huanucoensis
hylaea
lanata
lanianuligera
melanostele
mirabilis
nana
ritteri
ruficeps
senilis
superba
×*Espostocactus*
mirabilis
Eulychnia
ritteri
Haageocereus
acranthus ssp. acranthus
acranthus ssp. olowinskianus
albispinus
australis
chalaensis
chryseus
decumbens
fulvus
icensis
icosagonoides
lanugispinus
pacalaensis
platinospinus
pluriflorus
pseudomelanostele ssp. aureispinus
pseudomelanostele ssp. carminiflorus
pseudomelanostele ssp. pseudomelanostele
pseudomelanostele ssp. turbidus
pseudoversicolor
subtilispinus
tenuis
versicolor
vulpes
zangalensis
×*Haagespostoa*
×*albisetata*
climaxantha
Hylocereus
microcladus
peruvianus
Lasiocereus
fulvus
rupicola
Lepismium
brevispinum
micranthum
Matucana
aurantiaca ssp. aurantiaca

(Matucana)
- aurantiaca ssp. currundayensis
- aureiflora
- *comacephala*
- formosa
- fruticosa
- haynei ssp. haynei
- haynei ssp. herzogiana
- haynei ssp. hystrix
- haynei ssp. myriacantha
- *huagalensis*
- intertexta
- krahnii
- madisoniorum
- oreodoxa
- paucicostata
- polzii
- pujupatii
- ritteri
- tuberculata
- weberbaueri

Melocactus
- bellavistensis ssp. bellavistensis
- bellavistensis ssp. onychacanthus
- peruvianus

Mila
- caespitosa
- *colorea*
- nealeana
- pugionifera

Neoraimondia
- arequipensis ssp. arequipensis
- arequipensis ssp. roseiflora

Neowerdermannia
- chilensis ssp. peruviana

Opuntia
- *apurimacensis*
- *blancii*
- boliviana
- *bradleyi*
- brasiliensis
- corotilla
- *crassicylindrica*
- cylindrica
- *dactylifera*
- floccosa
- *fulvicoma*
- *galerasensis*
- *guatinensis*
- *hirschii*
- ignescens
- inaequilateralis
- infesta
- lagopus
- *mistiensis*
- pachypus
- pentlandii
- *punta-caillan*
- *pyrrhacantha*
- quitensis

(Opuntia)
- rossiana
- soehrensii
- sphaerica
- subulata
- *tumida*
- tunicata
- *unguispina*
- *yanganucensis*
- *zehnderi*

Oreocereus
- celsianus
- doelzianus
- hempelianus
- *leucotrichus*
- *ritteri*
- *tacnaensis*
- *varicolor*

Oroya
- *borchersii*
- peruviana

Pereskia
- aculeata
- horrida ssp. horrida
- horrida ssp. rauhii

Pilosocereus
- lanuginosus

Praecereus
- euchlorus ssp. amazonicus
- euchlorus ssp. diffusus
- euchlorus ssp. jaenensis

Pseudorhipsalis
- ramulosa

Pygmaeocereus
- *bieblii*
- bylesianus
- familiaris

Rauhocereus
- riosaniensis ssp. jaenensis
- riosaniensis ssp. riosaniensis

Rebutia
- neocumingii

Rhipsalis
- baccifera ssp. baccifera
- floccosa ssp. tucumanensis
- micrantha
- occidentalis

Selenicereus
- *megalanthus*
- wittii

Weberbauerocereus
- churinensis
- *cuzcoensis*
- johnsonii
- longicomus
- rauhii
- torataensis
- weberbaueri
- winterianus

PUERTO RICO (PR)

Harrisia
portoricensis
Hylocereus
trigonus
Leptocereus
grantianus
quadricostatus
Mammillaria
nivosa
Melocactus
intortus ssp. **intortus**
Opuntia
antillana
borinquensis
dillenii
(Opuntia)
moniliformis
repens
rubescens
triacantha
Pereskia
aculeata
Pilosocereus
royenii
Rhipsalis
baccifera ssp. **baccifera**
Selenicereus
grandiflorus
Stenocereus
fimbriatus

SURINAME (SR)

Cereus
hexagonus
Epiphyllum
phyllanthus
Hylocereus
lemairei
scandens
Melocactus
neryi
Pereskia
aculeata
Rhipsalis
baccifera ssp. **baccifera**
occidentalis
Selenicereus
setaceus

TRINIDAD & TOBAGO (TT)

Acanthocereus
tetragonus
Cereus
hexagonus
Epiphyllum
hookeri
Hylocereus
lemairei
Mammillaria
mammillaris
Melocactus
broadwayi
Opuntia
boldinghii
caracassana
caribaea
Pereskia
aculeata
Pilosocereus
royenii
Rhipsalis
baccifera ssp. **baccifera**
Stenocereus
griseus

UNITED STATES OF AMERICA (US)

Acanthocereus
tetragonus
Ariocarpus
♦fissuratus
Astrophytum
♦asterias
Bergerocactus
emoryi
Carnegiea
gigantea
Coryphantha
echinus

(Coryphantha)
macromeris ssp. **macromeris**
macromeris ssp. **runyonii**
nickelsiae
ramillosa
recurvata
robustispina ssp. **robustispina**
robustispina ssp. **scheeri**
robustispina ssp. **uncinata**
sulcata
Echinocactus
horizonthalonius
polycephalus ssp. **polycephalus**
polycephalus ssp. **xeranthemoides**
texensis
Echinocereus
apachensis
berlandieri
bonkerae
boyce-thompsonii
chisoensis
coccineus ssp. **coccineus**
dasyacanthus
engelmannii
engelmannii ssp. *decumbens*
enneacanthus ssp. **brevispinus**
enneacanthus ssp. **enneacanthus**
fasciculatus
fendleri ssp. **fendleri**
fendleri ssp. **rectispinus**
ledingii
mojavensis
nicholii
papillosus
pectinatus
pectinatus ssp. *wenigeri*
pentalophus ssp. **procumbens**
polyacanthus ssp. **polyacanthus**
poselgeri
pseudopectinatus
reichenbachii ssp. **baileyi**
reichenbachii ssp. **fitchii**
reichenbachii ssp. **perbellus**
reichenbachii ssp. **reichenbachii**
rigidissimus ssp. **rigidissimus**
×*roetteri*
russanthus
stramineus ssp. **stramineus**
triglochidiatus
viridiflorus ssp. **chloranthus**
viridiflorus ssp. *correllii*
viridiflorus ssp. **cylindricus**
viridiflorus ssp. **davisii**
viridiflorus ssp. **viridiflorus**
Epithelantha
bokei
micromeris ssp. **micromeris**
Escobaria
albicolumnaria
alversonii

(Escobaria)
dasyacantha ssp. **dasyacantha**
deserti
duncanii
emskoetteriana
guadalupensis
hesteri
♦**minima**
missouriensis ssp. **missouriensis**
orcuttii
organensis
robbinsorum
sandbergii
♦**sneedii** ssp. **leei**
♦**sneedii** ssp. **sneedii**
tuberculosa
villardii
vivipara
Ferocactus
cylindraceus ssp. **cylindraceus**
cylindraceus ssp. **lecontei**
eastwoodiae
emoryi ssp. **emoryi**
hamatacanthus ssp. **hamatacanthus**
hamatacanthus ssp. **sinuatus**
viridescens
wislizeni
Harrisia
aboriginum
fragrans
gracilis
simpsonii
Lophophora
williamsii
Mammillaria
dioica ssp. **dioica**
grahamii
heyderi ssp. **hemisphaerica**
heyderi ssp. **macdougalii**
heyderi ssp. **meiacantha**
lasiacantha ssp. **lasiacantha**
mainiae
pottsii
prolifera ssp. **texana**
sphaerica
tetrancistra
thornberi ssp. **thornberi**
wrightii ssp. **wilcoxii**
wrightii ssp. **wrightii**
Neolloydia
conoidea
Opuntia
abyssi
acanthocarpa
acanthocarpa ssp. *ganderi*
aciculata
aggeria
alcahes
ammophila
arbuscula

(Opuntia)
atrispina
aurea
aureispina
austrina
basilaris
bigelovii
brachyarthra
brachyclada
californica
camanchica
chisosensis
chlorotica
clavata
×***columbiana***
×***congesta***
×***curvispina***
cymochila
davisii
densispina
dillenii
echinocarpa
edwardsii
ellisiana
emoryi
engelmannii
erinacea
ficus-indica
×***fosbergii***
fragilis
fulgida
gosseliniana
grahamii
heacockiae
humifusa
imbricata
×***kelvinensis***
kleiniae
kunzei
laevis
leptocaulis
linguiformis
littoralis
longiareolata
macrocentra
macrorhiza
mamillata
martiniana
mojavensis
×***multigeniculata***
munzii
×***neoarbuscula***
×***occidentalis***
oricola
parishii
parryi
phaeacantha
polyacantha
prolifera
pulchella

(Opuntia)
pusilla
ramosissima
rufida
rutila
santa-rita
schottii
sphaerocarpa
×***spinosibacca***
spinosior
spinosissima
stricta
strigil
superbospina
tenuiflora
×***tetracantha***
treleasei
triacantha
trichophora
tunicata
turbinata
×***vaseyi***
versicolor
×***viridiflora***
×***vivipara***
whipplei
wigginsii
wolfei
wootonii

Pachycereus
schottii

Pediocactus
♦**bradyi**
♦***bradyi*** ssp. ***despainii***
♦***bradyi*** ssp. ***winkleri***
hermannii
♦**knowltonii**
nigrispinus
nigrispinus ssp. ***beastonii***
nigrispinus ssp. ***puebloensis***
♦**paradinei**
♦**peeblesianus**
♦**sileri**
simpsonii
simpsonii ssp. ***bensonii***
simpsonii ssp. ***idahoensis***
simpsonii ssp. ***indranus***
simpsonii ssp. ***robustior***

Peniocereus
greggii
striatus

Pereskia
aculeata

Pilosocereus
polygonus

Rhipsalis
baccifera ssp. **baccifera**

Sclerocactus
brevihamatus ssp. **brevihamatus**
brevihamatus ssp. **tobuschii**

(Sclerocactus)
♦**erectocentrus**
♦**glaucus**
intertextus
johnsonii
♦**mariposensis**
♦**mesae-verdae**
nyensis
♦**papyracanthus**
parviflorus ssp. ***havasupaiensis***
parviflorus ssp. ***intermedius***
parviflorus ssp. **parviflorus**
parviflorus ssp. ***terrae-canyonae***
polyancistrus
♦**pubispinus**
scheeri
sileri

(Sclerocactus)
spinosior
spinosior ssp. ***blainei***
uncinatus ssp. **crassihamatus**
uncinatus ssp. **wrightii**
warnockii
whipplei
whipplei ssp. ***busekii***
♦**wrightiae**
Selenicereus
spinulosus
Stenocereus
thurberi ssp. **thurberi**
Thelocactus
bicolor ssp. **bicolor**
bicolor ssp. **flavidispinus**
setispinus

URUGUAY (UY)

Cereus
aethiops
hildmannianus ssp. **uruguayanus**
Echinopsis
eyriesii
oxygona
rhodotricha
Epiphyllum
phyllanthus
Frailea
castanea
gracillima ssp. **gracillima**
phaeodisca
pseudopulcherrima
pumila
pygmaea ssp. **pygmaea**
Gymnocalycium
denudatum
hyptiacanthum
netrelianum
rauschii
schroederianum
uruguayense
Harrisia
tortuosa
Lepismium
lumbricoides
Opuntia
atrovirens
aurantiaca
canterae
megapotamica
monacantha

(Opuntia)
montevideensis
paraguayensis
Parodia
allosiphon
buiningii
concinna ssp. **concinna**
erinacea
erubescens
herteri
leninghausii
linkii
magnifica
mammulosa
mueller-melchersii
ssp. **mueller-melchersii**
neoarechavaletae
nothorauschii
ottonis ssp. **ottonis**
rutilans ssp. **veeniana**
scopa ssp. **scopa**
sellowii
tabularis ssp. **bommeljei**
tabularis ssp. **tabularis**
turbinata
turecekiana
werdermanniana
Pereskia
aculeata
nemorosa
Rhipsalis
cereuscula

VENEZUELA (VE)

Acanthocereus
  tetragonus
Cereus
  fricii
  hexagonus
  horrispinus
  mortensenii
  repandus
Disocactus
  amazonicus
Epiphyllum
  columbiense
  hookeri
  phyllanthus
Hylocereus
  estebanensis
  lemairei
Mammillaria
  columbiana ssp. **columbiana**
  mammillaris
Melocactus
  andinus
  curvispinus ssp. **caesius**
  curvispinus ssp. **curvispinus**
  mazelianus
  neryi
  schatzlii
Opuntia
  bisetosa
(Opuntia)
  boldinghii
  caracassana
  caribaea
  curassavica
  depauperata
  elatior
  lilae
  schumannii
Pereskia
  aculeata
  guamacho
Pilosocereus
  lanuginosus
Praecereus
  euchlorus ssp. **smithianus**
Pseudoacanthocereus
  sicariguensis
Pseudorhipsalis
  ramulosa
Rhipsalis
  baccifera ssp. **baccifera**
  floccosa ssp. **pittieri**
Selenicereus
  inermis
  wittii
Stenocereus
  griseus

VIRGIN ISLANDS (VI)

Hylocereus
  trigonus
Mammillaria
  nivosa
Melocactus
  intortus ssp. **intortus**
Opuntia
  antillana
  dillenii
(Opuntia)
  rubescens
  triacantha
Pereskia
  aculeata
Pilosocereus
  royenii
Selenicereus
  grandiflorus

OLD WORLD (OW)

Rhipsalis
  baccifera ssp. **erythrocarpa** *(Kenya, Tanzania)*
  baccifera ssp. **horrida** *(Madagascar)*
  baccifera ssp. **mauritiana** *(Tropical Africa, Sierra Leone and Ethiopia southwards to Angola and eastern South Africa, Madagascar, Comores, Mascarenes, Seychelles, Sri Lanka)*